DYNAMIC ANALYSIS OF OCEAN STRUCTURES

OCEAN TECHNOLOGY
Series Editor: John P. Craven

Department of Naval Architecture and Marine Engineering
Massachusetts Institute of Technology
Cambridge, Massachusetts

1969: OIL ON THE SEA
edited by David P. Hoult

1970: DYNAMIC ANALYSIS OF OCEAN STRUCTURES
by Bruce J. Muga and James F. Wilson

DYNAMIC ANALYSIS OF OCEAN STRUCTURES

Bruce J. Muga and James F. Wilson

Department of Civil Engineering
Duke University
Durham, North Carolina

PLENUM PRESS • NEW YORK–LONDON • 1970

Library of Congress Catalog Card Number 77-122021

SBN 306-30483-X

A Division of Plenum Publishing Corporation
227 West 17th Street, New York, N.Y. 10011

United Kingdom edition published by Plenum Press, London
A Division of Plenum Publishing Company, Ltd.
Donington House, 30 Norfolk Street, London W.C.2, England

Printed in the United States of America

FOREWORD

Improvements in the design process as applied to ocean structures have received intense interest in recent years. Part of this interest stems from the growing realization that design on a purely deterministic basis is inadequate for structures subject to random loads, which are best described by statistical (or probability) methods. This book is an attempt to bridge the gap between present design practices and available analytical techniques (which may be exploited to improve present procedures). The book itself is an outgrowth of a set of notes prepared for an intensive short course presented over the past three years by the Engineering Extension Division of the University of California at Los Angeles, California.

The ensuing presentation is composed of four parts. The material begins with a review of the physical environment (winds, surface gravity water waves and currents) for which engineering-type formulations are presented. Hindcasting and forecasting techniques using spectral concepts are included. Special problem areas are enumerated.

The interaction of structures and fluids is of major importance to the design process. The relevance of the dimensionless parameters (Froude, Reynolds and Strouhal numbers) to phenomenological occurrences of importance is illustrated. Methods of evaluating forces on fully restrained members, including circular cylinders are presented. This includes a discussion of methods of estimating values of inertial and drag coefficients for use in design. The importance of group effects (sheltering and solidification) is emphasized. Hydroelasticity (synchronization) is dealt with separately. Impact forces due to waves breaking on structures conclude the presentation on fluid-structure interaction.

Attention next turns to dynamic behavior, beginning with a complete review of statistical concepts needed to describe the excitation and response of structural elements. The stationary responses for single-degree of freedom and multiple degree of freedom systems are evaluated. One chapter is devoted to the dynamic behavior of structural materials with and without corrosion protection. In addition, a statistical theory for damage accumulation in metals under random type stresses is presented.

The final section is devoted to a discussion of the behavior of floating platforms. The equations for both free and moored-ship systems are derived. Linear and nonlinear responses are indicated. One chapter is devoted to the current topics of (i) double catenary relations, (ii) stresses induced in long lines by random loads, and (iii) bottom breakout forces.

This book is written primarily for the practicing engineer who has a need to refer to areas outside his own specialty or to those just beginning work in the design of structures for the ocean environment. The broad coverage would permit the book to serve also as a text for a course in Ocean Engineering at the senior or first-year graduate level.

To the extent possible, throughout the book, notation and terminology are those in wide usage among practicing engineers.

ACKNOWLEDGMENTS

I would like to thank the following persons whose influence and direction have helped significantly in shaping the senior author's career: Dr. John Brahtz, (formerly, Senior Staff Consultant, U. S. Naval Civil Engineering Laboratory, Port Hueneme, California), Mr. J. T. O'Brien, Director of Look Laboratory, Department of Oceanography, University of Hawaii, Honolulu, Hawaii, and Dr. Raymond J. Smith, Professor of Oceanography, U. S. Naval Postgraduate School, Monterrey, California. The numerous informal discussions with the above named, which extended over several years, were singularly rewarding and fruitful.

The excellent contributions (Chapters IX, X, XI and XIII) by my colleague, Dr. James F. Wilson, are appreciated. In addition, Chapter V was based largely on a set of notes prepared and furnished by Dr. Wilson for use herein. His attention to detail and clarity of expression is especially noteworthy.

Finally, I would like to express my appreciation to Professor R. F. Probstein for suggesting that these notes be made available in permanent form, and to Mr. Robert Ubell and Plenum Press for their patience and cooperation.

CONTENTS

PART I
ENVIRONMENT, CONSIDERATIONS FOR DESIGN

The first essential step in the design of any structure or hardware component requires estimation of the loads to which the facility may be subjected during its intended service life. In order to estimate these loads, an adequate description of the environment (in which the facility is intended to function) is an absolute necessity. This section is concerned with the description (both character and intensity) of this ocean environment which may be expected during the proposed service life of the facility.

All external forces acting on a structure can be classified into actions and reactions. Actions are defined herein as those forces which tend to move the structure and reactions are defined as those forces which resist the movement. Although the structure must be in equilibrium under the influence of the external forces, there are internal effects such as stresses and deflections which are important and must be considered. However, the treatment in these notes is generally limited to a rigid body considerstion.

All forces tending to move a structure (i.e., the action forces) arise from winds, currents (including the presence of ice), surface gravity and elastic water waves, seismic (ground motion) sources, and collision with foreign objects. Our primary interest in Part I is with winds, currents (excluding ice) and surface gravity water waves. Collision and impact type loadings will be considered in Parts II, III and IV.

Chapter I treats wind, and wind loadings while Chapter II treats current and current loadings. Surface gravity water waves, including the estimation or prediction of their intensities, are described in Chapters III, IV and V.

Chapter I

WIND AND WIND LOADINGS

Although forces induced by winds account for only about 15% of the total forces acting on most offshore structures during periods of intense wind, water wave and current excitation, it is useful to review many of the salient features of wind loadings for two reasons. First, there may be some structures for which the wind loadings dominate the loading pattern and second, many of the wind effects are also present in the case of surface water waves and thus a knowledge of the wind effects serve as a proper introduction to surface water wave excitation.

1.1 WIND FORCES

The general wind force equation appropriate to bodies exposed to a wind of uniform velocity, U is:

$$F = \frac{1}{2} C \frac{w}{g} U^2 A_{\perp} \tag{1.1}$$

where F is the component force in pounds
w is the specific weight of air in pounds/(foot)^3
g is the gravitational constant-32.2 feet/(second)^2
U is the uniform velocity in feet/second
C is the dimensionless drag (C_D) or lift (C_L) coefficient
$A_{\perp}$ is the projected area in (feet)^2.

The total force is usually described in terms of two components, drag and lift, acting orthogonally. The determination of the force by Equation 1.1 requires information on the specific weight of air, wind velocity and appropriate drag coefficients.

By substituting $C_D = 1.5$, $w = 0.0765$ pounds/(foot)^3, into Equation 1.1, the well-known Weather Bureau (ESSA) wind force equation (for flat surfaces) can be obtained, which is

$$F = 0.004\ U^2 A$$

where, U now has the units of miles/hour, and A is the area of the flat surface in (feet)^2.

1.2 SPECIFIC WEIGHT OF AIR

It has been observed on numerous occasions that in hurricanes, typhoons and other intense storms, the air-water interface is not clearly delineated. As a matter of fact, there are tremendous quantities of wind-borne water. Calculations of the change in density of air due to the measured rainfall show little variation, however. The present practice in most design offices do not account for this effect. Thus, the density of air is usually taken to be about 0.08 pounds/(feet)^3 which corresponds to the temperature of freezing water.

1.3 WIND VELOCITY RECORDS

At any given location, it has been found that the wind velocity profiles fluctuate with time both in shape and thus intensity, as well as direction. However, the velocity averaged over a suitable time period at one elevation has been found to be related to that at other elevations. Thus, from wind velocities measured at only one elevation, average velocities at other elevations can be calculated.

The time period selected for averaging should correspond to the minimum response time of the structure, which is usually of the order of a few seconds. In practice, however, velocities are measured over some longer interval and a correction is applied to account for gusts or smaller durations. Various time intervals of measurement are used (e.g., 1 minute, 1 hour, etc) but frequently the average velocity of each mile of wind passing the recorder is reported.

A facsimile of a wind measurement at a single point taken off the north coast of San Clemente Island with a cup type anemometer is shown in Figure 1.1. A spectral analysis of this record indicated a dominant "gust" period of around 75 seconds. The anemometer was located about 40 feet above mean water level, and with virtually no obstructions from any direction within the vicinity.

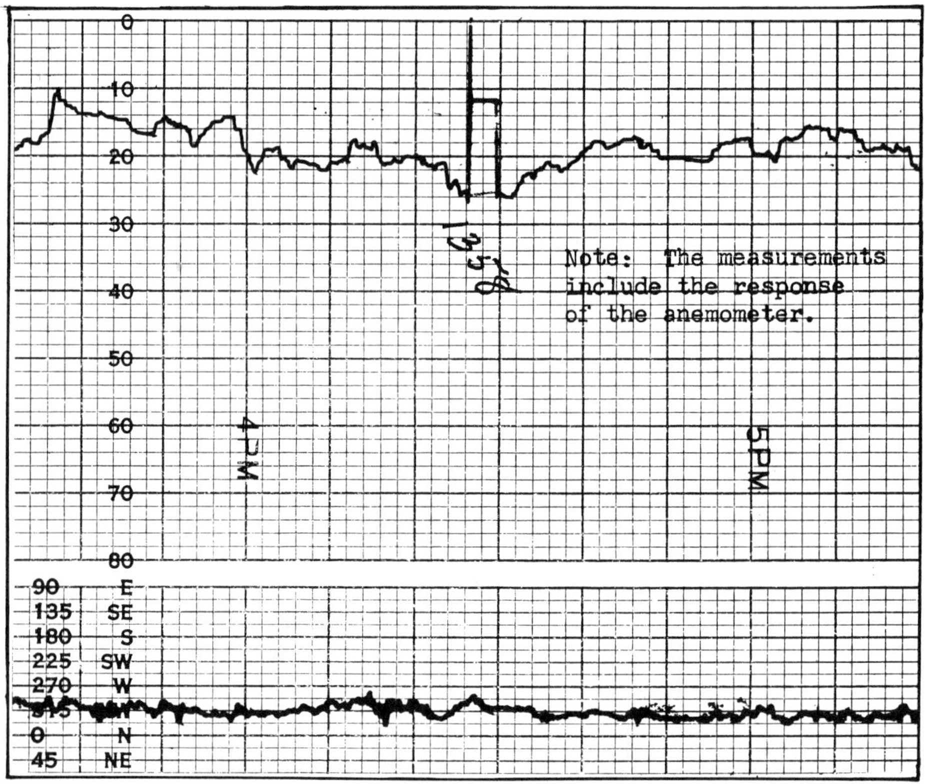

Figure 1.1. Example of Wind Velocity Record

1.4 DESIGN WIND VELOCITIES AND RECURRENCE INTERVAL

The wind velocity selected for design is usually that which will occur on the average only once in so many R years. But since this is only a statistical probability, it is entirely possible that a greater wind velocity will occur during the design life of the structure. Nevertheless, the value selected provides some measure of the risk being incurred.

The details of actual practice varies, but where loss of life is involved, the recurrence interval R is usually taken to be on the order of 100 years. In view of the fact that wind records for most locations of interest never approach anything near 100 years and most likely will be much shorter (say 10 or 15 years) a statistical method must be employed to estimate the wind velocity corresponding to the value of R.

Among the many methods available for estimating wind velocities corresponding to a specified recurrence interval, the method outline by H.C.S. Thom (1968) is especially suitable for locations where the recording period is not extensive.

Although this is hardly a handbook, it seems useful to outline at least the major steps involved in obtaining the wind speeds corresponding to the recurrence interval R. The steps are:

(1) The records for the locality are examined and arranged in descending order of extreme wind speeds, one for each year of record. The interval selected for the measurements can be of any length, but the length must be constant and all measurements must correspond to the same height or adjusted to a common height.

(2) The natural logarithms of the annual extreme wind speeds are then fitted to a Fisher-Tippet Type II distribution by simple order statistics. The latter is a two parameter distribution defined by

$$F(u_2) = \exp\left[-\left(\frac{u_2}{\beta_2}\right)^{-\gamma}\right]$$

where u_2 is the extreme wind speed, β_2 and γ are the scale and shape parameters and F is the distribution function or the probability that an extreme value is less than u_2. (The details of the fitting procedure are given in the previous reference.)

(3) The recurrence interval R is obtained from the relationship $R = \frac{1}{1-F(u_2)}$. An example of this relation is shown in Figure 1.2.

(4) From Figure 1.2, the maximum wind velocity can be estimated for any recurrence interval. Thom (1968) also indicates the confidence levels which can be placed on these estimates.

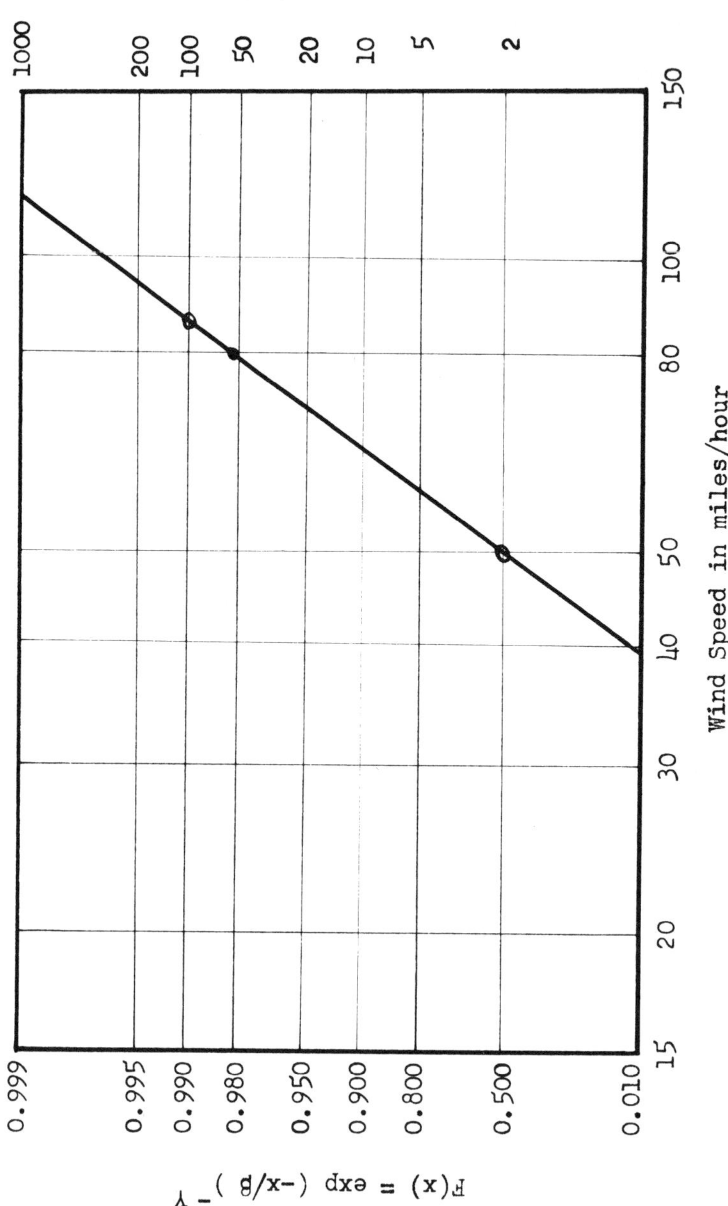

Figure 1.2 Example of Recurrence Interval Relationship

1.5 WIND VELOCITY PROFILES

It is generally accepted that the influence of height on wind velocity has the following general form:

$$\frac{U_z}{U_H} = \left(\frac{z}{H}\right)^{\frac{1}{n}}$$

where U_z is the wind velocity at height z
U_H is the wind velocity at reference height H

Sherlock (1953) proposed a reference elevation of 30 feet above the ground which has gained wide acceptance. The exponent has been found to vary from about 3 to 10, depending on the surface roughness. The final report of the ASCE Task Committee on Wind Forces* provides recommendations that specify variations of velocity with height in two different regions; (1) inland regions and (2) coastal regions. In this report, the term coastal was taken to extend at least 30 miles inland from a well-defined coastline. These variations are related to the basic wind velocity which is defined as the fastest mile of wind at a specified return period referenced to a height of 30 feet, in level, unobstructed terrain.

Application of the power law expressions are usually restricted to certain limiting elevations. Above this limiting elevation, the slope of the velocity profiles become almost vertical due to the decaying influence of the surface roughness. Above this height, the velocity is assumed to be constant and to depend mainly on the pressure gradient, earth's rotation and wind directional (curvature) effects. This velocity is called the gradient velocity, V_g.

Davenport (1960) states that the lower elevation limit of the gradient velocity varies with the exponent, n, because the surface roughness induces greater turbulence which in turn is effective to greater heights. However, it is to be noted that most structures, up to the present, extend to only a percentage of these limiting elevations (as determined from the intense storms of interest). The point to be emphasized is that because of surface roughness, the gradient wind velocity exists at a higher level (say 1000 feet) over inland areas than for coastal areas (say 600 feet). In the absence of any meaningful measurements, it is prudent to assume that the gradient wind velocity exists at an even lower level (say 500 feet) for marine areas (because of reduced surface roughness).

---

*"Wind Forces on Structures," Paper #3269, Transactions, American Society of Civil Engineers, Volume II, pp. 1124-1198.

Table 1.1 Variations of Wind Velocity with Height

Height zone in feet	Basic Wind Velocity (mph)									
	60	67	75	80	85	90	95	100	115	130
	a) For Inland Areas									
0-50	60	70	75	80	85	90	95	100		
50-150	70	80	90	95	100	105	110	120		
150-400	80	90	100	110	115	125	130	140		
400-700	90	100	115	120	130	135	145	150		
700-1000	100	110	125	130	140	145	155	160		
(V_G)-1000	105	115	130	135	145	150	160	165		
	b) For Coastal Areas									
0-50	60	70	75	80	85	90	95	100	115	130
50-150	85	95	100	105	110	115	120	125	140	150
150-400	115	125	130	135	140	145	150	155	170	180
400-600	140	150	160	165	170	175	180	185	190	195
(V_G) 600	150	160	165	170	175	180	185	190	195	200
	c) For Marine Areas									
0-50	130	140	140	145	150	150	155	160	160	165
50-150	145	155	160	165	165	170	175	180	185	190
150-300	160	170	175	180	185	190	195	200	205	210
300-500	175	185	190	195	200	205	210	215	220	225
(V_G) 500	180	190	195	200	205	210	215	220	225	230

This means that the velocity profile over marine areas is much steeper than that for inland areas.

Now, if the design gradient velocity for a nearby inland or coastal station is available, application of a suitable power law formula to the gradient wind velocity will provide the wind velocity profile. To illustrate the procedure, Table 1-1 is presented; parts a and b are taken from the previously cited reference, while part c represents an extension of the basic concepts to marine areas. Thom (1968) states that water fetches have a marked effect on extreme wind speeds which may be as much as 30 mph greater than those occurring a short distance inland. Thus, the gradient velocities shown in part c are those corresponding to the coastal stations increased by 30 mph. These gradient velocities have then been distributed to the height zones using the 1/7 power law.

If the only available records near an ocean site are from an inland station, the suggested procedure is to increase the basic wind speed by a fixed percentage before entering Table 1.1.

1.6 NATURE OF GUSTS

Gusts are defined as the instantaneous departure of the wind speed from the long term average or mean wind speed. Huss (1946) presents data on the relation between gusts and the average wind speed. The minimum gust period (or length) to be effective is related to the natural period of vibration (or characteristic length) of the proposed structure. Gust periods of shorter duration may be important for the design of individual components but the entire exposed area of the structure is not subject to this intense loading instantaneously. Sherlock (1947) outlines a procedure and provides charts to be used in estimating the design velocity for structures likely to be affected by gusts. Briefly, the procedure is as follows:

(1) Determine the minimum natural period (or periods) of oscillation of the complete structure.

(2) Determine the mean (or long term) wind speed corresponding to the specified recurrence interval as indicated in the previous sections.

(3) From the height of the proposed structure, and from (1) and (2), determine the appropriate gust factor. The gust factor is defined as the ratio of the instaneous speed to the mean speed. Figures 19, 20, 21, and 22 of the paper by Sherlock (1947) are particularly useful in obtaining the appropriate gust

factors. However, these factors were obtained from data collected over inland weather stations and thus the results should be adjusted for use in marine areas.

(4) From the gust factors so obtained, and from the mean wind speed, the design speed may be determined.

Although primary interest centers on wind variations in the horizontal direction, gusts may have finite lengths in the vertical direction as well appearing as vertically moving gusts. These transverse variations are usually of little or no consequence but it is well to keep in mind that this behavior is characterized by the appearance of moving lines of action of the induced loads.

Usually the gust factor at the higher elevations is less than that at the lower elevations. This may be easily seen by considering that:

$$G.F. = \frac{V_{gust}}{V_{Average}} = \frac{V_{Average} + \Delta V}{V_{Average}}$$

where, V, is the excess velocity of the gust over the average velocity, $V_{Average}$. As the height increases, $V_{Average}$ increases and hence the gust factor decreases.

Sherlock expresses this decrease in the gust factor, based on a 5-minute averaging, as

$$G_z = G_{30} \left(\frac{z}{30}\right)^{-0.0625}$$

where, G_z is the gust factor at height z
G_{30} is the gust factor at height 30.

1.7 EFFECTS OF GUSTS

In theory, dynamic loads of significance can be placed on a structure by a single gust or a sequence of gusts occurring at the natural frequency of the structure. However, the latter is somewhat remote since records obtained in open exposed locations indicate that gusts tend to be almost completely independent of one another. Also, there is a time-lag between the arrival of the leading edge of gust and its influence over the entire structure. Thus, for fixed ocean structures, the dynamic loads induced by gusts are usually unimportant. This is not the case for floating moored structures, however, since their natural periods may coincide with the gust periods.

The loads induced by gusts (of a given intensity and frequency) depend upon the natural frequency and damping of a proposed structure, as well as the size (exposed area). Fortunately, as size increases, the loads averaged over the entire structure area decreases. Also, as the intensity increases, the gust loads increase. In a recent paper, Vellozzi and Cohen (1968) present an approximate method for calculating gust response factors that account for the influence of those parameters. Several examples of application of the method to conventional land based structures are given. The procedures presented therein can be modified for application to structures located in marine areas or the procedure appearing in Part III of this nomograph can be used directly.

1.8 DRAG AND LIFT COEFFICIENTS - SIMPLE MEMBERS

Drag (C_D) and lift (C_L) coefficients for a large number of structural members of infinite length have been extensively tabulated. As used herein, the drag and lift coefficients are related to the aerodynamic force effect acting in the direction of the wind and orthogonal to the wind, respectively. These coefficients (sometimes referred to as shape factors) are usually given in terms of the Reynolds Number, where the characteristic length is some dimension normal to the free field wind direction.

The tabulations by Cohen and Perrin (1957) and by Hoerner (1965) are particularly noteworthy. For finite length members, the coefficients as given by Cohen and Perrin (1957) should be reduced by a factor, k, which varies with the slenderness ratio, λ. The slenderness ratio, λ, is defined as the ratio of the length of the member to the characteristic length dimension normal to the free field wind direction. Thus, $C_D = C_D^{\infty} \cdot k$ and $C_L = C_L^{\infty} \cdot k$, where C_D^{∞} and C_L^{∞} are the coefficients appropriate for infinite length, smooth surface members, and k is obtained from:

$$k = 1 - \frac{\lambda_o}{\lambda}(1 - k_o) \qquad \text{for} \quad \lambda > \lambda_o$$

$$k = k_o \qquad \text{for} \quad \lambda < \lambda_o$$

Values of the reference length, λ_o, and the reference factor, k_o, are provided by Cohen and Perrin (1957). Although these relationships are empirical and thus not strictly accurate, they are quite adequate for design purposes. For rough surfaces, values of C_D^∞ and C_L^∞ should be increased slightly.

1.9 DRAG AND LIFT COEFFICIENTS - TRUSSES

Solidification effects (in a plane normal to the prevailing wind direction) cause the drag and lift coefficients for trusses to be different than that obtained from the sum of the individual members. These coefficients can be conveniently related to the solidity ratio and slenderness ratio (computed on the basis of the overall dimensions of the truss). The solidity ratio, ϕ, is defined as the ratio of the solid area to the enclosed area.

On the basis of wind-tunnel model studies, Pagon (1934) obtained values of $C_D = 1.8\phi$ in the range $0 < \phi < 0.6$. For greater values of ϕ, C_D approaches a value of 2.0. Cohen and Perrin (1957) show that the factor, k, (in the expression $C_D = C_D^\infty \cdot k$) varies with the slenderness ratio of the truss for specified values of ϕ. It is to be emphasized that this data, however, applies only to sharp-edged truss members. For rounded or pipe members, these values (of k) should be reduced by an appropriate percentage. Finally, Hoerner (1965) presents additional information on the drag coefficients of various steel structures (flat plates, beam sections, bridge girders, girder models, and tower models.)

In summary, the effect of solidification is to increase the total drag force.

1.10 DRAG AND LIFT COEFFICIENTS - THREE-DIMENSIONAL EFFECTS (SHELTERING)

One member placed behind another with respect to the wind direction, will result in less force induced by the wind on the leeward member than on the windward member. This effect is variously known as sheltering or shielding, and it has been particularly well documented in a series of experiments described by Biermann and Herrnstein (1933).

The degree of shielding depends upon the solidity of the windward member of the group of members and the distance between members (i.e., spacing). The solidity is expressed, as before, as the ratio of the solid area to the enclosed area. The spacing ratio is defined as the ratio of the distance between members to the average characteristic length of the members. Hoerner (1965) provides some information on a pair of bridge girders which shows that for a solidity ratio of 30%, the drag coefficient on the leeward girder was 0.7 whereas that on the windward girder was 1.7. The spacing details are not given. Biggs (1953) shows that for a pair of Warren trusses (solidity ratio = 0.25), the force on the windward truss compared to the first varied from 11% for a spacing ratio of 0.27 to 52% at a spacing ratio of 1.22 when the wind was normal to the trusses. For oblique angles of incidence, the variation was found to be even greater.

Finally, Pagon (1934) provides the following relation for estimating the drag coefficient for structures consisting of two parallel trusses:

$$C_D = \phi \left[\sqrt{\frac{1.7}{\phi}} + \operatorname{Log}_{10} \frac{e}{h} \right]$$

where e/h is the spacing ratio. This coefficient is applicable to the total enclosed area of a projection of the structure onto a plane perpendicular to the wind direction.

For design purposes, the effect of shielding is quite significant for spacing ratios of up to 2.0 for simple structural members. As a matter of fact, at very close spacings, the force in the leeward member may even become negative. For spacing ratios from 2.0 to 9.0, a linear increase in the drag coefficients may be assumed from a value of 1.0 to a value of 2.0. Above spacing ratios of 9.0, the effect of shielding is negligible.

For orientations other than normal, the effect of shielding is not as significant. However, in this case, solidification effects come into play, especially for low values (i.e., < 2.0) of the spacing ratio.

1.11 REFERENCES

1. Biermann, D., and Herrnstein, W. H. (1933). "The Interference Between Struts in Various Combinations," NACA Report 468.

2. Biggs, J. M. (1953). "Wind Loads on Truss Bridges," Proceedings, American Society of Civil Engineers, V. 79, Separate Paper No. 201, p. 19, July.

3. Cohen, E., and Perrin, H. (1957). "Design of Multi-Level Guyed Towers: Wind Loading," Proceedings, American Society of Civil Engineers, V. 83, Paper 1355, September.

4. Davenport, A. G. (1960). "A Rationale for the Determination of Design Wind Velocities," Journal of the Structural Division, American Society of Civil Engineers, Vol. 86, No. ST5, Proceedings Paper 2476, May.

5. Hoerner, S. F. (1965). Fluid-Dynamic Drag, Published by the Author, New York City.

6. Huss, P. O. (1946). "Relation Between Gusts and Average Wind Speeds," D. Gugenheim Airship Institute, Akron, Ohio, Report No. 140.

7. Pagon, W. W. (1934). "Drag Coefficients for Structures Studied in Wind-Tunnel Mod Studies," Engineering News-Record, Vol. 113, No. 15, McGraw-Hill, Inc., New York, New York, October 11, pp. 456-458.

8. Sherlock, R. H. (1947). "Gust Factors for the Design of Buildings," Publications, Eighth Volume, International Association for Bridge and Structural Engineering, Zurich, Switzerland, pp. 207-236.

9. Sherlock, R. H. (1953). "Variations of Wind Velocity and Gusts with Height," Transactions, American Society of Civil Engineers, Vol. 118, pp. 463-508.

10. Thom, H. C. S. (1968). "New Distributions of Extreme Winds in the United States," Journal of the Structural Division, Proceedings, American Society of Civil Engineers, Paper 6038, Vol. 94, No. ST7, July.

11. Vellozzi, J., and Cohen, E. (1968). "Gust Response Factors," Journal of the Structural Division, Proceedings, American Society of Civil Engineers, Paper 5980, Vol. 94, No. ST6, June.

Chapter II

CURRENTS AND CURRENT LOADINGS

Currents induce additional forces (drag and lift) on ocean structures that on occasion may be of far greater magnitude than is initially suspected. This occurs when the currents and surface gravity water waves have components which coincide directionally. Thus, since the drag force is usually taken to be proportional to the square (rather than the first power) of the fluid particle velocity, the resulting force, due in part to the coupling (cross-multiplication products) terms, may be considerably enlarged. Currents also have another effect, namely that of refraction of surface waves which occurs at the boundary of an ocean current and still water. The end result is a change in the 'apparent' wave lengths and direction of the incident waves. The change in wave lengths also appears as a change in wave steepness. For details, the reader is referred to the paper by Johnson (1947).

2.1. NATURE OF CURRENTS

Several classifications of currents appear in the literature. Pierson and Neumann (1966) classify currents according to the dominant driving mechanism. The following current groups have been identified: (i) wind driven currents, (ii) slope currents and (iii) thermohaline circulations. In addition to these major groups, a fourth group might include those arising from local geophysical features. Finally, three important subgroups of currents are those arising from the surface gravity wave motion, particularly in the nearshore zone, tidal currents and storm surge, the latter two being special cases of the slope currents.

Wind drift currents, which are one type of wind driven currents, are considered to be due to the drag of the wind over the water surface. For most engineering design purposes, they are not very important but the mathematical treatment by Eckman is significant since this analysis explains the shift in current direction with depth and the deflection angle between the wind vector and the current vector. The appropriate equations (including derivations) as given by Pierson and Neumann (1966) is straightforward. One interesting finding is that the results seem to be in agreement with the observations of drifting icebergs--an observation which led Eckman to the formulation of the problem in the first place.

The problem of wind-driven currents in a limited depth ocean such as that overlying the continental shelf has been studied by Bretschneider (1967). The method requires the use of coefficients which presumably have been obtained by observation or experiment. Some empirical data is given as obtained from inland lakes. Although a great deal more work remains to be done, the significant feature is that the preliminary conclusions are not in disagreement with our notions of current systems as given, for example, by Eckman.

Slope currents result from changes in mean water surface elevations relative to constant geopotential surfaces. Changes in mean water surface elevations may be due to various sources (e.g., mass transport from wind drift currents.) The situation has been well depicted by Pierson and Neumann (1966). Eckman has also treated the problem of slope currents (under some limiting assumptions) and finds that the vertical distribution of velocity is similar to the logarithmic spiral given by the solution for wind drift currents, with some minor exceptions.

Using the results given by Eckman, Harris (1959) developed a method for predicting storm surge. The basic equation relating the slope β of the sea surface to the wind stress is

$$\beta = \lambda \frac{\tau}{\rho g d}$$

where $\lambda = 1$ for deep water

$\lambda = \frac{3}{2}$ for shallow water.

Continuing with these fundamental concepts, Bretschneider (1967) has proposed a method for determining storm surge over the continental shelf. Use of the empirical equations are illustrated by some example calculations.

In summary, of interest to the designer is that horizontal current profiles generally decay with depth (in the vertical plane) in deep water, while in shallow water the decay is not nearly as pronounced. Although interest centers primarily on horizontal currents, occasions may arise where vertical currents are also important and thus require investigation.

2.2 INTERPRETATION OF MODEL TEST DATA FOR CURRENTS

The present day procedures for the analysis of model test data of the drag of ship-shape vessels assumes that the total resistance consists of two parts, namely, frictional and residuary resistance. The frictional resistance is specified by the dimensionless Reynold's number and is calculated to be that for a flat plate of the same length and submerged hull area. No account is taken of the three-dimensional hull form. The residuary resistance is assumed to be due primarily to wave making and thus dependent upon the dimensionless Froude number. Model tests are usually conducted over the range of Froude numbers of the prototype vessel. The frictional drag is then calculated from boundary layer theory (two-dimensional) appropriate to an 'equivalent' flat plate and subtracted from the total measured resistance to yield the residuary resistance. The residuary resistance is then scaled according to the Froude model laws, on the assumption that residuary resistance is the same as the 'wave drag'. This scaled up residuary resistance is then added to the calculated frictional resistance of the prototype to obtain the total drag.

The procedure described above, while apparently successful in determining drag for streamlined surface vessels moving forward in otherwise still water, contains some shortcomings which may be significant when applied to any other circumstances.

For example, the frictional resistance does not take into account the three dimensional hull form and there is no accounting for the interaction of the wave field and the hull form. Also, the residuary resistance includes a form resistance which is apparently scaled according to the Froude number. Such deficiencies are most likely to exaggerate the difference between model tests and prototype behavior for ocean structures which are neither streamlined nor always oriented with the currents approaching from head on. Thus, the interpretation and use of model test data of ocean structures must be approached with care.

In the case of moored-ship model tests in relatively shallow water, one additional caution that must be observed is the correct simulation of the current profile. On occasion in the past, model tests have been conducted by moving the model through a still water tank at a speed which matches the magnitude of the current being studied. The measured resistances were then compared with those obtained by restraining the model in a current of equal magnitude and direction. The results were surprising in that the comparisons were poor. The reason for disagreement is probably due to the differing current profiles existing between the keel and the tank bottom. The analogous profiles are those resulting from Couette flow and Poiseuille flows, respectively.

2.3 DRAG AND LIFT COEFFICIENTS

There is a great deal of information on drag of ship-shaped vessels moving at finite forward speeds as well as drag of ship appendages. For example, Hoerner (1965) and Korvin-Kroukovsky (1961) cite numerous references. However, for non-streamlined forms, which are typical of many ocean structures and particularly for vessels moored in orientations where the currents approach at some direction other than from head-on, not much information is available in the published literature. Presumably, most of the information developed thus far is still proprietary. Hoerner (1965) presents some information on barge shapes, but again this is for head-on currents.

2.4 REFERENCES

1. Bretschneider, C. L. (1967).

(a) "Estimating Wind-driven Currents over the Continental Shelf," Fundamentals of Ocean Engineering-Part I, Ocean Industry, Vol. 2, No. 6, June.

(b) "How to Calculate Storm Surges Over the Continental Shelf," Fundamentals of Ocean Engineering-Part 2A, Vol. 2 No. 7, July.

(c) Ibid., Part 2B, No. 8, August.

(d) "Calculating Storm Surge Criteria for the Continental Shelf," Fundamentals of Ocean Engineering-Part 5, Vol. 2 No. 12, December.

(e) Ibid., Part 6, Vol. 3, No. 1, January, 1968.

2. Hoerner, S. F. (1965). Fluid Dynamic Drag, Published by the Author, New York, New York.

3. Johnson, J. W. (1947). "Refraction of Surface Water Waves by Currents," Transactions, American Geophysical Union, Vol. 28, No. 6, pp. 867-874, December.

4. Korvin-Kroukovsky, B. V. (1961). Theory of Seakeeping, Society of Naval Architects and Marine Engineers, New York.

5. Neumann, G., and Pierson, W. J., (1966). Principles of Physical Oceanography, Prentice-Hall, Inc., Englewood Cliffs, New Jersey.

Chapter III

SURFACE GRAVITY WATER WAVES

Of all of the forces induced by the ocean environment on structures, those due to surface gravity waves are the most important and at the same time the most difficult to determine. The critical evaluations to be made are: (i) what is the likelihood (or probability) of the occurrence of waves of a given magnitude, frequency (of the wave), and duration (of this wave intensity) at a given location during the proposed life of the structure; (ii) how can this 'time history' of waves be interpreted as a 'time history' of forces (or loads) acting on the structure; and, finally, (iii) what are the effects of the force history on the structure (i.e., the behavioral response of the structure). The first question is largely discussed in Chapter V of Part I. Part II is devoted to a discussion of the second question, and Part III treats the final question in a general manner. The behavioral response of specific structures are discussed in Part IV. The remainder of this chapter and the next chapter deal with the description of ocean water gravity waves and various methods of analyzing these waves.

Our primary interest is not in surface water wave theory alone, but rather in the process of selecting the appropriate theory and evaluating the hydrodynamic flow field and the latters effect on the proposed structure.

3.1 INTRODUCTION TO SURFACE GRAVITY WATER WAVES

Before entering into a detailed discussion of wave types, it seems appropriate to make a few preliminary observations of the general behavior of wave motions in all kinds of elastic media. All waves have two distinguishing features. First, like most other motions of fluids, waves have the property of transmitting energy through time and distance. Second, however, unlike most other fluid motions, waves carry out this transmission of energy with little or no permanent transport of the medium in which they propagate. Waves, as we know, occur in many different forms and they are characterized by the same fundamental equation which is an expression for the propagation of a disturbance, A, through a medium with velocity, c, or

$$c^2 \nabla^2 A = \frac{\partial^2 A}{\partial^2 t} \tag{3.1.1}$$

which is a second order linear partial differential equation. The disturbance quantity, A, may be the perturbation velocity of the particles, the particle displacements, potential functions or, in fluids, the pressure.

To the uninitiated, the large number of 'wave theories' discussed in the literature all arranged according to various classification schemes which overlap and merge into one another, would appear to be about as unordered as actual ocean waves are to the first-time observer. As a matter of fact, there are several classification schemes of wave theories which serve various useful purposes but when applied indiscriminately can be misleading. The problem is simply the lack of a unified comprehensive classification scheme of water wave theories. To illustrate, we have rotational and irrotational, long crested and short crested, finite amplitude and infinite amplitude, periodic and aperiodic and nonperiodic wave theories. The latter presumably include the oscillatory and translatory wave classification scheme. In addition, we have a numerical wave theory which includes at least two evaluation procedures.

It is illuminating to begin with the identification of surface gravity water waves as one of two broad categories of waves (i.e., longitudinal or transverse). Longitudinal waves are those where the particle motions are restricted to the direction of the wave advance whereas transverse waves have a component of particle motion which is in a direction normal to the direction of wave advance. Sound waves constitute an example of small amplitude longitudinal (compression-expansion) type waves and water hammer waves are examples of large amplitude longitudinal type waves. Water surface gravity waves are of the transverse type and these are the waves with which we shall be concerned. Transverse waves are generally associated with an interface between media of different densities, such as a liquid and a gas. Waves generated by deformations in the earth's crust are well known. These seismic waves contain examples of both types; (i.e., P (primary) waves are of the longitudinal type and S (shear or secondary) waves are of the transverse type.) From a physical point of view, water gravity waves can be considered as being free (swell) or forced (sea). Swell waves are former sea waves which have moved beyond the influence of the generating wind.

Measurements of the time varying surface water elevation at any given locale may appear completely disordered with no apparent form because many different wave forms from different directions are likely to be superposed. As a matter of fact, the sea surface may be considered as the resulting combination of an infinite number of infinitely small sine waves of various amplitudes, frequencies and directions. It is instructive to review the characteristics of some classical wave forms and to precede this with notation, nomenclature and definitions which have received wide acceptance and are in general use.

3.2 DEFINITIONS, NOTATION AND NOMENCLATURE

Consider first a single wave or a train of identical waves which propagate(s) without any change of properties past a given point. That is, one wave as it passes a given point looks exactly like all other waves as they pass this given point and further, the same wave at any other point looks like it did at the given point. Such a wave is shown in the sketch below.

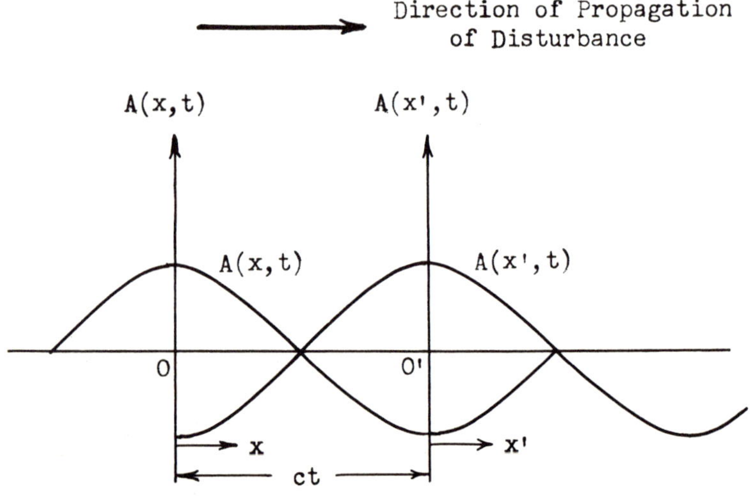

We denote the disturbance quantity, **A**, by **A**(x,t) since it is a function of time, t, and distance, x. We call the propagation velocity (phase velocity or wave celerity) by the symbol, c, and note that after any time, t, the disturbance quantity, **A**, looks exactly like it did at time $t = 0$ but with the origin displaced by the amount, ct. With reference to the origin, o, on the x-axis, this requires that $x = x' + ct$. Since $\mathbf{A}(x,t) = f(x')$ for arbitrary t, we can obtain a general expression for **A** as

$$\mathbf{A}(x,t) = f(x \pm ct) \tag{3.2.1}$$

where the + indicates a wave traveling in the negative x direction and the - indicates a wave traveling in the positive x direction. To illustrate this idea, let us consider the time varying surface elevation η for a simple harmonic sinusoidal wave as shown below.

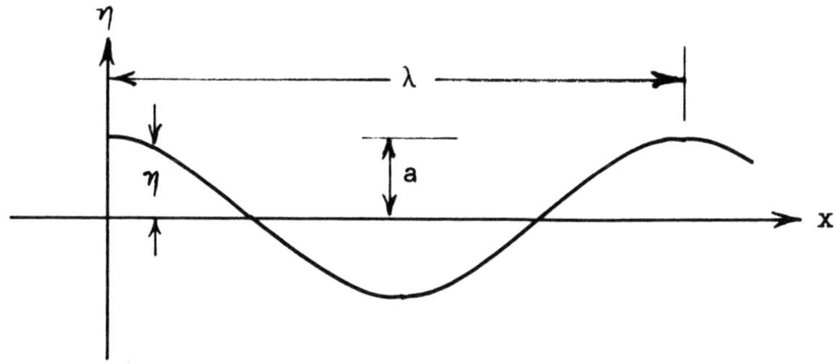

If a denotes the amplitude, then at time $t = 0$, the water surface elevation η is given by

$$\eta_{t=0} = a \cos kx \tag{3.2.2}$$

Here k is simply a parameter which converts the distance x into the proper units (i.e., angular measurement). It is the 'wave number' and is equal to $2\pi/\lambda$, where λ is the length (along the x-axis) where the wave form begins to repeat itself. The dimension λ is known as the wave length. In terms of the results obtained as indicated by Equation (3.2.1), the general expression for the water surface elevation for a sinusoidal wave moving in the positive x direction is

$$\eta(x,t) = a \cos k (x - ct) \tag{3.2.3}$$

It is easily seen that kc is a parameter which converts the time, t, into the proper units just as before. Thus, kc is called the wave frequency and is equal to $2\pi/T$, where in analogous fashion, T is the 'length' (along the imaginary t-axis), where the wave form begins to repeat itself. This can be easily seen by setting $x = 0$ in Equation (3.2.3) and noting the wave form along an imaginary t-axis. The quantity $2\pi/T$ is called the angular frequency and the symbols σ and ω are used interchangeably for this purpose. The symbol, T, is the well-known wave period. Since $k = 2\pi/\lambda$ and $kc = 2\pi/T$, it is seen immediately that $c = \lambda/T$ which is a fundamental relationship for all waves. Finally, Equation (3.2.3) can be rewritten as

$$\eta(x,t) = a \cos (kx - \sigma t - \epsilon) \tag{3.2.4}$$

where ϵ is the phase displacement and has been added to generalize the notation. Equation (3.2.4), which is the general expression for the water surface elevation of a progressive harmonic wave traveling in the $+x$ direction can be depicted graphically in a three-dimensional space-time sketch. (This exercise is left to the reader).

3.3 BASIC CONCEPTS

The motion of all fluids (including the waters of the ocean) are governed by four basic physical laws, or concepts, each of which is independent of the nature of the fluid. These concepts are those of:

(1) Continuity (conservation of mass).

(2) Newton's second law of motion (conservation of momentum).

(3) First law of thermodynamics.

(4) Second law of thermodynamics.

The concepts of the conservation of mass and momentum seem obvious but let us examine how these basic relations are developed for fluid wave motions. For this purpose, we assume a wave form propagating in the positive x direction as shown in the sketch below. Directing our attention to an elemental volume ($\Delta x \Delta y \Delta z$)

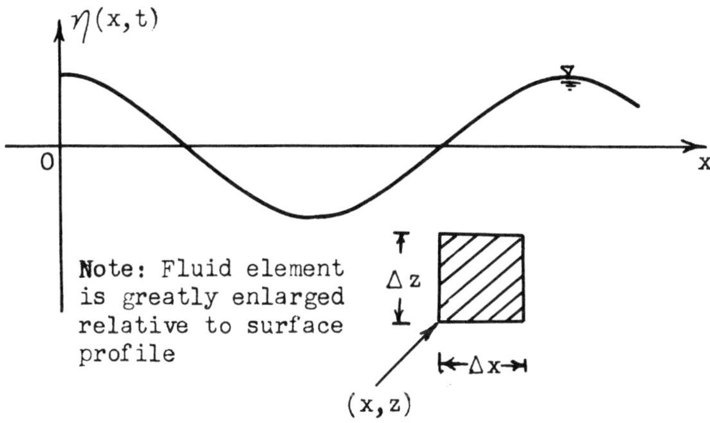

of fluid which is large as compared with the volume of a molecule, we are interested in the point-to-point variation of the fluid properties. For example, these properties can be pressure, density, viscosity, surface tension, temperature, thermal conductivity, internal energy, entropy and/or velocity. This elemental volume of fluid is located at point x, y, z where the fluid velocities are u and w in the x and z directions respectively. Assuming that the variation in a given fluid property per unit volume is denoted by A' and that no variation takes place in the y direction, then the disturbance quantity/unit volume entering the cuboid on the left vertical face is $A' \Delta y \Delta z u$/unit time and that leaving the right vertical face is $\Delta y \Delta z \left[A'u + \Delta x \;\partial(A'u)/\partial x\right]$. The net gain in A' due to motion in the x direction is, therefore, $\Delta x \Delta y \Delta z \;\partial(A'u)/\partial x$. Similarly, the net gain due to motion in the z direction is $\Delta x \Delta y \Delta z \;\partial(A'w)/\partial z$. In addition, there is a net change in A'/unit time which is given by $\Delta x \Delta y \Delta z \;\partial A'/\partial t$ so that the total rate of increase in A' with time and distance is

$$\left[\frac{\partial A'}{\partial t} + \frac{\partial(A'u)}{\partial x} + \frac{\partial(A'w)}{\partial z}\right] \Delta x \Delta y \Delta z \qquad (3.3.1)$$

Again, note that there is no change in the y direction because of our two-dimensional restriction. This result, as given above, can now be employed to obtain the continuity and momentum relationships directly.

Continuity - If we let A' be equal to the mass/unit volume or the density ρ, we obtain from Equation (3.3.1) the rate of increase of mass as

$$\left[\frac{\partial \rho}{\partial t} + \frac{\partial(\rho u)}{\partial x} + \frac{\partial(\rho w)}{\partial z}\right] \Delta x \Delta y \Delta z$$

Since matter cannot be created or destroyed, then

$$\frac{\partial \rho}{\partial t} + \frac{\partial(\rho u)}{\partial x} + \frac{\partial(\rho w)}{\partial z} = 0 \qquad (3.3.2)$$

If the fluid (i.e., sea water) is homogeneous and incompressible, that is, of constant density, then $\partial\rho/\partial t = 0$ and we obtain the equation of continuity which reduces to

$$\frac{\partial u}{\partial x} + \frac{\partial w}{\partial z} = 0 \qquad (3.3.3)$$

Momentum - If we next let the disturbance quantity A' per unit volume be equal to the product ρu (or momentum), we obtain by substitution in Equation (3.3.1) the momentum change per unit time in the x direction, which is

$$\left[\frac{\partial(\rho u)}{\partial t} + \frac{\partial(\rho u^2)}{\partial x} + \frac{\partial(\rho uw)}{\partial}\right] \Delta x \Delta y \Delta z$$

Similarly, with $A' = \rho w$, we can obtain the rate of change of momentum in the z direction which is

$$\left[\frac{\partial(\rho w)}{\partial t} + \frac{\partial(uw)}{\partial x} + \frac{\partial(\rho w^2)}{\partial z}\right] \Delta x \Delta y \Delta z$$

The momentum theorem states that the rate of change of momentum within a control volume plus the net flux of momentum through the surfaces of control volume is equal to the sum of all external forces acting on the control volume. The latter may be investigated by considering the forces acting on the elemental volume as shown in the sketch on the next page. The forces acting in the x direction are the pressures p and $p + \Delta p$ incident on the

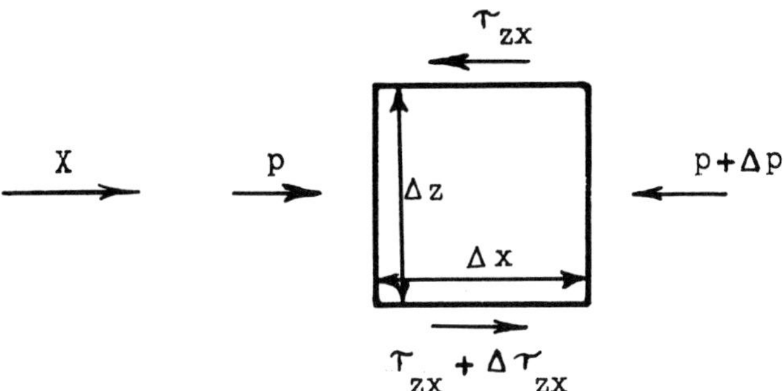

left and right vertical faces, respectively, and the shear stresses τ_{zx} and $\tau_{zx} + \Delta\tau_{zx}$ which act on faces normal to the z axis. In addition, there may be external body forces, X, per unit volume acting on the elemental body. Summing up the forces in the x direction, we obtain

$$X\rho - \frac{\partial p}{\partial x} + \frac{\partial \tau_{zx}}{\partial z} = \frac{\partial(\rho u)}{\partial t} + \frac{\partial(\rho uw)}{\partial z} + \frac{\partial(\rho u^2)}{\partial x} \qquad (3.3.4)$$

If we assume that ρ is constant and at the same time divide through by ρ, we obtain

$$X - \frac{1}{\rho}\frac{\partial p}{\partial x} + \frac{\partial \tau_{zx}}{\partial z} = \frac{\partial u}{\partial t} + u\frac{\partial w}{\partial z} + w\frac{\partial u}{\partial z} + u\frac{\partial u}{\partial x} + u\frac{\partial u}{\partial x} \qquad (3.3.5)$$

From continuity, noting that $u(\frac{\partial u}{\partial x} + \frac{\partial w}{\partial z}) = 0$, we obtain

$$X - \frac{1}{\rho}\frac{\partial p}{\partial x} + \frac{\partial \tau_{zx}}{\partial z} = \frac{\partial u}{\partial t} + u\frac{\partial u}{\partial x} + w\frac{\partial u}{\partial z} = \frac{D(u)}{Dt} \qquad (3.3.6)$$

where the operator

$$\frac{D}{Dt} = \frac{\partial}{\partial t} + u\frac{\partial}{\partial x} + w\frac{\partial}{\partial z} .$$

Equation (3.3.6) is a statement of Newton's second law of motion. In a similar manner, we can obtain the appropriate expression for the z direction, which is

$$Z - \frac{1}{\rho}\frac{\partial p}{\partial x} + \frac{\partial \tau_{xz}}{\partial x} = \frac{\partial w}{\partial t} + w\frac{\partial w}{\partial z} + u\frac{\partial w}{\partial x} = \frac{D(w)}{Dt} \qquad (3.3.7)$$

If it is assumed that the fluid cannot transmit shear stresses, (i.e., fluid is inviscid), and if all external force fields are eliminated, except the gravitational field, then Equations (3.3.6) and (3.3.7) simplify to

$$\frac{Du}{Dt} = -\frac{1}{\rho}\frac{\partial p}{\partial x} \quad \text{and} \quad \frac{Dw}{Dt} = -g - \frac{1}{\rho}\frac{\partial p}{\partial z} \qquad (3.3.8)$$

Equation (3.3.8) is the equation of motion due to Euler and appropriate to ideal fluids.

Equation (3.3.8) was derived on the basis that the flow was assumed to be irrotational (i.e., $\partial u/\partial z - \partial w/\partial x = 0$). Substitution of the irrotational condition in both equations leads to the Bernoulli equation as follows. Thus, rewriting Equations (3.3.8),

$$\frac{\partial u}{\partial t} + u\frac{\partial u}{\partial x} + w\frac{\partial u}{\partial z} = -\frac{1}{\rho}\frac{\partial p}{\partial x} \qquad (3.3.9a)$$

and

$$\frac{\partial w}{\partial t} + u\frac{\partial w}{\partial x} + w\frac{\partial w}{\partial z} = -g - \frac{1}{\rho}\frac{\partial p}{\partial z} \qquad (3.3.9b)$$

After substitution of $\partial u/\partial z = \partial w/\partial x$, these equations become

$$\frac{\partial u}{\partial t} + u\frac{\partial u}{\partial x} + w\frac{\partial w}{\partial x} = -\frac{1}{\rho}\frac{\partial p}{\partial x} \qquad (3.3.10a)$$

and

$$\frac{\partial w}{\partial t} + u\frac{\partial u}{\partial z} + w\frac{\partial w}{\partial z} = -g - \frac{1}{\rho}\frac{\partial p}{\partial z} \qquad (3.3.10b)$$

Noting that $u\frac{\partial u}{\partial t}$ may be written as $\frac{1}{2}\frac{\partial u^2}{\partial x}$ and similarly for $w\frac{\partial w}{\partial z}$, we obtain

$$\frac{\partial u}{\partial t} + \frac{1}{2}\frac{\partial u^2}{\partial x} + \frac{1}{2}\frac{\partial w^2}{\partial x} = -\frac{1}{\rho}\frac{\partial p}{\partial x} \qquad (3.3.11a)$$

and

$$\frac{\partial w}{\partial t} + \frac{1}{2}\frac{\partial u^2}{\partial z} + \frac{1}{2}\frac{\partial w^2}{\partial z} = -g - \frac{1}{\rho}\frac{\partial p}{\partial z} \qquad (3.3.11b)$$

Assuming the flow to be steady or that $\partial u/\partial t = 0 = \partial w/\partial t$, Equations (3.3.11) become

$$\frac{1}{2}\frac{\partial(u^2+w^2)}{\partial x} = -\frac{1}{\rho}\frac{\partial p}{\partial x} \qquad (3.3.12a)$$

and

$$\frac{1}{2}\frac{\partial(u^2+w^2)}{\partial z} = -g - \frac{1}{\rho}\frac{\partial p}{\partial z} \qquad (3.3.12b)$$

Integrating, we have

$$\frac{1}{2}(u^2+w^2) + \frac{p}{\rho} = F_1(z) \qquad (3.3.13a)$$

and

$$\frac{1}{2}(u^2+w^2) + \frac{p}{\rho} + gz = F_2(z) \qquad (3.3.13b)$$

By inspection, $F_1(z) - gz = F_2(z) =$ constant, so that final form of the steady-state Bernoulli equation is

$$\frac{1}{2}(u^2+w^2) + \frac{p}{\rho} + gz = \text{constant} \qquad (3.3.14)$$

Noting that u and w are the velocity components, Equation (3.3.14) reduces to

$$\frac{1}{2}(V^2) + \frac{p}{\rho} + gz = \text{constant}$$

or

$$\frac{V^2}{2g} + \frac{p}{w} + z = \text{constant} \qquad (3.3.15)$$

It is important to note that the assumption of an ideal fluid is a requirement for deriving the Bernoulli equation. However, the restriction of steady state conditions is not a strict limitation on the Bernoulli equation. This limitation may be lifted by the introduction of a potential function ϕ, which satisfies the Cauchy-Reiman equations and which results in the following form of the Bernoulli equation (integrated) under the assumption of irrotationality and constant fluid density:

$$-\frac{\partial\phi}{\partial t} + \frac{1}{2}(u^2+w^2) + \frac{p}{\rho} + gz = 0 \qquad (3.3.16)$$

3.4 CLASSIFICATION OF WAVES - PART I

Turning now to discussion of specific wave types, although measurements of the fluctuating water level at a point in the ocean represent a complex assemblage of wave types superposed without any apparent order, it is useful to classify waves as being either long-crested or short-crested. The essential distinction is that long-crested waves are waves whose crests extend infinitely far in the direction normal to the direction of wave propagation and that the crests coincide with a level surface. This permits the wave profile to be represented in a two-dimensional vertical plane in the direction of wave propagation since any vertical plane is identical to any other parallel vertical plane. On the other hand, short-crested waves are waves whose crests do not coincide with a 'level' surface. Thus, the crests may be considered to be of finite length. Although not extensively discussed in the literature, there is a theory of short-crested waves which was developed by

Jeffreys (1924) and to which Fuchs (1952) has made important contributions. We will not be concerned with short-crested waves except to point out the following salient features:

(i) In a sense, nearly all waves generated in the ocean are short-crested since crest lengths are finite.

(ii) As the short-crested waves propagate, they tend to become long-crested since there is a flow of energy along the crests (normal to the direction of wave advance).

(iii) The wave celerity, c, is greater for short-crested waves than for long-crested waves of the same wave length λ in the same water depth.

Waves may be further classified as being either periodic (oscillatory), aperiodic, or translatory, as shown in Table 3.4.1. Note that short-crested waves are considered as belonging to the class of periodic waves. All other wave types listed in the table are long-crested waves. Oscillatory or periodic waves are those for which there is no net transport of the fluid or at least the transport is nil. Translatory waves, however, involve a transport of the fluid particles in the direction of wave propagation. In addition to the solitary wave, the most important translatory waves are those generated by the tides, floods and seismic effects.

The recently developed numerical wave theories describe surface disturbances which can belong to either class depending on the boundary conditions. For this reason, they have been classified as aperiodic.

Table 3.4.1 Classification of Wave Theories

Periodic	Aperiodic	Translatory
(a) Short-Crested Waves	(a) Numerical Wave Theories	(a) Solitary Waves
(b) Airy Linear Waves		(b) Various Long Waves including Tidal Waves and Flood Waves
(c) Gerstner-Rankine (Trochoidal) Waves		
(d) Stokes Finite Amplitude Waves		
(e) Cnoidal Waves		

Initially, we want to consider the simple harmonic, periodic type of wave whose surface profile is given by $\eta(x,t) = a\cos(kx-\omega t)$. This type of wave can exist only for small amplitudes unless rotation or vorticity is present in the fluid wave motion. As the amplitude increases and if the particle orbits remain circular the surface profile approaches a trochoid. Next, if rotation is excluded and if the particle orbits are no longer required to be circular then there must be a net displacement or transport of the fluid particles. The latter are known as Stokian waves which are not exactly oscillatory waves but for which the assumption of no transport does not result in serious departure with measurements. Finally, to complete our discussion of periodic waves, we will briefly look at cnoidal waves.

Among the translatory type of waves, we will consider only the solitary wave but we will treat one of the numerical wave theories in substantial detail in a later section.

Airy Linear Waves: The small amplitude linear waves theory, otherwise known as the classical wave theory, has been developed by Airy, Laplace and others and is widely known as the Airy wave theory. The essential idea or restriction is that the amplitude of the surface disturbance must be small relative to the length of the wave or the depth of water in which the wave is propagating. Since the conditions of continuity and zero vorticity must be satisfied, we can write down the appropriate differential equations.

Continuity: $$\frac{\partial u}{\partial x} + \frac{\partial w}{\partial z} = 0 \qquad (3.4.1)$$

Irrotation: $$\frac{\partial u}{\partial x} - \frac{\partial w}{\partial x} = 0 \qquad (3.4.2)$$

Motion, (Eulerian): $$\frac{\partial u}{\partial t} + \cancelto{0}{u\frac{\partial u}{\partial x}} + \cancelto{0}{w\frac{\partial u}{\partial z}} = -\frac{1}{\rho}\frac{\partial p}{\partial x} \qquad (3.4.3)$$

$$\frac{\partial w}{\partial t} + \cancelto{0}{u\frac{\partial w}{\partial x}} + \cancelto{0}{w\frac{\partial w}{\partial z}} = -g - \frac{1}{\rho}\frac{\partial p}{\partial z} \qquad (3.4.4)$$

In the above equations, we have assumed the Coriolis acceleration to be insignificant and since the amplitude is small relative to depth or wave length, we can reason that the velocity head $\frac{1}{2}(u^2+w^2)$

is also small relative to the pressure. Thus, we can drop the convective acceleration terms in the preceeding formulation. Referring to the sketch below, and recalling that $\eta(t)$ is small, we can write the boundary conditions as

$$w = \frac{\partial \eta}{\partial t}, \quad p = p_a \quad \text{for} \quad z = 0$$

$$w = 0 \qquad\qquad \text{for} \quad z = -d$$

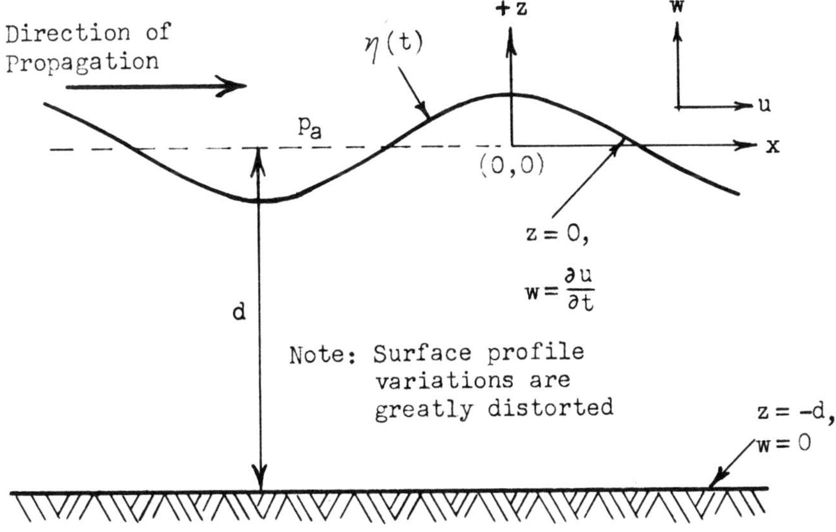

Solutions must satisfy Equations (3.4.1) through (3.4.2) in the interior region of the fluid while the boundary conditions specify conditions at the free surface and the bottom. One possible solution is given by the expression for the surface configuration for simple harmonic waves. That is

$$\eta(x,t) = a \cos(kx - \omega t)$$

where

$$u = a\omega \frac{\cosh k(z+d)}{\sinh kd} \cos(kx - \omega t) \qquad (3.4.5)$$

$$w = a\omega \frac{\sinh k(z+d)}{\sinh kd} \sin(kx - \omega t) \tag{3.4.6}$$

$$p = p_a - \rho gz + \rho ga \frac{\cosh k(z+d)}{\cosh kd} \cos(kx - \omega t) \tag{3.4.7}$$

$$\omega^2 = gk \tanh kd \tag{3.4.8}$$

That this is one possible solution is easily proved by substituting the appropriate expressions in Equations (3.3.1) through (3.3.4).

Noting that $c = \frac{\lambda}{T} = \frac{\omega}{k}$, we can rewrite Equation (3.4.8) to obtain

$$c^2 = \frac{g}{k} \tanh kd \tag{3.4.9}$$

The above expressions reduce to rather simple forms when the argument kd approaches large or small values. Often it is necessary to evaluate λ from Equation (3.4.8). This is easily done as illustrated by the subroutine given in Table 3.4.2.

Deep Water (Short Waves): If d/λ is greater than 1/2 than $kd > \pi$ and tank $kd \approx 1$ so that $c^2 = g/k = g/2\pi \cdot \lambda$. Since $c = \lambda/T$, then

$$\lambda = \frac{g}{2\pi} T^2 = 5.12\ T^2$$

which is a familiar relation for deep water waves. Also, for deep water waves, relations for particle velocity and pressure (Equations (3.4.5), (3.4.6), and (3.4.7)) simplify to

$$u = a\omega e^{kz} \cos(kx - \omega t) \tag{3.4.10}$$

$$w = a\omega e^{kz} \sin(kx - \omega t) \tag{3.4.11}$$

$$p - p_a = -\rho gz + \rho gae^{kz} \cos(kx - \omega t) \tag{3.4.12}$$

Since $\vec{v}=\vec{u}+\vec{w}$, then $v = a\omega e^{kz}$ and we observe that the particle orbits are closed circles of radius a and frequency ω. The orbital motions are independent of the phase angle of the waves and decrease in diameter with depth, as shown below.

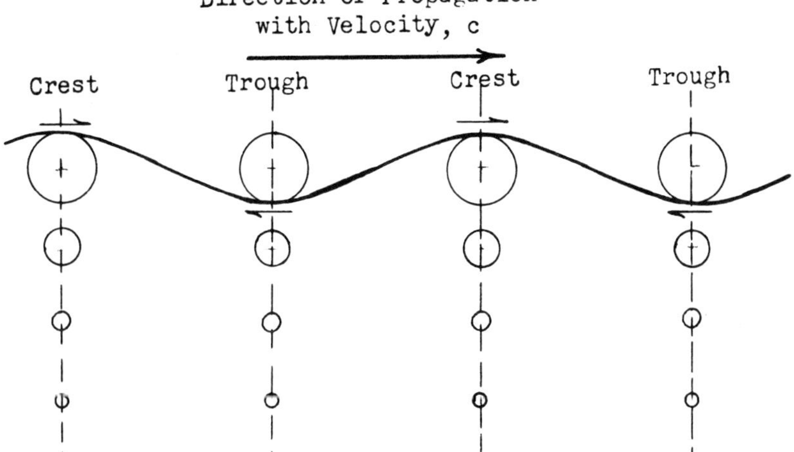

Schematic Representation of Particle Paths for Deep Water Small Amplitude Waves

Another important feature of deep water waves is the decay of the time varying pressure (second term on r.h.s. of Equation (3.4.12)). That is to say, that at a reasonable depth the second term vanishes and the pressure is very nearly equal to the hydrostatic pressure ρgz.

<u>Shallow Water (Long) Waves</u>: On the other hand, if $d/\lambda < 1/20$ then $kd < \pi/10$ and the $\tanh kd \approx kd$, so that $c^2 = gd$ or $c = \sqrt{gd}$ which is referred to in the literature as the Lagrangian wave velocity. The important feature of these waves is that the celerity is independent of wave length or period and thus they are non-dispersive. The equations for particle velocity and pressure corresponding to Equations (3.4.5), (3.4.6) and (3.4.7) become

$$u = \frac{ac}{d} \cos (kx - \omega t) \qquad (3.4.13)$$

$$w = \frac{ac}{d} k (z + d) \sin (kx - \omega t) \qquad (3.4.14)$$

$$p - p_a = \rho gz + \rho ga \cos (kx - \omega t) \qquad (3.4.15)$$

```
C             TABLE 3.4.2 - ILLUSTRATIVE SUBROUTINE

C       CALCULATE K FROM W**2 = GK TANH KD
C       WITH W,G,AND D AND TOLERANCE(TOL) KNOWN

        SIGMAG=(W**2)/G
        LAMBDA=SIGMAG
        TERC=1.4
10      TRIALK=0.5*(LAMBDA+TERC)
        TANHKD=TANHF(D*TRIALK)
        IF((TERC-LAMBDA)/TRIALK-TOL) 20,20,15
15      IF(TRIALK*TANHKD-SIGMA) 18,20,25
18      LAMBDA=TRIALK
        GO TO 10
25      TERC=TRIALK
        GO TO 10
20      END
```

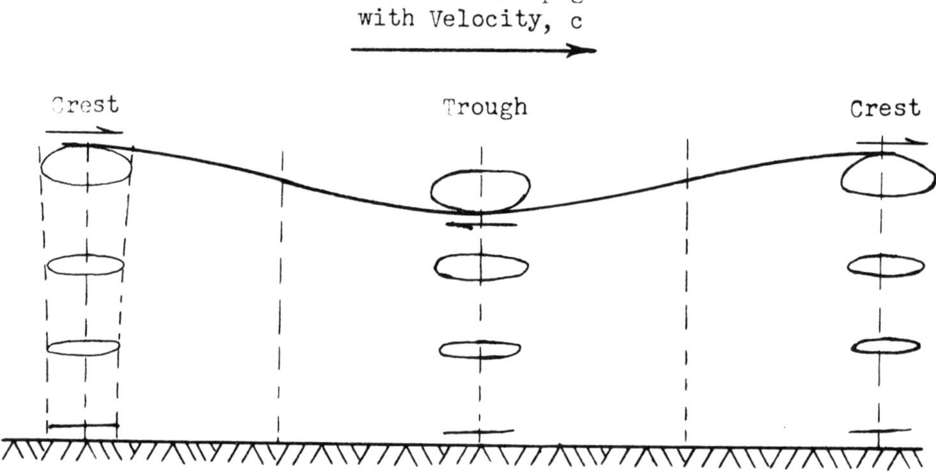

Schematic Representation of Particle Paths for Shallow Water Small Amplitude Waves

Thus u is independent of depth, but w is not so that the particle paths are ellipses with the long dimension horizontal and fixed and the axis of motion in the vertical direction decreasing with depth as shown in the foregoing sketch. The time-varying component of the pressure is also independent of depth. The horizontal axis of the particle paths have a diameter of length $2a/kd$ and the vertical axis diameter is of length $2a(z+d)/d$. The reference by Ippen, et. al. (1966) cites a paper by Ursell (1953) which shows that the shallow water wave equations are valid only if a/d and $a\lambda^2/d^3$ are both small. Thus, if $d/\lambda < 1/20$, then a/d must be $<< (1/20)^2$. Therefore, one is cautioned about using these relations for swell of moderate size in shallow water.

Finally, to terminate our discussion of small amplitude waves, Wiegel (1954) has prepared tables of functions to facilitate the determination of all of the characteristics of interest, (Gravity Waves, Tables of Function). Wilson has also prepared a useful kymogram (appearing in numerous publications) which depicts some of the characteristics of small amplitude waves.

Gerstner-Rankine (Trochoidal) Waves: Most treatments of water waves as ideal-fluid motions are limited to small amplitude waves in order to avoid the non-linear effects which appear with considerations of finite amplitudes. An exact large-amplitude wave theory for deep water conditions does exist, however, and it is illuminating to review the important features of trochoidal waves as an introduction to finite amplitude wave theory. For waves of finite amplitude, the surface configuration no longer remains sinusoidal if the particle paths remain circular. This is easily illustrated by the sketch shown on the next page. The sketch depicts the path traveled by a point, P, on a circle as the circle is rotated on the underside of a horizontal line. Note that the circumference of the circle is equal to the wave length. If the point, P, is moved from very near the center of the circle to a point near the periphery, several features are immediately obvious. First, the orbital path of the particles remains circular independent of the location of point, P. Second, as the point, P, is moved closer to the center of the circle, the surface profile traced by the point is almost indistinguishable from a sinusoidal profile. Third, as the point, P, is moved from the center to the periphery, the area between the line traced by the center of the circle and the path traced by the point, P, becomes greater beneath the center line and less above the center line. This has the effect of requiring the relocation of a new mean water line (MWL) if the areas are to remain equal. Further, as the point, P, is moved further and further away from the center, then the MWL moves further and further away from the line traced by the center of the circle. Fourth, the point, C, which lies at the intersection of the surface profile and the MWL is closer to the crest than the trough. If P were near the center, this point would move to a point equidistant

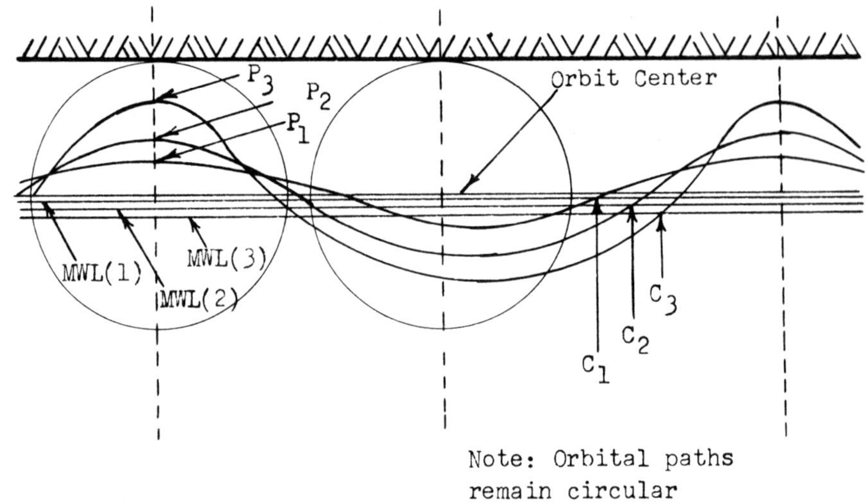

Sketch Illustrating Important Features
of Trochoidal Wave

from the crests and troughs.

The relations for particle velocity, wave celerity, pressure, and surface profile are widely published in the pertinent literature (see for example, Korvin-Kroukovsky, 1961) and need not be repeated herein. In summary, the characterizing features of the Gerstner-Rankine wave are (i) the trochoidal shape of the surface profile, and (ii) the condition that the vorticity is not zero. The expressions for the particle velocity and pressure fields are similar to but not identical to the expressions for simple harmonic waves. Finally, the particle orbits are circular and thus no mass transport or net displacement of the particles is predicted, a fact which is not in agreement with observations. These latter observations, in fact, gave impetus to the development of finite amplitude wave theory.

Stokian Waves (Finite Amplitude, Irrotational): For small amplitude waves, it was noticed that (i) the particle orbits were circular (closed), (ii) the shape of the surface profile was sinusoidal, and (iii) the flow was irrotational. As a consequence of (i), there was no net displacement of the particle, a fact that was noted to be in disagreement with observations. If the amplitude of the surface disturbance is not required to be small relative to the length of the wave or depth of water, then the basic differential equations to be satisfied are

$$\text{Continuity:} \quad \frac{\partial u}{\partial x} + \frac{\partial w}{\partial z} = 0 \tag{3.4.16}$$

$$\text{Irrotation:} \quad \frac{\partial u}{\partial x} - \frac{\partial w}{\partial x} = 0 \tag{3.4.17}$$

$$\text{Motion (Eulerian):} \quad \frac{\partial u}{\partial t} + u\frac{\partial u}{\partial x} + w\frac{\partial w}{\partial z} = -\frac{1}{\rho}\frac{\partial p}{\partial x} \tag{3.4.18}$$

$$\frac{\partial w}{\partial t} + u\frac{\partial w}{\partial x} + w\frac{\partial w}{\partial z} = -g - \frac{1}{\rho}\frac{\partial p}{\partial z} \tag{3.4.19}$$

For finite amplitude waves, the convective acceleration terms cannot be set equal to zero as was done for small amplitude waves. The boundary condition on the free surface is non-linear and is given by

$$\frac{\partial p}{\partial t} + u\frac{\partial p}{\partial x} + w\frac{\partial p}{\partial z} = 0 \tag{3.4.20}$$

where the free surface is defined by $p(x, \eta, t) = p_a$ = constant. The search for a solution which satisfied the boundary conditions given above is stated to be a matter of some difficulty. Stokes employed a successive approximation technique wherein the solution is approached via a series of alternating harmonic functions.

Wiegel (1964) notes that in view of the continuity requirement, a velocity potential function ϕ exists which is a function of x, z and t. Moreover, Equations (3.4.18) and (3.4.19) can be rewritten in terms of the integrated form of Bernoulli's equation for unsteady flow which then appears as

$$\frac{p}{\rho} = gz + \frac{\partial\phi}{\partial t} - \frac{1}{2}\left[\left(\frac{\partial\phi}{\partial t}\right)^2 + \left(\frac{\partial\phi}{\partial z}\right)^2\right] \qquad (3.4.18a) \text{ and } (3.4.19a)$$

where

$$u = \frac{\partial\phi}{\partial x}$$

and

$$w = \frac{\partial\phi}{\partial y}$$

In view of the requirement of irrotation, this velocity potential ϕ must satisfy Laplace's equation, or

$$\frac{\partial^2\phi}{\partial x^2} + \frac{\partial^2\phi}{\partial y^2} = 0$$

It is noted that the results given previously for small amplitude waves constitute in effect a first order approximation to the above equations where

$$\phi = -ac\,\frac{\cosh k(z+d)}{\sinh kd}\,\sin(kx - \omega t)$$

For deep water, this reduces to

$$\phi = -ace^{kz}\sin(kx - \omega t).$$

The second order correction must contain a term with argument 2 (kx $-\omega t$). The appropriate expressions for the second order approximations are widely published in the literature and summarized below:

$$\phi = -\, ac \frac{\cosh k(z+d)}{\sinh kd} \sin (kx - \omega t)$$

$$- \frac{3}{8} ka^2 c \frac{\cosh 2k(z+d)}{(\sinh kd)^4} \sin (kx - \omega t) \quad (3.4.21)$$

$$\eta = a \cos (kx - \omega t)$$

$$+ \frac{1}{4} ka^2 \frac{(2+\cosh 2kd)\cosh kd}{(\sinh kd)^3} \cos 2(kx - \omega t)$$

(3.4.22)

$$c^2 = \frac{g}{k} (\tanh kd) \quad (3.4.23$$

$$u = a\omega \frac{\cosh k(z+d)}{\sinh kd} \cos (kx - \omega t)$$

$$+ \frac{3}{4} (ak)^2 c \frac{\cosh 2k (z+d)}{(\sinh kd)^4} \cos(kx - \omega t) \quad (3.4.24)$$

$$w = a\omega \frac{\sinh k(z+d)}{\sinh kd} \sin(kx - \omega t)$$

$$+ \frac{3}{4}(ak)^2 c \frac{\sinh 2k(z-d)}{(\sinh kd)^4} \sin 2(kx-\omega t) \qquad (3.4.25)$$

$$p - p_a = -\rho g z - \frac{\rho(a\omega)^2}{2}\left[\frac{\sinh k(z+d)}{\sinh kd}\right] + \rho g a \frac{\cosh k(z+d)}{\cosh kd}\cos(kx-\omega t)$$

$$+ \frac{\rho (a\omega)^2}{4(\sinh(kd))^2}\left[\frac{3\cosh 2k(z+d)}{(\sinh kd)^2} - 1\right] \cos 2(kx-\omega t) \qquad (3.4.26)$$

For deep or shallow water the above expressions reduce to simpler forms in much the same fashion as those for the Airy linear wave theory. Results for higher order approximations have also been evaluated. For example, Skjelbreia (1959) presents results for Stokes third order approximation and Skjelbreia and Hendrickson (1962) presents the results for Stokes fifth order approximation. The important features to notice and which can be developed from Equations (3.4.21) through (3.4.26) are as follows:

(1) The surface profile is no longer sinusoidal but approaches an elongated or stretched-out trochoid with increasing amplitude.

(2) By substitution of the appropriate phase angles (0° at crest and 180° in trough) in the arguments of Equation (3.4.22), it can be shown that the crest amplitude is greater than the trough amplitude, as was the case for the trochoidal wave.

(3) The forward speed of the particle at the top of the orbit is greater than the backward speed at the bottom of the orbit. When this effect is superposed on the wave speed, it is seen that the particle moves forward over a longer time interval than it moves backward. This results in a non-circular and non-closing particle orbit.

(4) There is a net forward displacement of the particles per cycle which is in agreement with observations. However, the net transport velocity is not $\bar{u}$, since u is zero if averaged over one wave cycle. (This is because the velocity field is non-uniform). The net transport velocity is given by another expression.

(5) The wave celerity, c, to the second order approximation for finite amplitude waves is the same as that given for Airy linear (small amplitude) waves. The higher approximations show that the wave celerity wave length, and particle velocities are slightly greater than those predicted by small amplitude theory.

To sum up our review of Stokian waves, we note that the particle paths are as indicated in the sketch below; that is, the paths are similar to those for small amplitude waves except that they are not closed circles. There is a net transport per wave cycle.

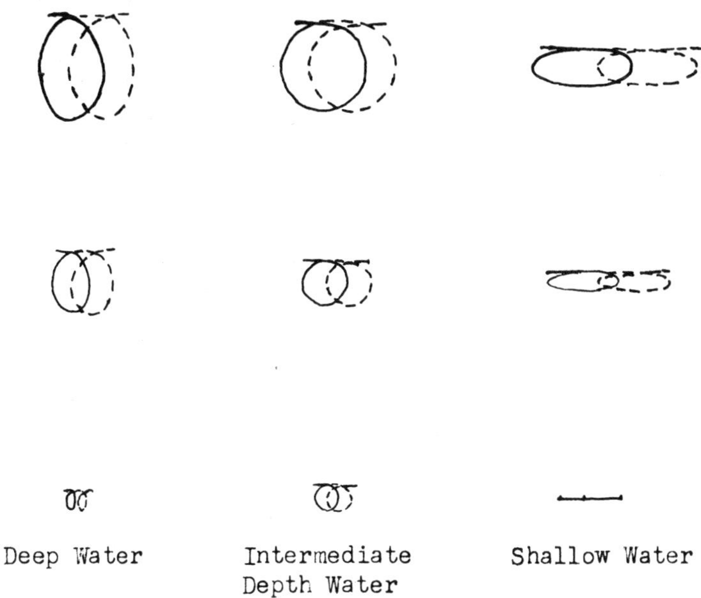

Deep Water | Intermediate Depth Water | Shallow Water

Sketch Illustrating Particle Orbits for Finite Amplitude Waves

The matter of long waves of finite amplitude has been studied by Stokes (1847) and Ursell (1953). The important findings (corresponding to the limits of the Airy linear wave theory in shallow water) are that the quantity $2a(\lambda^2/d^3)$ must be very small for the second order approximation to be valid. Since d/λ must be of the order of 1/20 in shallow water, then a/d must also be small. In order to treat finite amplitude waves in shallow water, then higher order approximations are required. In principle, any wave can be treated in the Stokes manner by inclusion of a large number of harmonic functions in the series representation but this requirement makes the Stokes theory unsuitable and inconvenient. For these particular waves, it is more convenient to employ the cnoidal wave theory which is discussed next, or the solitary wave theory.

Cnoidal Waves: One final long-crested periodic wave is the cnoidal wave first studied by Korteweg and de Vries (1895) and more recently by Masch and Wiegel (1961). The wave parameters are given in terms of elliptic cosine functions from which the term "cnoidal" arises. Although tables have been published by the Council on Wave Research (Masch and Wiegel, 1961) to aid in application of the theory, it is still difficult to use relative to the other wave theories. It represents a second order approximation to the wave equations for long waves based on the assumption that the square of the inclination of the water surface (wave steepness) is small relative to unity. One important feature is that the waves are periodic; thus, the particle paths are orbital being almost closed for small values of H/d where its limiting form is the sinusoidal wave. For values of H/d approaching 0.78, the period becomes infinite and the particle paths degenerate effectively into the translatory type given by solitary waves which is its other limiting form. The range of application of the cnoidal wave theory is from $0.01 \leq H/d \leq 0.78$ and for d/λ ratios less than 1/8, it describes the progression of the periodic waves more accurately than does the theory for Stokes waves. Thus, the cnoidal wave theory seems to bridge the gap between periodic and translatory (solitary) type waves. The pertinent equations are far too complex to present here and the reader is referred to the appropriate references for details on use of the cnoidal theory.

In using the theory, one must be careful to avoid using values of the cnoidal (elliptic cosine) function in the region where these values increase with decreasing values of the argument k, as shown in the sketch on the following page. The reason is because as the argument approaches a value of 1, the cnoidal function yields results comparable to those given by the solitary wave theory. On the other hand, in order for the cnoidal theory to yield results comparable to the Airy linear wave theory, then the cnoidal function should approach a value of zero. For this to occur, an approximation $Cn(k)$ must be made in the region as indicated by the sketch. (It is to be noted that tabulated values of the argument allow for complex numbers but that application to surface gravity water waves restricts values of the argument to real numbers.)

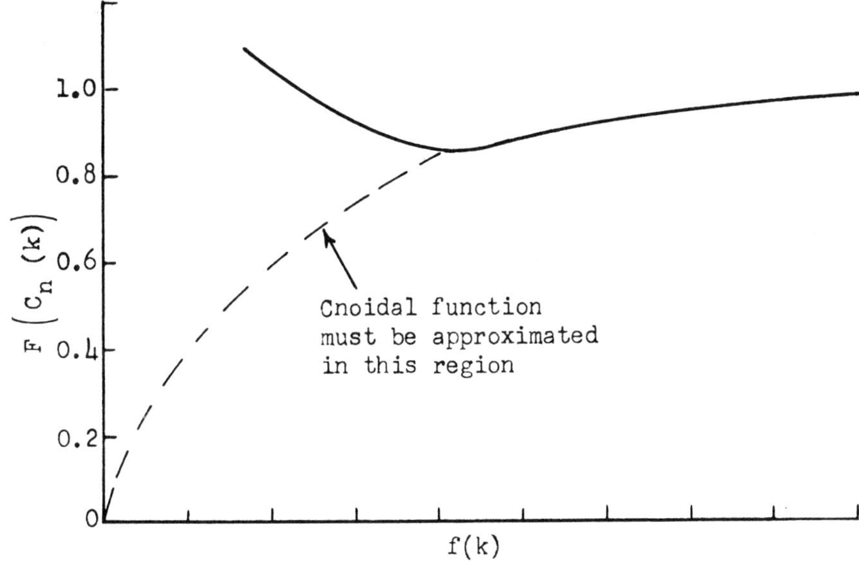

Sketch Illustrating Region Where Cnoidal Function Must be Approximated

Finally, it should be noted that with the availability of large memory high-speed computers along with well developed programs, the effort required to manipulate a large number of complex terms for evaluation of wave parameters via high order solutions of either the Stokian or cnoical wave theory has diminished to a great extent.

This terminates our discussion of periodic waves of which one (Airy infinite small amplitude) was found to be linear while the other three (Stokian finite amplitude, Gerstner-Rankine(Trochoidal) and Cnoidal waves) were found to be non-linear. The Gerstner-Rankine wave was found to be rotational while the remaining types were found to be irrotational. The pressure distribution was found to be hydrostatic for shallow water Airy linear waves and although not illustrated, it is hydrostatic for the cnoidal waves. In all other cases, the pressure distribution is non-hydrostatic. It was noted that the particle orbits are closed circles or ellipses for the Airy linear and Gerstner-Rankine waves and thus no mass transport is present while a small amount of mass transport is present in the Stokian and cnoidal waves. Finally, the solutions for Airy linear and Gerstner-Rankine waves are exact while only the Stokian finite amplitude and cnoidal waves represent only approximate solutions approached via power series.

3.5 CLASSIFICATION OF WAVES - PART II

Our discussion now turns to translatory waves of which the solitary and long waves (tidal, tsunami, flood) are the best known.

Solitary Waves: The solitary wave was first studied in the laboratory by Russell in the 1840's. It is a single, shallow-water wave of apparently permanent form which can travel considerable distances with little attenuation. It can be regarded as a limiting form of periodic shallow-water waves of finite height if the period or wave length is stretched out indefinitely as the relative height H/d is held constant. Thus, in the vicinity of the crest, it has the bell shape shown below.

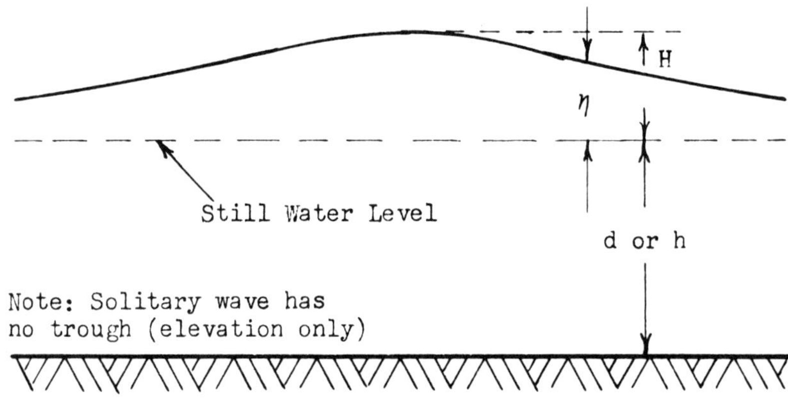

Definition Sketch for Solitary Wave

Russell (1845) found that the wave celerity was given by $c = \sqrt{g(H + h)}$. The usual method of generating solitary waves in the laboratory is by the addition of a finite volume of water at one end of the tank. Most analyses of the solitary wave use the technique of superposing the wave celerity $-c$ onto the field of motion thereby reducing the moving solitary wave to a stationary wave (i.e., the origin of the coordinate axes translates with a velocity $-c$). On this basis, the conditions to be satisfied are those of continuity and zero vorticity. Hence, a field function ϕ must exist which satisfies Laplace's equation along with the boundary condition at the free surface as follows:

$$(u - c)\frac{\partial p}{\partial x} + w\frac{\partial p}{\partial z} = 0 \qquad (3.5.1)$$

$$p\,(x, \eta, t) = p_a \qquad (3.5.2)$$

where p must satisfy the following special form of Bernoulli's equation:

$$\frac{p}{\rho} - uc + \frac{1}{2}(u^2 + w^2) + gz = \text{constant}$$

If the moving horizontal axis is denoted by ξ, where $\xi = x - ct$, then the potential function ϕ and the surface water elevation η and vertical particle velocity w must vanish for large ξ. For these conditions, and provided H/d is less than approximately 0.7, Lamb (1945) gives the following solutions for ϕ and η.

$$\phi = -ch\frac{N}{M}\left[\frac{\sinh M\frac{\xi}{h}}{\cos M\left(1 + \frac{z}{h}\right) + \cosh M\frac{\xi}{h}}\right] \qquad (3.5.3)$$

$$\eta = H\left[\operatorname{sech}\sqrt{\frac{3H}{4(h + H)}}\,\frac{\xi}{h}\right]^2 \qquad (3.5.4)$$

where M and N are dimensionless parameters given by

$$N = \frac{2}{3}\sin^2\left[M\left(1 + \frac{2}{3}\frac{H}{h}\right)\right] \qquad (3.5.5)$$

$$\frac{H}{h} = \frac{N}{M} \tan \left[\frac{1}{2} M \left(1 + \frac{H}{h}\right)\right] \qquad (3.5.6)$$

The particle velocity components u and w are obtained directly from Equation (3.5.3). It is to be noted that η for sinusoidal waves is periodic and changes sign whereas for the solitary wave η is always positive. Thus, the solitary wave is a wave of elevation only. Also, the particle movement is purely translatory describing arcs of parabolas as shown below.

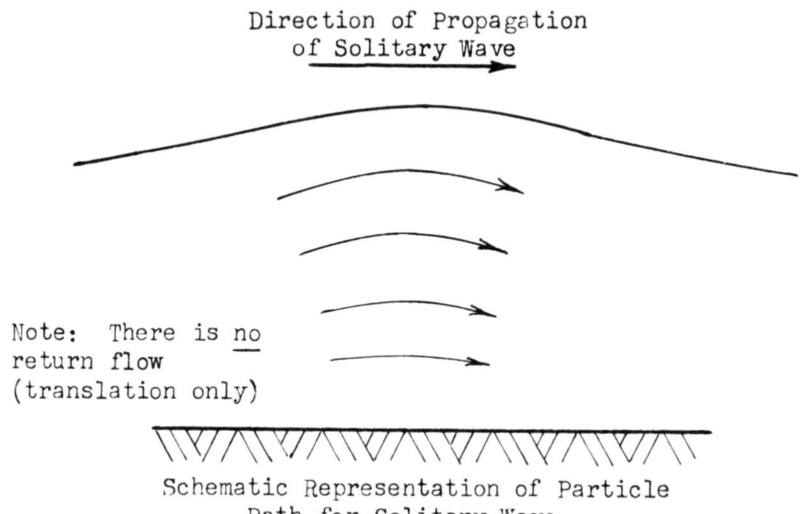

Schematic Representation of Particle Path for Solitary Wave

It has been noted in the literature that the maximum value of H/d is equal to 0.78 which occurs when u = c. Also, the expressions for particle velocity and surface configuration are compatible with those for small amplitude wave theory when H/h approaches zero. Munk (1949) has summarized the work on solitary waves and proposed a modification which is applicable to the problem of periodic waves. The reader should consult the pertinent references for details.

Long Waves: Long waves consist of tidal (including bores), tsunami, and flood waves. Although the governing differential equations to be satisfied are somewhat different for each class of wave, the essential feature is that these waves are present only in very shallow water. The appropriate equations are:

$$\text{Continuity:} \quad \frac{\partial \eta}{\partial t} + \frac{\partial}{\partial x}\left[u\,(d + \eta)\right] = 0 \qquad (3.5.7)$$

Motion, (Eulerian): $\frac{\partial u}{\partial t} + u\frac{\partial u}{\partial x} + g\frac{\partial \eta}{\partial x} = 0$ (3.5.8)

In deriving Equation (3.5.7) from the integrated form of the continuity relation, u has been taken as approximately constant in the vertical plane. In formulating Equation (3.5.8) the assumption has been made that the vertical acceleration of the particles has been taken as negative. The bottom is located at $z = -d$, where d is a function of x (but may be taken as constant). These are non-linear partial differential equations in the independent variables x and t and they must be solved simultaneously to find the temporal and spacial development of some wave form. The effect of amplitude on wave celerity must be included. For simplicity of analysis, this is approximated as

$$c = \sqrt{g\,(d + \eta)} \tag{3.5.9}$$

Interest usually centers on the manner in which the wave form is modified as it propagates over a free surface. Solutions may be approached via the method of characteristics which has been extensively developed for the analysis of analogous problems in gas dynamics.

Solution of Equations (3.5.7), (3.5.8) along with (3.5.9) by the method of characteristics requires that the first two be placed in standard form in the following manner. Adding the terms $\partial d/\partial t$ and $\partial d/\partial x$ to both sides of Equations (3.5.7) and (3.5.8), respectively, we obtain

$$\frac{\partial}{\partial t}(\eta + d) + \frac{\partial}{\partial x}\left[u\,(\eta + d)\right] = \frac{\partial d}{\partial t} \tag{3.5.7a}$$

$$\frac{\partial u}{\partial t} + u\frac{\partial u}{\partial x} + g\frac{\partial}{\partial x}(\eta + d) = g\,\frac{\partial d}{\partial x} \tag{3.5.8a}$$

Taking the partial derivative of the square of the wave celerity c^2, first with respect to the variable x and then with respect to the variable t, we obtain

$$2c \frac{\partial c}{\partial x} = g \frac{\partial}{\partial x} (\eta + d) \tag{3.5.9a}$$

$$2c \frac{\partial c}{\partial t} = g \frac{\partial}{\partial t} (\eta + d) \tag{3.5.9b}$$

Multiplying Equation (3.5.7a) by g/c and then alternatively adding and subtracting Equations (3.5.7a) and (3.5.8a), and finally making the substitutions indicated by Equations (3.5.9a) and (3.5.9b), the following equations result:

$$\frac{\partial}{\partial t} (u + 2c) + (u + c) \frac{\partial}{\partial x} (u + 2c) = g \frac{\partial d}{\partial x} \tag{3.5.10a}$$

$$\frac{\partial}{\partial t} (u - 2c) + (u - c) \frac{\partial}{\partial x} (u - 2c) = g \frac{\partial d}{\partial x} \tag{3.5.10b}$$

or

$$\left[\frac{\partial}{\partial t} + (u \pm c) \frac{\partial}{\partial x} \right] (u \pm 2c) = g \frac{\partial d}{\partial x} \tag{3.5.11}$$

As shown by Stoker (1957), when the bottom is a plane, its slope $\partial d/\partial x$ is a constant, say m/g, and Equation (3.5.11) appears as

$$\left[\frac{\partial}{\partial t} + (u \pm c) \frac{\partial}{\partial x} \right] (u \pm 2c - mt) = 0 \tag{3.5.12}$$

The quantities $u \pm 2c - mt$ are known as the Riemann invariants which remain constant for a point moving through the fluid with velocity $(u \pm c)$. In the $x - t$ plane, two sets of curves C_+ and C_-, termed characteristics occur and are defined by the equations

$$C_+ \; : \; \frac{dx}{dt} = u + c$$

$$C_- \; : \; \frac{dx}{dt} = u - c$$

The Riemann invariants are separately constant along the characteristic curves but the latter are not usually straight lines since the wave celerity c varies with depth d and η, as indicated by Equation (3.5.9). The solution by the method of characteristics is well documented in the literature. The solution to a given problem may be found graphically in the xt plane by using small increments in Δx and Δt for which the characteristics are taken as straight lines.

The application of the method of characteristics is well illustrated by the simple case of a wave propagating in a channel of uniform depth, where $u < c$. At some initial point the wave profile $\eta(x_i, t_i)$ is known and it is required to find the profile $\eta(x,t)$ as the wave passes some point downstream where $x_n = x_o - \Delta x$. It is assumed that the increment is small enough that changes in celerity are insignificant. Therefore, the characteristics are straight lines sloping away from the point x_i, t_i, as shown by the sketch below.

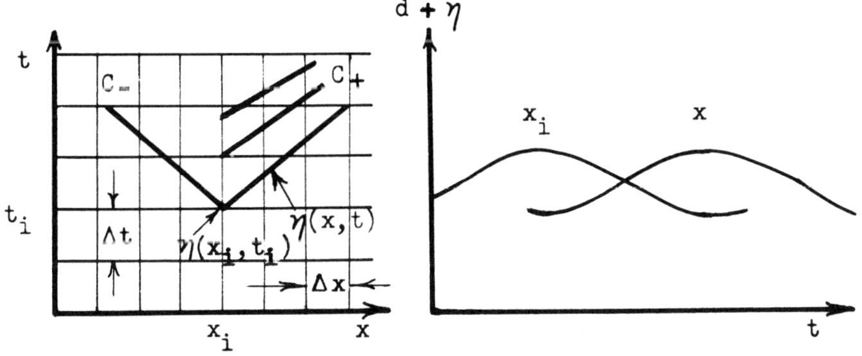

Graphical Portrayal of Method of Characteristics

The C_- characteristics progress upstream in this case since u is less than C thus only the C_+ characteristics are significant. Since the Riemann invariant $u + 2c$ is constant along the C_+ characteristic, surface elevations remain constant being evaluated at intersections of the C_+ characteristic and x_n lines. Because the characteristics have different slopes, the wave changes shape as it progresses.

3.6 CLASSIFICATION OF WAVES - PART III

In addition to the classical wave theories discussed thus far and which fit neatly into the periodic or translatory type of wave, there is another group of wave theories which, for want of a better description, might best be called aperiodic waves. These are the numerical wave theories of rather recent origin, at least compared to the classical wave theories.

As pointed out by Dean (1967) the analytical wave theories discussed up to now satisfy the Nth order boundary conditions on the $(N - 1)^{th}$ order wave form. Two numerical wave theories have been developed which permit the free surface boundary condition for the Nth order solution to be applied to the Nth order water surface. One is termed the stream function theory developed by Dean (1965) and the other is simply a numerical wave theory as developed by Chappelear (1961).

Stream Function Theory: To illustrate briefly the stream function theory, the differential equation to be satisfied within the interior of the wave is Laplace's equation since the fluid is incompressible and irrotational, or in two dimensions

$$\frac{\partial^2 \phi}{\partial x^2} + \frac{\partial^2 \phi}{\partial z^2} = 0, \qquad \frac{\partial^2 \psi}{\partial x^2} + \frac{\partial^2 \psi}{\partial z^2} = 0$$

where

$$u = -\frac{\partial \psi}{\partial z} = -\frac{\partial \phi}{\partial x}$$

$$w = \frac{\partial \psi}{\partial x} = -\frac{\partial \phi}{\partial z}$$

The boundary conditions (see sketch below) which must be satisfied are:

(a) At the bottom where $z = -d$, $w = \frac{\partial\psi}{\partial x} = -\frac{\partial\phi}{\partial z} = 0$ since the boundary is impermeable.

(b) At the free surface, $z = \eta$,

$$\frac{\partial\eta}{\partial t} + u\frac{\partial\eta}{\partial x} = w$$

which states that particles on the free surface remain on the surface.

(c) Also, on the free surface at $z = \eta$, the pressure is uniform meaning that Bernoulli's equation for unsteady flow must be satisfied or

$$\eta + \frac{1}{2g}\left[u^2 + w^2\right] - \frac{1}{g}\,\phi_t = \text{constant.}$$

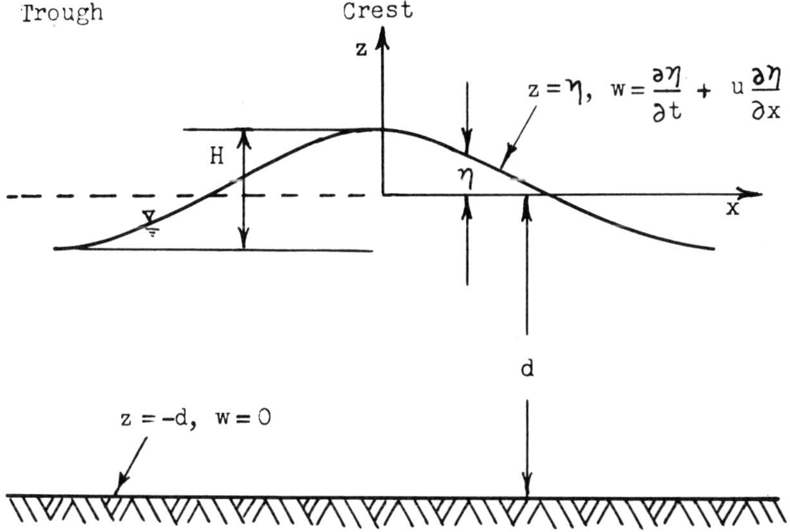

Dean (1967) shows that if the wave can be assumed to propagate without change in form, then a uniform velocity field c (equal to the wave celerity) can be imposed on the field of motion reducing boundary conditions (b) and (c) to the following:

$$\frac{\partial \eta}{\partial t} = \frac{w}{u - c}$$

and

$$\eta + \frac{1}{2g}\left[(u-c)^2 + w^2\right] = \text{constant}$$

which is Bernoulli's equation for steady state conditions.

Numerical Wave Theory: The procedure adopted by Chappelear (1961) requires an iterative process to obtain coefficients satisfying the parameters d/gT^2 and H/gT^2. The boundary conditions to be satisfied are essentially the same as those given by Dean; i.e., the wave profile is a streamline and Bernoulli's equation holds on the surface. (See the definition sketch presented below.)

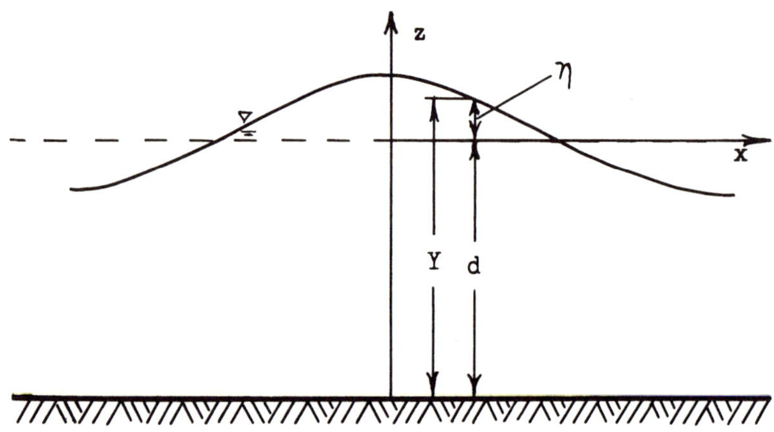

The hydrodynamic flow field is determined by evaluation of the coefficients A_j and B_j such that

$$Y\ (x) = d + \eta = \sum A_j \cos jx \tag{3.6.1}$$

where

$$\begin{bmatrix} u \\ w \end{bmatrix} = \sum_{j=0}^{\eta} B_j \begin{matrix} \cos jx & \cos jy \\ \sin jx & \sinh jy \end{matrix} \tag{3.6.2}$$

and the following boundary conditions are satisfied:

$$u\left[x, Y(x)\right] = \sqrt{\frac{2c - Y(x)}{1 + \left[\frac{dY}{x}/dx\right]^2}} \tag{3.6.3}$$

$$\frac{dY(x)}{dx} = -\frac{w\left[x,\ Y(x)\right]}{u\left[x,\ Y(x)\right]} \tag{3.6.4}$$

First, an estimate is made of the constants

$$\sum_{j=0} A_j \quad \text{and} \quad c$$

and an initial estimate of A_j $(j = 0, 1, 2, \ldots n)$ from any other suitable wave theory. Then the procedure is started by computing $Y(x)$ and $d(Y)/dx$ along an arbitrary network of x_k $(k = 1, 2\ldots)$ spaced along the wave profile between the wave crest and trough. These values are then substituted into Equation (3.6.3) to give

$u\ (x_k, Y(x_k))$ at each network point. With these values of u, a least squares fit of the B_j in Equation (3.6.2) is made and w is computed with the resulting values. Finally, u and w are substituted into Equation (3.6.4) which requires that the wave surface be a streamline. The last operation yields $d(Y(x))/dx$ at each x_k. A least squares fit of the A_j values to the $d(Y(x))/dx$ provides a new estimate of the A_j, and the iteration is carried to the next cycle. Eventually the changes in the A_j and B_j values are below a specified level and the iteration is halted. The parameters

$$\frac{d}{gT^2} = \frac{A_o B_o^2}{2\pi^2}$$

and

$$\frac{H}{gT^2} = \frac{B_o^2}{\pi^2}(A_2 + A_2 + A_5 \quad + \quad \ldots.)$$

are calculated and compared with the desired values. If the departure is too large, a new sequence of iterations is initiated and the whole process repeated. Eventually a solution is obtained which has the particular values of d/gT^2 and H/gT^2 which are under study.

3.7 BREAKING WAVES AND WAVE THEORY SELECTION

In the preceeding material, the reader has become acquainted with the vast amount of effort and references pertaining to the development of theories describing single discrete "waves" or a succession of waves, each of which looks exactly like one another as they progress past a single point. Unfortunately ocean waves never conform to this idealization. As a matter of fact, observations of ocean waves show them to be generally unpredictable both spatially and temporally. Thus, the design engineer is usually faced with two problems. On the one hand, he must determine the largest wave force likely to be induced on a structure with a given probability and, on the other, he must somehow determine what the time history of forces are likely to be during an intense period of wave motion.

The former requirement may be greatly assisted by use of a wave theory selection chart which depicts the domains of the various wave theories. These charts have evolved on the premise that for any given water depth, a particular gravity wave form is explicitly described by a wave height, H, and period, T. From dimensional analysis considerations, two dimensionless parameters H/gT^2 and d/gT^2 can be formed from these wave characteristics. For a given value of d/gT^2, there is a maximum value of H/gT^2 determined by the height H_m at which a wave breaks. The ratio of a specific H/gT^2 and the maximum value H_m/gT^2 is the relative height H/H_m. Alternatively the Ursell parameter ($\eta_o L^2/d^3$) can be used in lieu of the relative height H/H_m. In any event, values of d/gT^2 and H/H_m or $\eta_o L^2/d^3$ can be used to determine which theory may be used for a solution of a given wave fluid flow property.

Unfortunately, there is no unanimity of opinion as to the range of validity of the various wave theories. Wilson (1957) and Dean (1967) have proposed two theoritical criteria resulting in delineation of the various domains shown in Figures 3.7.1 and 3.7.2, respectively. Dean's (1967) study focused on the validity of wave theories in the region near breaking. He found that for near-breaking waves the stream function numerical theory (carried to the fifth order) provided the best fit for $d/T^2 > 0.08$ ft/sec. With respect to the analytical theories only, the best boundary condition fit was provided by the Stokes fifth-order theory in deep water and the cnoidal and Airy wave theories in intermediate and shallow water depths. To reiterate, the problem is complicated by the lack of common basis upon which to evaluate the various wave theories. In deep water, the Stokes theory, if carried to a sufficiently high order of approximation, should be sufficient to describe almost any conceivable wave type. Dean found that in shallow water the boundary condition fit was not an appropriate measure of wave theory validity unless the associated errors were quite small. All hydrodynamically possible non-breaking waves are shown in these two figures. Actually there is some overlapping of the regions but in these regions, the different theories yield approximately similar results. The decision on which theory is to be used requires some intuition and experience or judgement. Generally, the higher order wave theories predicts higher drag force but not necessarily higher inertia force. To be specific, if drag force dominates the loading pattern, then a higher order theory is sometimes desirable.

Usually, uncertainties in the establishment of design wave parameters far outweigh the relatively minor differences given by the various wave theories. There is some variation in the value of the breaking wave parameter H/d, or determination of the maximum wave steepness. The criterion for maximum wave

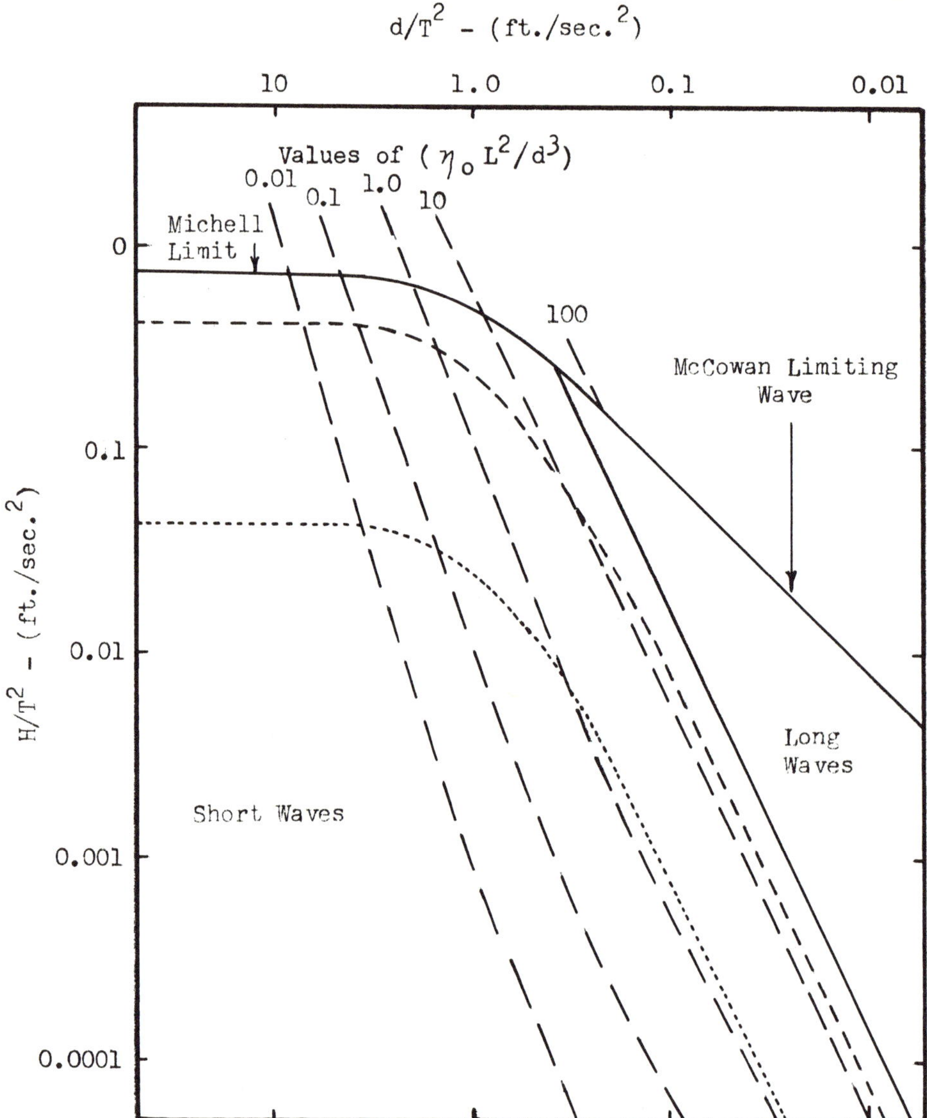

Figure 3.7.1 Representation of the zones of validity of various wave theories. (After Wilson, 1957)

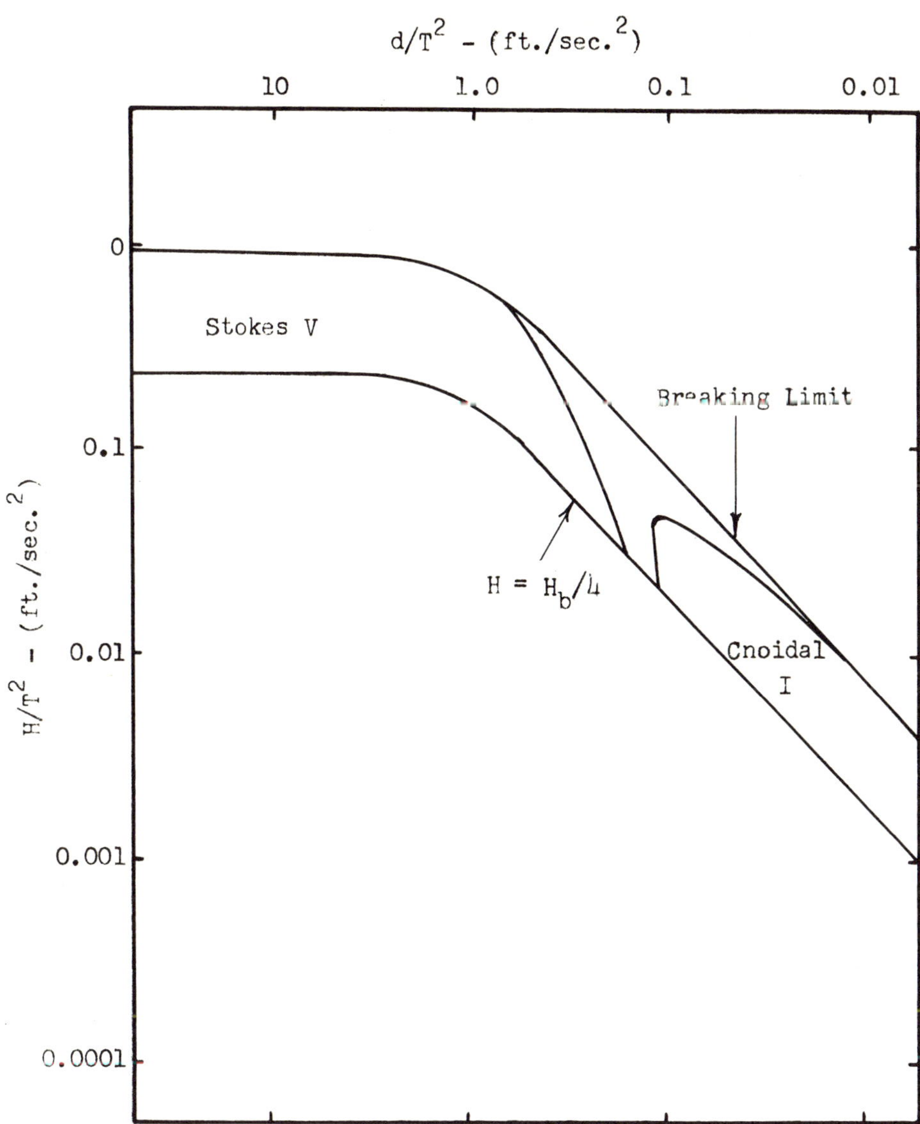

Figure 3.7.2 Regions of validity of wave theories. (After Dean, 1967)

steepness adopted by Stokes and used almost exclusively by others is that "waves break when the particle velocity u at the crest exceeds the wave celerity c". For finite amplitude waves in deep water, this occurs when the angle between tangents to the water surface profile at the crest is 120°. The maximum wave steepness is thus 0.1418 or $H/T^2 = 0.875$ (based on wave length for finite amplitude waves). Michell (1893) and Havelock (1918) also obtained essentially the same result independently. When this criterion is applied to the solitary wave, McCowan (1894) obtained a value of $H/d = 0.78$. Chappelear (1959) has obtained a maximum value of $H/d = 0.87$ for the numerical theory and Laitone (1962) has obtained a minimum value of 0.73 for the cnoidal theory.

3.8 WAVES AS RANDOM PROCESSES

The problem of determining the time history of loading of an ocean structure can be accomplished by means of computations in the time domain representation. This requires some knowledge of information theory. Suppose we have a surface wave record (a wiggly line) as shown in the sketch below. The record can be

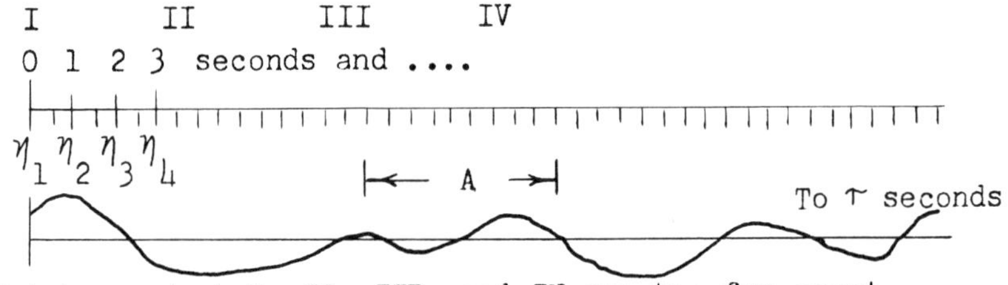

Points read at I, II, III, and IV are too far apart.
One hundred points over interval A are far too many.
The one (1) second marks suggest a proper reading interval.

marked off in equally spaced time intervals such that the sequence of points, when connected by a smooth curve, reproduces the important details of the record. One can readily see that there is an 'optimum' spacing which is neither too fine nor too coarse, and which reproduces the important information in the record. This 'optimum' sampling interval is equal to one-half of the shortest period in the record and is alternatively known (when converted to frequency) as the "Nyquist frequency". What one does is to inspect the record for the highest frequency (or shortest period) and then to read off the record at equally spaced time intervals. The result is a sequence of numbers, $\eta_1, \eta_2, \eta_3 \ldots \eta_N$. For example, if we want to reproduce a 60 minute ocean wave record at a two-second interval, then we would need 1800 numbers. Thus N in this case would be 1800. There is a theorem in information theory that permits a finite record of any length to be written in terms of a Fourier series. In this case, the Fourier series would consist of 1800 terms. In other words, it would be possible to write down a sum of sines and cosines that will reproduce this record exactly at all 1800 points and approximately at intermediate points. Thus, the record of η versus t can be represented in terms of a Fourier series in the form

$$\eta(t) = \sum_{n=1}^{n=\frac{N}{2}} a_n \cos \frac{2\pi nt}{\tau} + \sum_{n=1}^{n=\frac{N}{2}} b_n \sin \frac{2\pi nt}{\tau} \tag{3.8.1}$$

where $2\pi/\tau$ is the frequency corresponding to the record length.

It is seen that the Fourier series will contain all of the harmonics of τ, namely $\tau/1$, $\tau/2$, $\tau/3$ . . . τ/N. If a zero mean is assumed (i.e., $a_o = 0$), then the periods in the record will be 3600, 1800, 900, 450, and so forth down to 4 seconds. Essentially, the series representation, Equation (3.8.1) is a description of an apparent record of finite length, and the discrete frequencies $n(2\pi/\tau)$ contained in the series do not necessarily represent the actual periods even if they exist. As a matter of fact, there may be nothing in the record with a period of 4 seconds. This just means that the a's and b's corresponding to a 4 second period are zero. The coefficients in Equation (3.8.1) can be evaluated from the record of duration τ by the relations

$$a_n = \frac{1}{\tau} \int_{(\tau)} \eta(t) \cos \frac{2\pi nt}{\tau} \, dt \tag{3.8.2}$$

$$b_n = \frac{1}{\tau} \int_{(\tau)} \eta(t) \sin \frac{2\pi nt}{\tau} \, dt \tag{3.8.3}$$

One can show that the total wave energy is proportional to the average of the squares of η (i.e., $\overline{\eta^2}$) and is equal to the sum of the energy content of each of the individual wave components. To do this, use is made of the Parseval theorem which states that for zero mean (i.e., $a_o = 0$), then

$$\frac{1}{\tau}\sum_{n=1}^{n=\frac{N}{2}}\left[a_n^2 + b_n^2\right] = \frac{1}{\tau}\int_{(\tau)}\left[\eta(t)\right]^2 dt \qquad (3.8.4)$$

The right-hand side of Equation (3.8.4) corresponds to the variance and it is seen to be related in some way to the sum of the squares of the Fourier coefficients.

If we define a spectral function $S(\omega)$ such that it has the property that

$$S(\omega)\, d\omega = \psi \qquad (3.8.5)$$

we see that the spectrum is the resolution of the variance, ψ, of $\eta(t)$ into its frequency components.

The character of the function $S(\omega)$ depends, of course, upon the surface wave elevation record $\eta(t)$. As an example, $S(\omega)$, for unidirectional swell, is a narrow banded function centered about the dominant frequency. On the other hand, sea (i.e., waves recorded in the generating area) likely will be characterized by a broad banded spectral function, $S(\omega)$.

Our interest in wave spectra arises from the need to select a suitable or representative loading history (acting on the proposed structure) corresponding to the "design wave spectra". The latter is usually furnished by a specialist as obtained from hindcast or forecast studies. Obviously, it is desirable to select a loading history which represents as nearly as possible the maximum loads or worst possible combination of loads to which the structure will be subjected. It is also obvious that there are many different surface water wave elevation histories which are characterized by the same spectrum. In other words, a one to one correspondence between the surface water wave elevation record and the spectrum does not exist. However, all records which yield identical spectra do have the same statistical properties although details in the records may be widely varying. The suggested procedure, then, is to select a number of water wave elevation records corresponding to the "design spectrum" and to test the proposed structure under these conditions.

It is to be emphasized that the spectrum (as described above, which is based on the concept of superposition) as a means of describing the sea is adequate only for "linear" seas. Various efforts have been made to include non-linear phenomena into the spectral description, but as of this date (1970) these efforts have not been completely successful or widely accepted. (See, for example, Longuet-Higgins (1963)).

For the present, one may consider the linear spectral models to provide adequate descriptions of the sea surface, at least for design purposes.

The important point of this section is that the mathematical model used to represent the sea surface should conform as close as possible to the actual waves on the sea surface and that there are two ways to accomplish this. It may be done in a purely deterministic manner by employing a sufficient number of surface water wave records which result from a mechanical process of harmonic analysis. Or the same wave records may be described by a spectrum. The spectrum description has at least one obvious advantage which is that only 60 or so numbers describe it (the spectral function) completely in the frequency domain. This compares with (1800 x n) or so numbers to describe a representative wave record by a Fourier series. In this case, n is the number of wave records to be considered. It is quite apparent, however, that the Fourier series representation is not affected by non-linear effects which is not quite the case for the spectral representation.

For an introduction to the subject of waves as random processes (and spectral representation) the reader is referred to the well-known article by Munk (1950), and for a more extensive treatment, the works of Kinsman (1965) and Pierson and Neumann (1966) are recommended.

3.9 SELECTION OF DESIGN WAVE SPECTRA

The selection of the wave spectra to be used in evaluating the design of a particular structure depends on, among other things, the risk criteria adopted by the owner of the structure. That is, the owner of the structure must decide what risk can be economically justified for a particular structure located at a given site. This presumes, of course, that all other conditions are satisfied. The most economical overall design might be based on a storm occurring on the average once in 20 years in one case, and once in 100 years in another. A rather complete discussion of risk criteria is given by Borgman (1963). More recently, the reference by Freudenthal and Gaither (1969), presents information in a somewhat different vein on the subject of design criteria. The wave spectra eventually selected for design purposes should be based on the probability of occurrence of "waves" of a given intensity at the site in question. This information is normally developed from hindcast studies which are discussed in Chapter IV. In elementary terms, the procedure is to determine the paths of all known storms which might have caused waves of interest at the site. From the data available on each of the storms (which usually begins with a knowledge of the wind field), estimates can be made of the intensity, duration and direction of the wind generated waves which reach the site. From this meteorological information, then, the "length", "period", and "heights" of waves at the site can be determined. The above approach is not the way it is done in practice but it is the easiest way to conceive what is being done.

In practice, the way to forecast (and/or hindcast) waves is to forecast (and/or hindcast) the spectrum of waves. From the spectrum, one can obtain many of the properties of waves such as heights, average period and average wave length. Thus, one begins with a deep water wave spectrum at a location which is not necessarily the same location as the particular site being investigated. Then, by means of various transformations (which depend on the distance direction, hydrography, spectral and storm properties) one can determine the wave spectrum to be expected at the site under consideration. The well-known work by Pierson, Neumann and James (1955) is highly recommended for an introduction to the subject.

Once the wave spectra (at the site) has been selected for evaluation of a proposed design, one can then generate a number of deterministic (but representative) surface water elevation histories all of which yield the same spectra. Obviously, this is a job for a high speed digital computer, but to my knowledge it has not been done although "in principle" at least, it can be done. It has been done by the "trial and correct" approach, for example, at the Netherlands Ship Model Basin at Wageningen. Here random waves are generated by the snake type wave generators following a preset signal. The time varying surface water elevation at a point is sensed and recorded on an analog tape. A spectral analysis is made of the recording and then depending on the results of the analysis, the signal to be followed by the wave generators is adjusted until the sensed signal yields a spectrum closely matching that (one) desired.

In summary then, selection of the "design wave spectra" is dependent upon (i) risk criteria and (ii) results of the hindcasting studies. In the absence of hindcasting studies, one must be guided by precedence and experience at the site under consideration as well as that acquired at other similar locations.

3.10 REFERENCES

1. Borgman, L. E. (1963). "Risk Criteria," Paper 3607, Journal of the Waterways and Harbors Division, Proceedings, American Society of Civil Engineers, Vol. 89, No. WW3, pages 1-35, August.

2. Chappelear, J. E. (1959). "On the Theory of the Highest Waves," Beach Erosion Board, TM No. 116, July.

3. Chappelear, J. E. (1961). "Direct Numerical Calculation of Wave Properties," Journal of Geophysical Research, AGU, Vol. 66, No. 2, February.

4. Dean, R. G. (1965). "Stream Function Representation of Nonlinear Ocean Waves," Journal of Geophysical Research, Vol. 70, No. 18, September.

5. Dean, R. G. (1967). "Relative Validities of Water Wave Theories," Proceedings of the First Conference on Civil Engineering in the Oceans, American Society of Civil Engineers, San Francisco, California.

6. Freudenthal, A. M. and Gaither, W. S. (1969). "Design Criteria for Fixed Offshore Structures, Preprints, First Offshore Technology Conference, Houston, Texas.

7. Fuchs, R. A. (1952). "On the Theory of Short-Crested Oscillatory Waves," Gravity Waves, NBS Circular 521, November.

8. Havelock, T. H. (1918). "Periodic Irrotational Waves of Finite Height," Proceedings, Royal Society, Vol. 95, Series A.

9. Ippen, A. T., et. al. (1966). Estuary and Coastline Hydrodynamics, McGraw-Hill Book Company.

10. Jeffreys, H. (1924). "On Water Waves Near the Coast," Phil. Mag., Series 6, Vol. 48.

11. Kinsman, B. (1965). Wind Waves, Their Generation and Propagation on the Ocean Surface, Prentice-Hall, Inc.

12. Korteweg, D. J. and de Vries, G. (1895). "On the Change of Form of Long Waves Advancing in a Rectangular Canal, and on a New Type of Long Stationary Waves," Phil. Mag., Series 5, Vol. 39.

13. Korvin-Kroukovsky, B. V. (1961). Theory of Seakeeping, The Society of Naval Architects and Marine Engineers.

14. Laitone, E. V. (1962). "Limiting Conditions for Cnoidal and Stokes' Waves," Journal of Geophysical Research, AGU, Vol. 67, Series 4.

15. Lamb, H. (1945). Hydrodynamics, Dover Publications.

16. Longuet-Higgins, M. S. (1963). "The Effect of Non-Linearities on Statistical Distributions in the Theory of Sea Waves," Journal of Fluid Mechanics, Vol. 17, Part 3.

17. Masch, Frank D., and Wiegel, R. L. (1961). "Cnoidal Waves: Tables of Functions," Council on Wave Research, The Engineering Foundation, Berkeley, California.

18. McCowan, J. (1894). "On the Highest Wave of Permanent Type," London, Edinburgh and Dublin Phil. Mag. and Journal of Science, Vol. 38.

19. Michell, J. H. (1893). "On the Highest Waves in Water," Phil. Mag., Vol. 36, Series 5.

20. Munk, W. H. (1949). "The Solitary Wave and Its Application to Surf Problems," Annals, New York Academy of Sciences, Vol. 51.

21. Munk, W. H. (1950). "Origin and Generation of Waves," Proceedings, First Conference on Coastal Engineering, Council on Wave Research, Berkeley, California.

22. Pierson, W. J., Jr., Neumann, G., and James, R. W. (1955). "Practical Methods for Observing and Forecasting Ocean Waves by Means of Wave Spectra and Statistics," U. S. Navy Hydrographic Office, Pub. No. 603, 284 pp. (Reprinted 1960).

23. Pierson, W. J., Jr., and Neumann, G. (1966). Principles of Physidal Oceanography, Prentice-Hall, Inc.

24. Russell, J. S. (1845). "Report of the Committee on Waves," Meeting of the British Association for Advancement of Science.

25. Skjelbreia, Lars (1959). "Gravity Waves, Stokes' Third Order Approximation; Tables of Functions," Council on Wave Research, The Engineering Foundation, Berkeley, California.

26. Skjelbreia, Lars and Hendrickson, J. (1962). Fifth Order Gravity Wave Theory and Tables of Functions, National Engineering Science Company.

27. Stoker, J. J. (1957). Water Waves, Interscience Publishers, Inc., New York.

28. Stokes, G. G. (1847). "On the Theory of Oscillatory Waves," Transactions, Cambridge Phil. Soc., Vol. 8.

29. Ursell, F. (1953). "The Long Wave Paradox in the Theory of Gravity Waves," Proceedings, Cambridge Phil. Soc., Vol. 49, Part 4.

30. Wiegel, R. L. (1954). "Gravity Waves, Tables of Functions," Council on Wave Research, The Engineering Foundation, Berkeley, California.

31. Wiegel, R. L. (1964). Oceanographical Engineering, Prentice-Hall, Inc.

32. Wilson, B. W. (1957). "Origin and Effects of Long Period Waves in Ports," XIX International Navigational Congress, Section II Comm., Vol. 1.

33. Wilson, B. W. (1957). "Results of Analysis of Wave Force Data-Confused Sea Conditions Round a 30" Test Pile, Gulf of Mexico," Texas A & M Research Foundation, TR No. 55-6 and TR55-7, July.

Chapter IV

WAVE FORECASTING AND HINDCASTING

The purpose of this chapter is not that of providing instruction on how to make wave forecasts and/or hindcasts, but rather to illustrate the basic concepts and procedures which are employed by specialists who regularly furnish forecasting services. The objective here is that of providing sufficient insight into the mechanics of wave forecasting and the underlying premises so that information contained in the forecast and/or hindcast can be used to maximum advantage.

In spite of the intense interest in developing improved techniques of wave forecasting in recent years, it is still a fact that forecasting (and hindcasting, as well) is more of an art than a science. This is because the features (or decisions) which exert the major influence on the results of a forecast are largely subjective. These features include those associated with the process of defining the wind field, (i.e., wind speeds and duration, fetch length and movement).

It should be clear from the material presented up to now that the way to forecast waves is to forecast the spectrum of waves. In addition, since forecasts involve estimation of the wind field and the waves generated therefrom, whereas hindcasts are based on a known wind field, the latter (hindcasts) are generally to be regarded as more accurate than forecasts. This point is often overlooked in many of the discussions of wave forecasting and hindcasting. Thus, wave forecasts are limited in accuracy by the accuracy of the wind field determination, a matter which is beyond the scope of our present interest. We are interested in wind forecasts and/or wind field determinations only insofar as they bear directly on the generation of waves.

Once the space-time wind field has been determined, either subjectively or rigorously, the next step is the determination of the wave spectrum resulting from the imposed wind field acting on the water surface. At present, there are essentially two methods employed for this purpose, both of which are thoroughly covered in the published literature. One is the wave spectra method, sometimes called the P-N-J method (after Pierson, Neumann and James), but to which Baer (1962) has contributed and developed to an advanced stage. The other is the significant wave method, sometimes called the S-M-B method (after Sverdrup, Munk and Bretschneider), and which leads naturally to the space-time wind field method described by Wilson (1961). As to the choice of method, two comments are appropriate. One is that less difference in forecasting results from the same investigator using different methods than from different investigators using the same method. The other is that because of the inaccuracies inherent in wind forecasts, the method employed however refined and sophisticated will give results that are comparable in accuracy only to that of the wind forecast. Thus, the expenditure of a great deal of effort and time is usually not warranted.

Finally, after the wave spectrum is determined, modification of the spectrum must be considered as the waves propagate out of the generating area. This requires some estimate of the effects of wave decay and following and opposing winds as well as the presence of ocean currents. For waves which are generated in shallow water, the effects of bottom friction, refraction, seiching and wave breaking must be taken into account. At the present state of development, very little has been done on wave generation in shallow water.

4.1 GENERATION OF WAVES BY WIND IN DEEP WATER

The generation of waves by wind is best introduced by considering the available theories which attempt to explain how energy is transferred from one fluid media (the air) to another (the ocean). Our picture of the process by which energy is transferred from the atmosphere to the ocean is still incomplete as of this date but at least three different concepts or mathematical models have been advanced to explain this mechanism.

(a) Kelvin (1887)-Helmholtz (1888): This concept, sometimes referred to as the Kelvin-Helmholtz instability criteria, employs a mathematical model consisting of two ideal fluids with a common surface. A complete discussion of this theory is presented by Bretschneider (1966) in the reference by Ippen et al.(1966). Only a brief summary of the important results are included herein. The upper fluid of density ρ_1 has a horizontal velocity U_1 and the lower fluid has a density ρ_2 and a horizontal velocity U_2. Under the small amplitude assumption, Helmholtz found that the oscillation induced by a disturbance on the water surface satisfied the following equation.

$$\rho_1 (U_1 - C)^2 + \rho_2 (U_2 - C)^2 = (\frac{g}{k}) (\rho_2 - \rho_1) \tag{4.1.1}$$

where C is the velocity at which the interface travels. (See the sketch below).

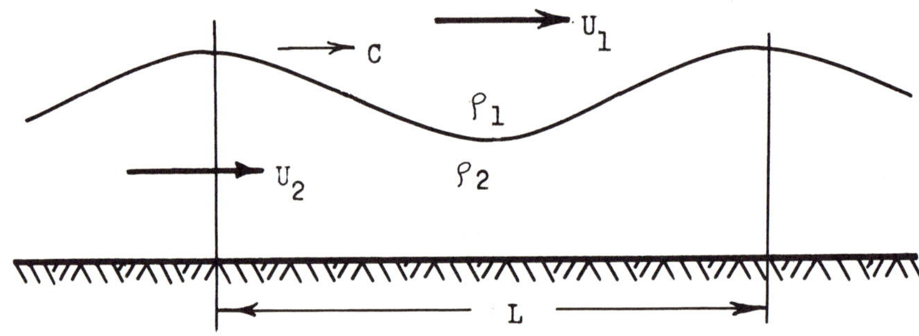

Equation (4.1.1) can be solved explicitly for C resulting in

$$C = \left[\frac{\rho_2 U_2 \quad \rho_1 U_1}{\rho_1 + \rho_2}\right] \pm \left[\frac{g}{k}\left(\frac{\rho_2 - \rho_1}{\rho_2 + \rho_1}\right) - \frac{\rho_1 \rho_2}{(\rho_1 + \rho_2)^2} (U_2 - U_1)^2\right]^{\frac{1}{2}} \tag{4.1.2}$$

or

$$C = \overline{U} + \overline{C} \qquad \text{where} \qquad \overline{U} = \frac{\rho_1 U_1 + \rho_2 U_2}{\rho_1 + \rho_2} \tag{4.1.3}$$

and

$$(\overline{C})^2 = C_o^2 - \frac{\rho_1 \rho_2}{(\rho_1 + \rho_2)^2} (U_2 - U_1)^2 \tag{4.1.4}$$

$$C_o^2 = \frac{g}{k}\left(\frac{1 - \rho_1/\rho_2}{1 + \rho_1/\rho_2}\right)$$

Bretschneider (1966) states that when $(C_o)^2$ is less than the rightmost term in Equation (4.1.4), then $\overline{C}$ is an imaginary quantity "which implies a condition of instability" leading to a continuous transfer of energy from the wind system to the water system and hence to the formation and growth of water waves. Substituting appropriate values of ρ_1/ρ_2 into Equation (4.1.4) one obtains the result that

$$\overline{U} > \frac{1 + \rho_1/\rho_2}{\sqrt{\rho_1/\rho_2}}\, C_o \qquad (4.1.5)$$

or, finally $C_o/U < 1/28$. The quantity C_o/U is referred to in the literature as wave age. This means that waves are unstable when their wave age is less than 1/28 and this instability results in a progressive increase in wave amplitude. By assuming the lowest velocity possible for capillary waves, Kelvin determined that $U > 12.5$ knots for gravity wave generation. The latter is known as the critical wind speed. You will note that the Kelvin-Helmholtz model presupposes an ideal fluid, hence no vorticity is present.

(b) Jeffreys (1925)-Munk (1955)-Miles (1957): The model proposed by Jeffreys and to which Munk and Miles have made contributions is based on the concept that the air pressure variations due to flow about wave profile is the dominant mechanism by which energy is transferred to the waves. Jeffreys considered only the normal pressures (stresses) but Munk took into account the tangential stresses as well.

Jeffreys (1925) model of the problem (see sketch on the next page) assumed that the pressure exerted by the wind of mean velocity U on an element of the wave surface was given by

$$p = s\,\rho_1\,(U_1 - C)^2\,\frac{\partial\eta}{\partial x}$$

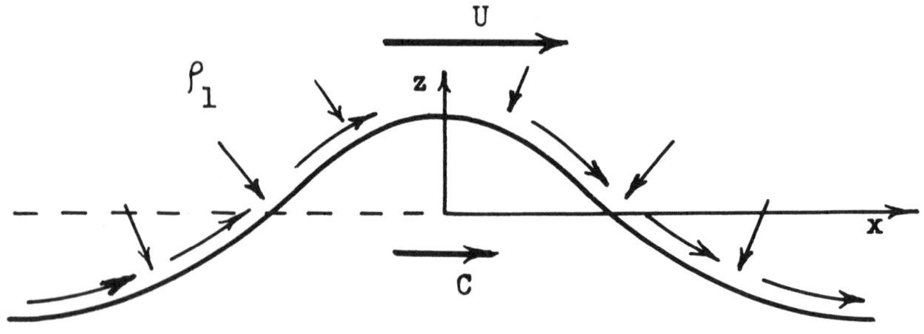

where

ρ_1 = air density,

$(U_1-C)^2$ is the velocity differential squared,

$\frac{\partial \eta}{\partial x}$ is the horizontal area exposed to the wind, and

s is a "sheltering" coefficient intended to account for the fact that part of the wave form is at various times "sheltered" by adjacent waves.

The horizontal component of this pressure, that is, the drag force or wind stress, is given by

$$\tau = s\,\rho_1 \left[(U-C)^2 \left(\frac{\partial \eta}{\partial x}\right)^2\right]_{\text{mean}} \tag{4.1.6}$$

Munk (1955) has noted that C is in reality the horizontal velocity of translation of a small element of water surface which he refers to as the "facet velocity", c. This velocity is related to the wave slope by

$$\frac{\partial \eta}{\partial t} = -\,c\,\frac{\partial \eta}{\partial x}\,.$$

Rewriting the previous expression for pressure and neglecting the square of C (assuming it to be small relative to U) the following relation is obtained:

$$p = s \rho_1 U(U-2C) \frac{\partial \eta}{\partial x} \qquad (4.1.7)$$

Making the appropriate substitution for C, we obtain

$$p = s \rho_1 U(U \frac{\partial \eta}{\partial x} + 2 \frac{\partial \eta}{\partial t}) \qquad (4.1.8)$$

The drag force (form drag) per unit area thus becomes

$$\tau = s \rho_1 U \left[U \frac{\partial \eta}{\partial x} \cdot \frac{\partial \eta}{\partial x} + 2 \frac{\partial \eta}{\partial x} \cdot \frac{\partial \eta}{\partial t} \right]_{mean} \qquad (4.1.9)$$

By averaging the quantities in the brackets by statistical methods and by assuming that the waves are propagating in one direction only, Munk (1955) arrived at the expression that

$$\tau = s \rho_1 U^2 \sigma^2$$

where σ^2 is the mean square slope of the water surface. The latter was taken to be proportional to U so that τ was found to be proportional to U^3. This latter finding seems to be in conflict with the original work of Neumann where τ is found to be proportional to $U^{1.5}$.

The important conclusions obtained from the work of Jeffreys and Munk are (1) the wave slopes are much more significant than elevations, and (2) the small high frequency waves (ripples) are thus very important in accounting for the transfer of energy from the wind system to the waves. Bretschneider (1966) states that based on the work of Jeffreys, the lowest wind speed for gravity wave generation is on the order of 2 knots. Miles (1957) has extended the work by Jeffreys and Munk by assuming the existence of a slightly viscous air in proximity to the sea surface. This results in the formation of a velocity gradient via an elegant mathematical development. The principle result is that the energy transfer is proportional to the curvature of the velocity profile at that point in the profile where the mean air speed is equal to the wave speed.

(c) Eckart (1953) - Phillips (1957): All of the foregoing developments have considered that the wind is constant in velocity in order to account for the transfer of energy. The concept proposed by Eckart (1953) and Phillips (1957) consider that fluctuations in the air pressure due to gusts are the primary cause of wave formation. Phillips theory is of a much more general nature than that postulated by Eckart. Eckart assumed a random distribution of gusts in space and time but specified the diameters and durations of the gusts and that the gusts moved with the wind velocity U. Phillips assumed that the dimensions and lifetime of gusts are also random. He states that "it is found that waves develop most rapidly by means of a resonance mechanism which occurs when a component of the surface pressure distribution moves at the same speed as the free surface wave with the same wave number.... The development of waves is .... considered in two stages in which the elapsed time (from the onset of a turbulent wind) is respectively less or greater than the time of development of the pressure fluctuations." In the initial stage of development, the most prominent waves are ripples of wave length corresponding to the lowest velocity for capillary waves. Phillips continues, "most of the growth of gravity waves occurs in the second or principle stage of development, which continues until the waves grow so high that non-linear effects become important. An expression for wave spectrum is derived from which the following result is obtained:

$$\overline{\eta^2} = \frac{\overline{p^2}\, t}{2(2\rho^2\, U_c g)^{\frac{1}{2}}}$$

where $\overline{\eta^2}$ is the mean square surface displacement, $\overline{p^2}$ is the mean square turbulent pressure on the water surface, t is the elapsed time, U_c is the convection speed of the surface pressure fluctuations, and ρ is the water density. Finally, Phillips found the energy spectrum of the sea surface to be

$$E(\omega) = \alpha g^2 \omega^{-5}$$

where α is a constant, ω the circular frequency, and g the acceleration of gravity. This result is in agreement with some of the more important empirical observations.

4.2 WIND-WAVE RELATIONSHIPS

Although the mechanics of wave generation in the laboratory under carefully controlled conditions are susceptible to rational analysis, actually very little is known about the detailed mechanics of wave generation by winds acting over the ocean surface. It is likely that the mechanics are so complex and the governing parameters so difficult to monitor that empirical procedures similar to and evolving from the present practices will continue to be the most useful approach to the forecasting (hindcasting) of waves. Two widely known semi-empirical methods have evolved for which the significant features are reviewed herein. The first to be discussed is the P-N-J method which grew out of Neumann's 1946-1952 work on wave generation by wind, an excellent summary of which is given in Appendix B of the reference by Korvin-Kroukovsky (1961). (The discussion by Kinsman (1965) is also illuminating). As noted by Korvin-Kroukovsky, this work is outstanding because (1) the experimental data covered a very wide range in wind and sea conditions as well as geographic locations, and (2) a deliberate effort was made to follow as rational an approach as possible.

(a) P-N-J Method: From an examination of a large number of wave records, Neumann found an empirical expression relating the observed wave height H, observed wave period T, and mean wind velocity U. This empirical equation, which represents an upper bound envelope curve for a large number of observations, is,

$$H = C \quad T^2 \exp\left[- \frac{g}{2\pi} \frac{T}{U} \right]^2 \tag{4.2.1}$$

where C is a constant. The reader may consult the reference by Pierson (1955) which clearly indicates this relation and which was obtained by plotting the logarithm of H/T^2 versus $(T/U)^2$. The observed wave height H is not the spectral height but rather is the equivalent height such that

$$[H]^2 = \left[H_{spectrum}\right]^2 dT \tag{4.2.2}$$

Thus, Neumann then assumed that the observable wave height squared, H^2, was in some way proportional to the potential energy averaged over a time period corresponding to the period band for H. Thus, Equation (4.2.1) then becomes

$$[H]^2 = C^2 T^4 \exp\left[-2 \frac{g}{2\pi} \frac{T}{U}\right]^2 \quad (4.2.3)$$

Next, it was noticed that as the period band ΔT (over which the observations were made) becomes smaller, then $[H]$ and $[H]^2$ must also become increasingly smaller. Kinsman (1965) states that the only way for this to happen is for C^2 to become infinitesimal with dT or

$$C^2 \cong C_1 dT \quad (4.2.4)$$

Rewriting Equation (4.2.3), we then have the result that

$$[A(\omega)]^2 d\omega = C_1 dT \cdot T^4 \exp\left[-2 \left(\frac{gT}{2\pi U}\right)^2\right] \quad (4.2.5)$$

The lefthand side of the above expression simply states that the area under the energy spectrum $[A(\omega)]^2$ in the interval $d\omega$ is proportional to the observable wave height squared, $[H]^2$. The symbol T has also acquired a new meaning, being no longer the observed period in the record but instead it is now the spectral period. Finally, the frequency symbol is substituted for its corresponding period symbol to obtain Neumann's spectrum.

$$[A(\omega)]^2 d\omega = C_2 \left(\frac{2}{\pi}\right) \omega^{-6} \exp\left[-2g^2 \omega^{-2} U^{-2}\right] d\omega \quad (4.2.6)$$

In Equation (4.2.6), C_2 is a constant having the value of 51.5 $(\text{feet/sec})^2$. Note that Equation (4.2.6) states that for a fully developed sea, the spectral density is proportional to the inverse 6th power of the frequency, ω. However, if one checks the dimensional units of both the left and righthand sides of the equation, it becomes clear that the relationship is not dimensionally correct. This leads us to the S-M-B method and the spectra derived by Bretschneider (1959).

(b) S-M-B Method: The S-M-B method evolved from the work of Sverdrup and Munk (1947) and Bretschneider (1959) approached it from dimensional analysis considerations. Dimensional analysis techniques serve two purposes. First, it serves as a check on the dimensional accuracy of an equation, for if an equation is not correct dimensionally, then it is not correct physically. Second, dimensional analysis provides a means of organizing physical research in problems for which a theoretical relationship among variables cannot be obtained. Buckingham's π theorem is one of the elements of dimensional theory which states that

(1) If A_1 is a function of a certain number of variables alone, then

$$A_1 = f(A_2, A_3, \ldots\ldots A_n), \text{ or}$$

$$f'(A_1, A_2, A_3, \ldots\ldots A_n) = 0, \text{ and}$$

(2) If each variable can be expressed by one or all of m number of dimensional units, then there will be a n-m number of dimensionless π terms in the functional equation and each term will have m+1 variables, and will be a constant, thus,

$$\phi(\pi_1, \pi_2, \pi_3, \ldots\ldots \pi_{n-m}) = 0$$

The basis for applying Buckingham's π theorem to the problem of wind-wave relationships is founded on the premise that a group of parameters such as wind velocity V, fetch length F, and wind duration t, will adequately describe the wind field. Similarly, one must be willing to accept only wave height and wave period as an adequate description of the wave conditions.

Bretschneider (in a number of publications) has presented the results of applying the π theorem to the problem under consideration for which the following relations are obtained:

$$\frac{C}{U} = \Psi_1 \left[\frac{gF}{U^2}, \frac{gt}{U}\right] \tag{4.2.7}$$

and

$$\frac{gH}{U^2} = \Psi_2 \left[\frac{gF}{U^2}, \frac{gT}{U}\right] \tag{4.2.8}$$

where C is the wave celerity and is equal to $\frac{g}{2\pi} T$ for deep deep water, F is the fetch length, and U is the wind speed. The problem then is to determine the nature of the functions ψ_1 and ψ_2 by the use of wave data. Essentially, what has been done is that a large amount of wind and wave data has been collected, placed in the form of the dimensionless parameters given above, and then plotted. Bretschneider (1968) states that most of the data collected was obtained from the Pacific Ocean but that there should be little difference in wave characteristics for the same wind field from one ocean to another. From this basically empirical approach, Bretschneider (1966) has obtained a relationship for the spectral density for fully developed seas corresponding to Equation (4.2.6), as follows:

$$\left[A(\omega)\right]^2 = \alpha g^2 \omega^{-5} \exp\left[-0.675 \frac{g}{U\omega F_2}\right]^4 \tag{4.2.9}$$

where α and F_2 are dimensional quantities defined as

$$\alpha = 3.437 \frac{F_1^2}{F_2^4}$$

$$F_1 = \frac{g\bar{H}}{U^2} = f_1\left(\frac{gF}{U^2}, \frac{gT}{U}\right)$$

$$F_2 = \frac{g\bar{T}}{2\pi U} = f_2\left(\frac{gF}{U^2}, \frac{gT}{U}\right)$$

The important feature to notice is that the spectral density is proportional to the inverse 5th power of the frequency, a fact which is now generally agreed to be proper.

Detailed step-by-step procedures for making a forecast by the P-N-J method are given by Pierson, Neumann and James (1955). Similar procedures for making a forecast by the S-M-B method are given by Bretschneider (1968). It should be recognized that both of the methods contain certain inherent deficiencies and both are undergoing continual revision.

One problem concerns the delineation of the fetch which has no clear-cut boundaries and furthermore the wind is not constant within the fetch. As one forecaster stated it, "there are no rectangular storms." Therefore, one major contribution would be a method whereby the subjectivity in defining the fetch could be eliminated. An additional problem is that the fetch is continually moving and moving sometimes with a variable direction. Wilson (1961) in extending the S-M-B method has attempted to take this factor into account. This and other deficiencies have been considered by a numerical model proposed and discussed by Baer (1962). It is an extension of the P-N-J method and holds the most promise for obtaining accurate forecasts. Unfortunately, use of a large digital computer is required for its application. This is not the case for the earlier elementary methods and, of course, the cost and accessibility to such a facility may be prohibitive factors.

4.3 TRANSFORMATION OF DEEP WATER WAVE SPECTRA

After the deep water wave spectra has been determined for the generating area, as described in the preceeding paragraphs, it is necessary to account for the various factors which affect its modification. These factors include the effects of:

(1) Decay into swell and subsequent modification of swell.

(2) Interaction of opposing and following wind and waves.

(3) Effects of ocean currents.

(4) Effects of breaking waves and surf.

(5) Effects of refraction and shoaling.

(6) Effects of bottom friction.

Of these factors, the decay into swell has received far the greatest attention. Again two methods of estimating decay into swell have been proposed. The S-M-B method is relatively simple to use and is essentially an empirical method based on a moderate amount of data. Bretschneider (1968) presents various charts and examples on how to use the method. The P-N-J method, on the other hand, is a rational approach based on the selective filtration of the generated wave train (as represented by its spectrum) as it passes through a given point in space. In either case, the pertinent references should be consulted.

Specific studies of items (2) through (6), particularly where simple sinusoidal waves are involved, have been reported in the literature. However, generally their effects on wave spectra have not received a great deal of study.

4.4 REFERENCES

1. Baer, Ledolph (1962). "An Experiment in Numerical Forecasting of Deep Water Ocean Waves," LSMC-801296, Lockheed Missile and Space Company, Sunnyvale, California, June.

2. Bretschneider, C. L. (1959). "Wave Variability and Wave Spectra for Wind-Generated Gravity Waves," Technical Memorandum No. 118, Beach Erosion Board, U. S. Army Corps of Engineers.

3. Bretschneider, C. L. (1966). "Wave Generation by Wind, Deep and Shallow Water," Chapter 3, _Estuary and Coastline Hydrodynamics_, edited by A. T. Ippen, _McGraw-Hill Book Company, Inc._

4. Bretschneider, C. L. (1968).

   (a) "Significant Waves and Wave Spectrum," Fundamentals of Ocean Engineering-Part 7, Ocean Industry, Vol. 3, No. 2, February.

   (b) "Decay of Ocean Waves," Fundamentals of Ocean Engineering-Part 8A, Ocean Industry, Vol. 3, No. 3, March.

   (c) "Decay of Ocean Waves," Fundamentals of Ocean Engineering-Part 8B, Ocean Industry, Vol. 3, No. 4, April.

5. Eckart, C. (1953). "The Generation of Wind Waves on a Water Surface," Journal of Applied Physics, Vol. 24, No. 12, December.

6. Helmholtz, H. (1888). SITZ.-Ber. Akad. Wiss Berlin.

7. Ippen, A. T. et. al. (1966). _Estuary and Coastline Hydrodynamics_, McGraw-Hill Book Company.

8. Jeffreys, H. (1925). "On the Formation of Water Waves by Wind," Proceedings, Royal Society of London, Series A, Vol. 107, April.

9. Kelvin, Lord (1887). "On the Waves Produced by a Single Impulse in Water of Any Depth, or in a Dispersive Medium," Proceedings, Royal Society of London, Series 80, Vol. 42.

10. Kinsman, Blair (1965). Wind Waves, Their Generation and Propagation on the Ocean Surface, Prentice-Hall, Inc., Englewood Cliffs, New Jersey.

11. Korvin-Kroukovsky, B. V. (1961). Theory of Seakeeping, Society of Naval Architects and Marine Engineers, New York.

12. Miles, J. W. (1957). "On the Generation of Surface Waves by Shear Flows," Journal of Fluid Mechanics, Vol. 3, Part 2, November.

13. Munk, W. H. (1955). "Wind Stress on Water: A Hypothesis," Quarterly Journal, Royal Meteorology Society, Vol. 81, No. 349, July.

14. Phillips, O. M. (1957). "On the Generation of Waves in Turbulent Wind," Journal of Fluid Mechanics, Vol. 2, Part 5, July.

15. Pierson, W. J., Jr. (1955). "Wind Generated Gravity Waves," Advances in Geophysics, Vol. 2, Academic Press, New York.

16. Pierson, W. J., Jr., Neumann, G., and James, R. W. (1955). Observing and Forecasting Ocean Waves by Means of Wave Spectra and Statistics, Hydrographic Publication No. 603, U. S. Department of the Navy.

17. Sverdrup, H. U., and Munk, W. H. (1947). Wind, Sea and Swell, Theory of Relations for Forecasting, Hydrographic Office Publication No. 601, U. S. Department of the Navy.

18. Wilson, B. W. (1961). "Deep Water Wave Generation by Moving Wind Systems," Proceedings, American Society of Civil Engineers, Vol. 87, (WW2), May.

PART II
STRUCTURE-FLUID INTERACTION PHENOMENOLOGY

In order to estimate the forces induced on structures immersed in fluid systems, a review of the pertinent fluid-obstruction interaction relationships can be quite illuminating. In this context, "obstruction" and "structure" are used interchangeably. This review is most easily approached via a brief look at the basic concepts of fluid mechanics, some of which were introduced in Part I but which are repeated in this section so that integrity of the presentation is maintained.

Before proceeding, however, recall that in order to design any structure, we must have some idea of the nature of and magnitude of the forces to be imposed on the structure. This requirement has been partially met by the discussion given in Part I where the complex nature of the ocean environment was described as quantitatively as possible. To complete the picture, we need now to acquire a firm grasp of the fundamentals of fluid mechanics. Thereafter, we can proceed hopefully to predict the forces of the wind and water environment on any proposed structure. We need to know not only estimates of the forces but also how far in error these estimates may be. In other words, we need to know the degree of reliability (or accuracy) of these predictions (of forces). Since the forces are extremely complex and the limited number of field measurements certainly verify this, predictions as to the behavior of the structure may be very difficult to achieve. Various techniques of statistical analysis, however, have been employed to characterize the behavior of structures excited similarly, and these techniques are discussed in Part III.

Returning to the ocean environment, which might also be said to be a fluid environment, one of the basic concepts of fluid mechanics is the idea of a continuum. This concept seems to be a natural premise since the engineer deals with matter on a scale which is large relative to the molecular size of the fluid. Thus, any discontinuities in the fluid are obscured by the magnitude of the fluid region being considered. Thus, physical properties such as density, viscosity, temperature, velocity, surface tension and pressure, which are used to characterize fluids, are assumed to vary continuously in the region of interest. Moreover, meaningful numbers can be attached to each one of these properties at every point in the region.

In Part I, we considered briefly the four basic physical laws which govern the motion of fluids and they are repeated here. They are:

(1) Continuity (conservation of mass).

(2) Newton's second law of motion (conservation of momentum).

(3) First law of thermodynamics.

(4) Second law of thermodynamics.

It is to be emphasized that these laws are independent of the nature of the fluid being considered. The last two laws have special applicability to high speed, compressible fluid flow; thus, we will not be concerned with them further.

In addition, there are subsidiary laws which characterize the fluid (or fluids) of interest. These subsidiary laws take various forms. For compressible fluids, for example, there is a perfect gas law, or an equation of state. On the other hand, for incompressible fluids there are various relationships between shearing stress and rate of strain. Some fluids follow the simple Newtonian assumption (which is really a first-order approximation) of a linear proportionality between shearing stress and rate of shear deformation. Hence, they are known as Newtonian fluids. This assumption is used along with the first two basic laws in deriving the well-known Navier-Stokes equations. These equations were introduced and employed in Part I. Thus, the Navier-Stokes equations are strictly limited to a special kind of fluid — a Newtonian fluid. The point to be made is that the engineer must recognize the assumptions and the limitations of the various equations (representing various combinations of the four basic laws and subsidiary relations) as they appear in the literature.

It is instructive also to look at various areas of fluid mechanics, both theoretical and experimental, in order to gain some insight into the interrelationships between the areas and at the same time provide some perspective to the very broad range of problems where a knowledge of fluid mechanics can be applied. For this purpose, Wilson (1968) has prepared the figure which appears on the next page. Note that ocean applications appear as invicid as well as viscous flows, and that the latter includes both laminar and turbulent occurrences.

I. Invicid flow (zero viscosity, ideal fluid, perfect fluid, potential flow)

- Classical hydrodynamic theory; ρ = const. ocean applications
- Compressible, subsonic flow; $\rho \neq$ const.

II. Viscous flow (external flow - around objects; internal flow - in pipes; partly confined flow - open channel)

- Laminar(Newtonian); boundary layer theory, both compressible and incompressible; the Navier-Stokes eqs.; aircraft; lubrication; ocean applications
- Turbulent; Navier-Stokes eqs. are modified; ocean applications
- Non-Newton; modern continuum mechanics; chemical industry

III. Gas Dynamics - aircraft and missile industry

- Subsonic - M close to 1 $\left(M = \frac{\text{fluid velocity}}{\text{sound velocity}}\right)$
- Transonic - $M < 1$ and $M > 1$ on the same air foil.
- Supersonic - $1 < M < 3$
- Hypersonic - $M > 3$
- Rarefied gases - no longer a continuum

IV. Electrically - conducting fluids

- Magnetohydrodynamics I and II above with Maxwell's field equations
- Plasma physics; III above with coupled electrical effects

Areas of Fluid Mechanics (After Wilson, 1968)

Before proceeding to a detailed discussion of the basic laws, a brief consideration of the historical development of fluid mechanics is in order. This is proper since the development of fluid mechanics has taken place over many hundreds of years and a great many individuals have contributed to the body of knowledge now known collectively as fluid mechanics. It is worthwhile to mention the readable and informative work by Rouse and Ince (1957) in this connection. Only some significant milestones are mentioned herein.

The earliest known contribution to fluid mechanics seems to be that due to Archimedes (287-212 B.C.) with his analysis of the buoyancy of submerged bodies. History is replete with many accounts of the Roman engineers who built large conveyance structures for water supply but whose understanding of the role of friction in fluid flow seems to have been limited. The basis for our understanding of modern fluid mechanics originated with Euler (in 1755) who extended Newton's particulate mechanics to continuous fluid matter. The equations of motion attributed to Euler, which neglect all viscous (frictional) contributions are:

$$\frac{\partial u}{\partial t} + u\frac{\partial u}{\partial x} + v\frac{\partial u}{\partial y} + w\frac{\partial u}{\partial z} = -\frac{1}{\rho}\frac{\partial p}{\partial x} + F_x$$

$$\frac{\partial v}{\partial t} + u\frac{\partial v}{\partial x} + v\frac{\partial v}{\partial y} + w\frac{\partial v}{\partial z} = -\frac{1}{\rho}\frac{\partial p}{\partial y} + F_y$$

$$\frac{\partial w}{\partial t} + u\frac{\partial w}{\partial x} + v\frac{\partial w}{\partial y} + w\frac{\partial w}{\partial z} = -\frac{1}{\rho}\frac{\partial p}{\partial z} + F_z$$

where F_x, F_y and F_z are the body forces and the equations are referred to a fixed Cartesian coordinate frame. For the case of steady flow, the temporal acceleration terms

$$\frac{\partial u}{\partial t}, \frac{\partial v}{\partial t} \text{ and } \frac{\partial w}{\partial t}$$

all vanish and if the only body force acting is the gravitational force (as, for example, $F_z = -g$), then the resulting equations can be integrated along a streamline. The result is the well-known Bernoulli equation, which is

$$\frac{V^2}{2g} + \frac{p}{\rho g} + z = \text{constant}$$

where V is the total or absolute velocity. In obtaining this result, the density ρ has been taken as constant but the Eulerian equations are not limited to a constant density. Euler's contributions were not advanced for a number of years during which efforts were directed along empirical lines especially toward developing the 'science of hydraulics.'

In 1827, however, Navier introduced viscous forces into the equations of motion and Stokes did it independently in 1845. As indicated earlier, these equations are restricted to flows in which the shearing stress is related to the velocity gradient by a simple constant of proportionality (i.e., μ, the absolute viscosity). For incompressible flow, these equations are:

$$\frac{\partial u}{\partial t} + u\frac{\partial u}{\partial x} + v\frac{\partial v}{\partial y} + w\frac{\partial w}{\partial z} = -\frac{1}{\rho}\frac{\partial p}{\partial x} + F_x + \frac{\mu}{\rho}\left(\frac{\partial^2 u}{\partial x^2} + \frac{\partial^2 u}{\partial y^2} + \frac{\partial^2 u}{\partial z^2}\right)$$

$$\frac{\partial v}{\partial t} + u\frac{\partial v}{\partial x} + v\frac{\partial v}{\partial y} + w\frac{\partial v}{\partial z} = -\frac{1}{\rho}\frac{\partial p}{\partial y} + F_y + \frac{\mu}{\rho}\left(\frac{\partial^2 v}{\partial x^2} + \frac{\partial^2 v}{\partial y^2} + \frac{\partial^2 v}{\partial z^2}\right)$$

$$\frac{\partial w}{\partial t} + u\frac{\partial w}{\partial x} + v\frac{\partial w}{\partial y} + w\frac{\partial w}{\partial z} = -\frac{1}{\rho}\frac{\partial p}{\partial z} + F_z + \frac{\mu}{\rho}\left(\frac{\partial^2 w}{\partial x^2} + \frac{\partial^2 w}{\partial y^2} + \frac{\partial^2 w}{\partial z^2}\right)$$

These equations differ from Euler's equations only by the term containing the absolute viscosity, μ.

The next major advance in fluid mechanics comes largely as a result of experimentation when Reynolds in 1883 recognized the distinction between laminar and turbulent flow. He eventually extended the Navier-Stokes equations of viscous motion to include the effects of turbulence.

Fluid mechanics made a giant leap forward with the introduction of the boundary layer concept by Prandtl in 1904. Essentially, he realized that the viscous terms in the Navier-Stokes equations dominated the flow pattern in a thin region close to a boundary surface and that outside of this layer, the flow pattern was nearly a potential flow. The modern era of fluid mechanics can be said to have begun with the boundary-layer concept, which brought theory and experiment into close agreement. Recent developments include (1) the incorporation of the subsidiary laws of electromagnetic theory into fluid dynamic theory, and (2) the application of high-speed computers (employing numerical methods) to the solution of fluid mechanics problems. Both of these contributions began in the 1950's. The latter development, which has been pursued most actively by Harlow and his associates with the Los Alamos Scientific Laboratory, has almost brought theory and experiment into complete agreement. The computer simulation of fluid flow problems such as is now possible with the Particle-in-Cell (and Marker-and-Cell) method, is of much greater consequence than is commonly realized and perhaps will be recognized as important a development as the boundary-layer concept in advancing the understanding of fluid flows. As is universally known, it is now possible to simulate flows with the aid of a computer and observe them as they develop just as one might do in the laboratory. This will be treated in some detail in this section. The basic concept will be discussed in Chapter V, and it will be shown how the well-known "Morison et. al." equation is derived. The Marker-and-Cell method will be introduced to the reader in Chapter VI. Chapter VII will then deal with inertial and viscous drag contributions to induced force. Fluid induced vibrations will be discussed in Chapter VIII, and finally, an example of the phenomenological occurrences of importance will be illustrated.

Chapter V

BASIC CONCEPTS

The basic concepts to be illustrated in this chapter are those of continuity and momentum. We must first define or at least make some distinctions between a system, control volume, and a property. A system is an arbitrary volume of mass particles which may be stationary or moving through space. The distinguishing feature of a system is that no mass may be exchanged through its boundaries. A control volume is also an arbitrary volume but in contrast with a system, mass as well as momentum may be exchanged across its boundaries which are referred to as control surfaces. Thus, the control volume may be fixed with respect to the fluid particle motion. The situation is depicted in Figure 5.1.1 below.

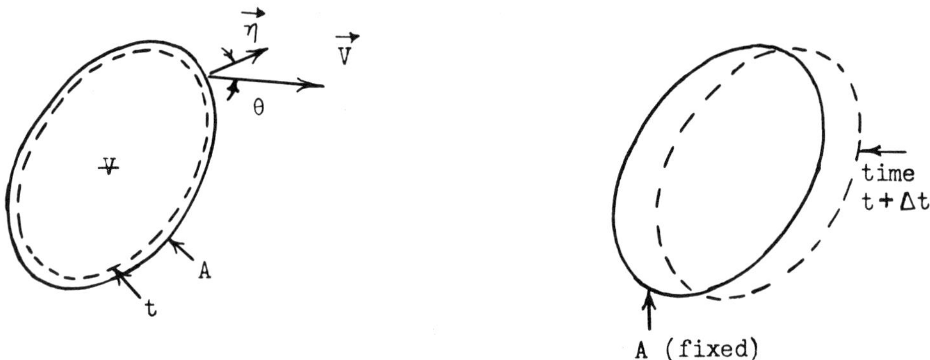

Figure 5.1.1 System of fixed set of mass particles moving through space.

At time t, we have the control surface A which encloses the control volume $\forall$. At time $t+\Delta t$, A, the control surface, has remained stationary while the system of fluid particles has moved to occupy a new volume.

Properties are the characterizing features which describe the state of the system. The fundamental properties with which we are all familiar are mass, length, time and temperature. Usually we will not be concerned with temperature. From these four fundamental properties, we can obtain all of the others, for example, velocity and acceleration. It is intuitively obvious that some of these properties are scalar whereas others are vectors.

5.1 CONSERVATION OF MASS

With this brief introduction, we can now express the concept of mass conservation in integral form as related to the control volume $\forall$. As illustrated by Figure 5.1.1, the mass having left $\forall$ (solid curves) during elapsed time Δt is Δm_{out}; the mass having entered $\forall$ during this time interval is Δm_{in}. At time t, the mass occupying $\forall$ is m_t, while at time $t+\Delta t$, the mass occupying $\forall$ is $m_{t+\Delta t}$.

Conservation of mass within the control volume $\forall$ is expressed by

$$m_t = m_{t+\Delta t} + \Delta m_{out} - \Delta m_{in} \tag{5.1.1}$$

Dividing by Δt, we have

$$\frac{m_{t+\Delta t} - m_t}{\Delta t} = \frac{\Delta m_{in}}{\Delta t} - \frac{\Delta m_{out}}{\Delta t} \tag{5.1.2}$$

At Δt becomes infinitely small, the

$$\frac{\partial}{\partial t}(\text{mass in } \forall) = \frac{\partial m}{\partial t} = \frac{\partial}{\partial t}\int_{\forall} \rho \, d\forall \tag{5.1.3}$$

where ρ is the mass density and $d\forall$ is an elemental volume of $\forall$. Now, since integration is carried out over a fixed volume $\forall$ then Equation (5.1.3) may be written as

$$\frac{\partial m}{\partial t} = \int_{\forall} \frac{\partial \rho}{\partial t}\, d\forall \tag{5.1.4}$$

In a corresponding fashion, the expression for the difference in mass flow rates across the boundaries of the control volume, $\forall$ may be expressed in integral form. Recall that the distance which a particle having velocity $\vec{v}$ traverses in a direction normal to an elemental area dA in time Δt is $(\vec{v}\cos\theta\,\Delta t)$. Here θ is the angle between the outward normal to dA and $\vec{v}$, as shown in Figure 5.1.1. Then,

$$\Delta m_{out} = (\rho \cdot \Delta\,\text{volume})_{out} = \int_{A_{out}} \rho\vec{v}\cos\theta\,\Delta t\, dA$$

Again, dividing by Δt and then letting Δt become infinitesimally small, we have

$$\lim_{\Delta t \to 0} \frac{\Delta m_{in}}{\Delta t} = \int_{A_{out}} \rho\vec{v}\cos\theta\, dA \tag{5.1.5}$$

The corresponding expression for $\Delta m_{in}/\Delta t$ is

$$\lim_{\Delta t \to 0} \frac{\Delta m_{in}}{\Delta t} = -\int_{A_{in}} \rho\vec{v}\cos\theta\, dA \tag{5.1.6}$$

where the negative sign is due to the fact that the $\cos\theta$ is negative for inflow and an outward positive normal to dA.

After the appropriate substitutions, Equation (5.1.2) becomes

$$\int_{\forall} \frac{\partial\rho}{\partial t}\, d\forall = -\int_{A_{in}} \rho\vec{v}\cos\theta\, dA - \int_{A_{out}} \rho\vec{v}\cos\theta\, dA \tag{5.1.7}$$

which usually appears in vector notation as

$$\int_{\forall} \frac{\partial\rho}{\partial t}\, d\forall = -\int_{A} \rho\vec{v}\cdot\vec{n}\, dA \tag{5.1.8}$$

The right-hand side of Equation (5.1.8) is interpreted to mean that integration is carried out over the entire control surface and $\vec{n}$ is the unit normal vector to dA taken positive in the outward sense. This concept contained in Equation (5.1.8) can also be expressed in differential form. To do this, use is made of Gauss' integral theorem which gives the relation between a volume integral and a surface integral as follows:

$$\int_{A} \rho\vec{v}\cdot\vec{n}\, dA = \int_{\forall} \nabla\cdot(\rho\vec{v})\, d\forall \tag{5.1.9}$$

where the "del" or "nabla" symbol stands for the operator. For example, in a fixed cartesian reference frame

$$\nabla = \vec{i}\frac{\partial}{\partial x} + \vec{j}\frac{\partial}{\partial y} + \vec{k}\frac{\partial}{\partial z}$$

Replacing the right-hand side of Equation (5.1.8) with its volume equivalent given by Equation (5.1.9), we have

$$\int_{\forall} \frac{\partial \rho}{\partial t}\, d\forall = -\int_{\forall} \nabla\cdot(\rho\vec{v})\, d\forall \qquad (5.1.10)$$

Both of these terms can be grouped under the same volume integral and one notices that the resulting integrand must be equal to zero for all values of the integral, the final result is given by

$$\nabla\cdot(\rho\vec{v}) + \frac{\partial \rho}{\partial t} = 0 \qquad (5.1.11)$$

Since the medium (ocean water) we are dealing with is incompressible, then $\partial\rho/\partial t$ and we have

$$\nabla\cdot\vec{v} = 0 \qquad (5.1.12)$$

5.2 CONSERVATION OF MOMENTUM

Directing out attention to the system of fluid mass particles given in Figure 5.1.1, and applying Newton's second law to the mass particles contained in the control volume $\forall$ at any time t, gives

$$\vec{F}_{\text{on } A} = \frac{d\vec{M}}{dt} = \frac{d}{dt}\int_{\forall} \rho\vec{v}d\forall \qquad (5.2.1)$$

Here, $\vec{F}$ represents the vector sum of all external forces acting on the control surface A, and $\vec{M}$ is the momentum vector (sum of the mass of each particle ($\rho d\forall$) times (velocity $\vec{v}$ of each particle)). Describing the fixed control volume $\forall$ in a fixed cartesian coordinate frame of reference (x, y, z) we can write the x-component of Equation(5.2.1), as

$$F_x = \frac{dM_x}{dt} = \frac{d}{dt}\int_{\forall} \rho v_x d\forall \tag{5.2.2}$$

which can be written as

$$F_x = \frac{dM_x}{dt} = \lim_{\Delta t \to 0} \frac{(M_x, t+\Delta t)_{sys} - (M_{x,t})_{sys}}{\Delta t} \tag{5.2.3}$$

We can select a time t, such that the x-component of momentum of the system of particles $(M_{xt})_{sys}$ is the same as that for the coincident control volume $(M_{xt})_A$. Thus,

$$(M_{xt})_{sys} = (M_{xt})_A \tag{5.2.4}$$

whence, upon substituting this result in Equation (5.2.3) gives

$$F_x = \lim_{\Delta t \to 0} \frac{(M_{x,t+\Delta t})_{sys} - (M_{xt})_A}{t} \tag{5.2.5}$$

Now, after some elapsed time Δt, the system of particles has been displaced from the control surface A denoted by the solid line and now occupies volume bounded by the dotted line. This system of particles has an x-momentum given by

$$(M_{x,t+\Delta t})_{sys} = (M_{x,t+\Delta t})_A + (\Delta M_x)_{out} - (\Delta M_x)_{in}$$

which, when substituted in Equation (5.2.5), results in

$$F_x = \lim_{\Delta t \to 0} \left\{ \frac{(M_{x,t+\Delta t})_A - (M_{x,t})_A}{\Delta t} + \frac{(\Delta M_x)_{out} - (\Delta M_x)_{in}}{\Delta t} \right\} \quad (5.2.6)$$

Equation (5.2.6) can be further simplified by noting that the first term (R.H.S.) of Equation (5.2.6) represents the time rate of change of momentum in the x-direction across the fixed control surface A. Recall that the mass that crosses an elemental section of A in time Δt is $\rho(|\vec{v}| \cos\theta \; \Delta t) dA$. Thus, the total change in momentum across dA in time Δt, is given by $\rho(|\vec{v}| \cos\theta \; \Delta t) dA \cdot \vec{v}$ and the x-component of this change in momentum is obviously $\rho(|\vec{v}| \cos\theta \; \Delta t) dA \cdot v_x$. This leads to the equality

$$(M_{x,t+\Delta t})_A - (M_{x,t})_A \quad \int_A \rho |\vec{v}| \cos\theta \; \Delta t \; dA \; v_x \quad (5.2.7)$$

Dividing through by Δt, the first term (R.H.S.) of Equation (5.2.6) becomes

$$\lim_{\Delta t \to 0} \left\{ \frac{(M_{x,t+\Delta t})_A - (M_{x,t})_A}{\Delta t} \right\} = \int_A \rho |\vec{v}| \cos\theta \; v_x \; dA$$

$$= \int_A \rho \, v_x \; \vec{v} \cdot \vec{n} \; dA \quad (5.2.8)$$

Here again, $\vec{v} \cdot \vec{n} \quad |\vec{v}| \cos\theta$, where $\vec{n}$ is the unit outward normal to dA, as illustrated in Figure 5.1.1.

The second term in Equation (5.2.6) is

$$\lim_{\Delta t \to 0} \left\{ \frac{(\Delta M_x)_{out} - (\Delta M_x)_{in}}{\Delta t} \right\}_{in\ A} = \left\{ \frac{\partial M_x}{\partial t} \right\}_{in\ A}$$

$$\lim_{\Delta t \to 0} \left\{ \frac{(\Delta M_x)_{out} - (\Delta M_x)_{in}}{\Delta t} \right\}_{in\ A} = \frac{\partial}{\partial t} \int_{\forall} \rho v_x \, d\forall$$

$$= \int_{\forall} \frac{\partial}{\partial t} (\rho v_x) \, d\forall \qquad (5.2.9)$$

Substituting Equation (5.2.8) and (5.2.9) into Equation (5.2.6), we obtain one form of the well-known momentum theorem which states that the sum of all the x directed external forces acting on a fixed control volume $\forall$, whose boundaries are the control surface A, is equal to the net flow rate of momentum through A plus the net rate of change of momentum within $\forall$.

$$F_x = \int_A \rho v_x (\vec{v} \cdot \vec{n}) \, dA + \int_{\forall} \frac{\partial}{\partial t} (\rho v_x) \, d\forall \qquad (5.2.10)$$

Corresponding to Equation (5.2.10) are equations for the y and z directed forces, which are precisely the same except that the subscripts x are changed to y for F_y and to z for F_z. Finally, by multiplying each of the three resulting equations by the unit vectors $(\vec{i}, \vec{j}, \vec{k})$, respectively, and adding them, we obtain the momentum theorem in vector form as

$$\vec{F} = \int_A \rho \vec{v} (\vec{v} \cdot \vec{n}) \, dA + \int_{\forall} \frac{\partial}{\partial t} (\rho \vec{v}) \, dA \qquad (5.2.11)$$

Equations (5.2.11) is valid for mass systems of fluids or solids; viscous or non-viscous, or compressible or incompressible fluids. Although Equation (5.2.11) was derived on the basis of a stationary reference frame and control volume, the theorem is valid for a control volume moving at constant velocity because Newton's laws are also valid in a reference frame moving at constant velocity.

We are now in a position to apply the momentum theorem just derived to find the external force F_T acting per unit length on a vertical cylinder fixed to the ocean floor. The derivation presented below is due to Wilson (1968) and is taken in its entirety from this reference.

"Suppose, at some instant of time t that the horizontal velocity distribution $u = u(y,t)$ is given. Physically reasonable profiles of $F_T = F_T(y,t)$ and $u = u(y,t)$ are sketched in Figure 5.2.1.

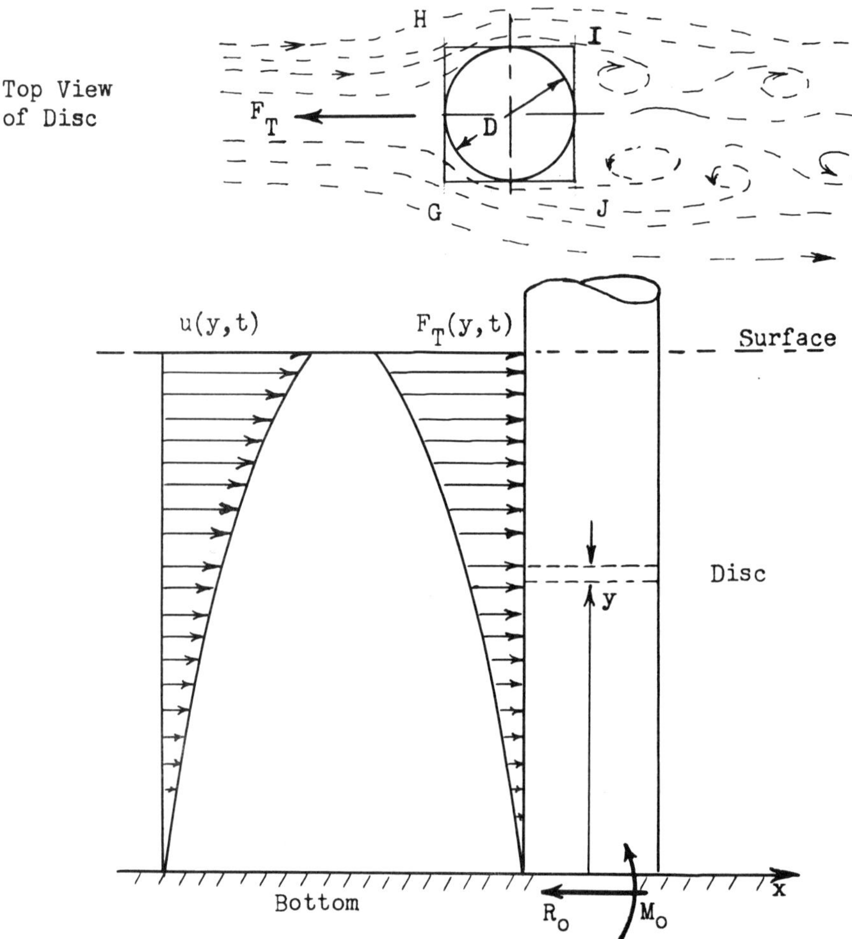

Figure 5.2.1 Illustration of momentum theorem applied to a "circular cylinder control volume".

We recall that F_T exists because the impinging water particles lose momentum to the obstruction, the cylinder. If we assume that the cylinder is entirely rigid, the reaction force R_o and moment M_o at the ocean bottom can be calculated from F_T by statics. Physically, the cylinder carries this distribution of F_T to the bottom through shear.

To find F_T, let us assume the following mathematical model. Isolate a small disc of the cylinder at position y and let the disc be of unit thickness. The horizontal force necessary to hold this disc stationary is F_T as shown in the top view in Figure 5.2.1. A control volume GHIJ and approximate fluid path lines around the cylinder are shown also. From Equation (5.2.10), we have

$$- F_T = \int_A \rho v_x (\vec{v} \cdot \vec{n}) \, dA + \int_{\forall} \frac{\partial}{\partial t} (\rho v_x) \, d\forall$$

Across area GH, we assume

$$\int_{GH} \rho v_x (\vec{v} \cdot \vec{n}) \, dA \approx \rho u (u \cos 180^\circ) D \cdot 1 = - \rho u^2 D$$

Across the areas HI, IJ, and JG, assume that $v_x \approx 0$. Also, assume that $v_x \approx u$ inside $\forall$. Thus,

$$\int_{\forall} \frac{\partial}{\partial t} (\rho v_x) \, d\forall \approx \rho \frac{du}{dt} \cdot D \cdot D \cdot 1$$

Thus, the external force acting on each unit disc of the cylinder is approximated by:

$$F_T \approx - \rho D^2 \frac{du}{dt} + \rho D u^2$$

We now pile all of the unit discs together to form our rigid cylinder. We know that both u and F_T are along $+x$. If u is in the opposite direction, F_T will change sign. To account for a direction change of F_T with a direction change in u, the quantity ρDu^2 is often written $\rho Du|u|$. Also since u slows down or decelerates in $\forall$, $du/dt = -a$, where a is a positive number.

Equality in the expression for F_T is achieved by introducing the empirical constants C_I and C_D, so that

$$F_T = C_I \rho \frac{\pi}{4} D^2 a + C_D \rho \frac{D}{2} u|u|$$

Experimental values of C_I and C_D depend on the type of wave theory used to calculate u and a, but generally are both between 1 and 2.5. These constants are usually correlated as a function of Reynold's Number which is based on a root mean square maximum flow velocity, averaged over depth. It should be emphasized that, if the exact distribution of $\vec{v}$ were known everywhere in the control volume surrounding the cylinder, then the integrals of the momentum theorem, and thus F_T, could be evaluated without the need for these experimental constants."

From this derivation, one notes that the form of F_T can be justified from the law of conservation of momentum for a control volume. Also, the form F_T given by the above derivation corresponds to that proposed by Morison et. al. (1950), which is expressed as

$$F_T = F_I + F_D$$

where F_I is the inertial component of wave force and F_D is the drag component of wave force. More specifically, F_I is the oscillating component of force due to acceleration of water particles and F_D is the 'steady state' drag component. These will be discussed more fully in Chapter VII.

5.3 CIRCULATION AND VORTICITY

In addition to the basic concepts of conservation of mass and momentum, an understanding of fluid-structure interactions is greatly assisted by a brief review of two additional topics. These topics are those of circulation and vorticity to be discussed in this section and of the stream function and velocity potential to be discussed in the following section.

If we select an arbitrary path, s, in the plane of a two-dimensional velocity field, the circulation c around this path is defined by

$$C = \int_S \vec{c} \cdot \vec{ds} \tag{5.3.1}$$

where $\vec{ds}$ is an element of arc of s, and $\vec{c}$ is the total velocity on s. The capital S on the integral sign indicates that the integral must be taken around the closed curve. The situation is depicted in Figure 5.3.1 shown below where a radial vector $\vec{r}$ describes the paths.

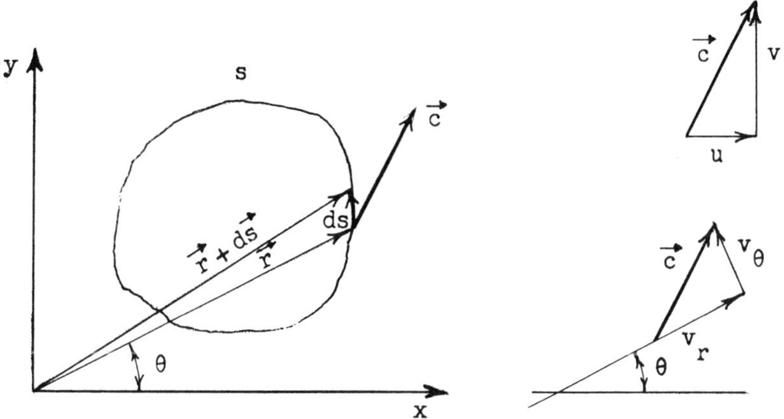

Figure 5.3.1 Definition sketch for circulation and vorticity.

As shown by the figure, an increment in $\vec{r}$ is $\vec{ds}$, which is an element of arc of s directed along s in the x-y plane. The vector $\vec{r}$, measured from a fixed set of cartesian coordinates is

$$\vec{r} = \vec{i}x + \vec{j}y$$

from which, by differentiation

$$\vec{ds} = \vec{dr} = \vec{i}dx + \vec{j}dy \qquad (5.3.2)$$

Recall that the velocity is given in terms of its component as

$$\vec{c} = \vec{i}u + \vec{j}v \qquad (5.3.3)$$

and that the scalar product of the latter two equations yields

$$\vec{c} \cdot \vec{ds} = (\vec{i}u + \vec{j}v) \cdot (\vec{i}dx + \vec{j}dy)$$

$$= udx + vdy$$

Substituting this last result into Equation (5.3.1), we have for plane circulation

$$C = \int_S (udx + vdy) \qquad (5.3.4)$$

We can change the line integral of Equation (5.3.4) into a surface integral by use of Stokes' theorem which, in the two-dimensional formulation, is

$$\int_S (udx + vdy) = \int_{\substack{\text{inside} \\ S}} \left(\frac{\partial v}{\partial x} - \frac{\partial u}{\partial y} \right) dxdy$$

Thus, the circulation around an arbitrary path s can be found if u and v are known everywhere inside the curve s. Making the appropriate substitution gives

$$C = \int_{\substack{\text{inside} \\ s}} \left(\frac{\partial v}{\partial x} - \frac{\partial u}{\partial y} \right) dxdy \tag{5.3.5}$$

In polar coordinates (shown in Figure 5.3.1), this result is

$$C = \int_{\substack{\text{inside} \\ s}} \left(\frac{1}{r} \frac{\partial (r v_\theta)}{\partial r} - \frac{1}{r} \frac{\partial v_r}{\partial \theta} \right) r dr d\theta \tag{5.3.6}$$

which can be obtained by a simple transformation of coordinates.

The integrand in Equations (5.3.5) and (5.3.6) is usually defined as the <u>vorticity</u> or <u>curl</u> of the velocity field in the xy plane. These equations clearly show the connection between the circulation and the vorticity. As an example, when frictional (or viscous) forces are negligible, the motion may be assumed to be approximately irrotational. That is,

$$\text{vorticity} = 0 = \text{curl } \vec{c} = \frac{\partial v}{\partial x} - \frac{\partial u}{\partial y} \tag{5.3.7}$$

Note that the vorticity could have been defined more generally (i.e., along a space curve instead of a plane curve), but the present definition is adequate for our purposes.

5.4 THE STREAM FUNCTION AND VELOCITY POTENTIAL

Suppose that we have a two-dimensional flow (plane motion) that is irrotational (i.e., free from vortices). In cartesian coordinates, this condition is given by Equation (5.3.7) or

$$\frac{\partial v}{\partial x} = \frac{\partial u}{\partial y} = 0 \tag{5.4.1}$$

A stream function $\psi(x,y)$ may be defined in terms of velocities as

$$u = \frac{\partial \psi}{\partial y} \quad ; \quad v = -\frac{\partial \psi}{\partial x} \tag{5.4.2}$$

If the definitions of Equation (5.4.2) are substituted into Equation (5.4.1), we obtain the well-known Laplace equation

$$\frac{\partial^2 \psi}{\partial x^2} + \frac{\partial^2 \psi}{\partial y^2} = 0 \tag{5.4.3}$$

Unique stable solutions to Laplace's equation can be found if either ψ or $\partial\psi/\partial n$ is specified around the closed boundary of the velocity field. The quantity $\partial\psi/\partial n$ is the derivative of ψ normal to the fluid boundary. The lines along which the function $\psi(x,y)$ is constant are called streamlines. Note that irrotationality is a requirement for their existence. The slope of the streamlines are obtained from the relation

$$d\psi = \frac{\partial \psi}{\partial x} dx + \frac{\partial \psi}{\partial y} dy = 0$$

or

$$\left.\frac{dy}{dx}\right|_{\psi=\text{constant}} \approx \frac{v}{u}$$

The equations corresponding to Equations (5.4.1), (5.4.2) and (5.4.3) in polar coordinates are respectively,

$$\frac{1}{r}\frac{\partial(rv_\theta)}{\partial r} - \frac{1}{r}\frac{\partial v_r}{\partial\theta} = 0 \qquad (5.4.4)$$

$$v_r = \frac{1}{r}\frac{\partial\psi}{\partial\theta} \quad ; \quad v_\theta = -\frac{\partial\psi}{\partial r} \qquad (5.4.5)$$

$$\frac{\partial^2\psi}{\partial r^2} + \frac{1}{r}\frac{\partial\psi}{\partial r} + \frac{1}{r^2}\frac{\partial^2\psi}{\partial\theta^2} = 0 \qquad (5.4.6)$$

Next, suppose that the fluid flow is incompressible. The continuity relation in cartesian coordinates is

$$\frac{\partial u}{\partial x} + \frac{\partial v}{\partial y} = 0 \qquad (5.4.7)$$

A velocity potential $\phi = \psi(x,y)$ can now be defined such that

$$u = \frac{\partial\phi}{\partial x} \quad , \quad v = \frac{\partial\phi}{\partial y} \qquad (5.4.8)$$

By substituting the defining Equation (5.4.8) into Equation (5.4.7), the result is again Laplace's equation, or

$$\frac{\partial^2\phi}{\partial x^2} + \frac{\partial^2\phi}{\partial y^2} = 0 \qquad (5.4.9)$$

Again, the equations in polar coordinates corresponding to Equations (5.4.7), (5.4.8) and (5.4.9) are, respectively,

$$\frac{\partial}{\partial r}(r\, v_r) + \frac{\partial v_\theta}{\partial \theta} = 0 \tag{5.4.10}$$

$$v_r = \frac{\partial \phi}{\partial r}, \qquad v_\theta = \frac{1}{r}\frac{\partial \phi}{\partial \theta} \tag{5.4.11}$$

$$\frac{1}{r}\frac{\partial}{\partial r}\left(r\frac{\partial \phi}{\partial r}\right) + \frac{1}{r^2}\frac{\partial^2 \phi}{\partial \theta^2} = 0 \tag{5.4.12}$$

This concludes our discussion of the basic concepts pertinent to fluid-structure interaction phenomenology.

5.5 REFERENCES

1. Morison, J. R., O'Brien, M. P., Johnson, J. W., and Shaaf, S. A. (1950). "The Force Exerted by Surface Waves on Piles," Transactions, American Petroleum Institute, Vol. 189, No. TP 2846.

2. Wilson, J. F. (1958). "Dynamic Analysis of Ocean Structures," Unpublished Lecture Notes, UCLA Short Course, Extension Division, University of California, Los Angeles, California.

Chapter VI

PARTICLE-IN-CELL METHOD

In Chapter V, it was stated that "if the exact distribution of $\vec{v}$ were known everywhere in the control volume surrounding the cylinder, then the integrals of the momentum theorem, and thus F_T could be evaluated without the need for the experimental constants." As a natural extension of this statement, it should be pointed out that there is a way for determining what amounts to the "exact distrubution" of $\vec{v}$. This approach, although not widely used, for a number of reasons which will be pointed out later, is the subject of this chapter. The original method (known as Particle-In-Cell (PIC)) was developed by Harlow (1955) and by Evans and Harlow (1957), and has been greatly extended by Welch, Harlow, Shannon and Daly (1966). The latter treatment is known as the Marker and Cell (MAC) method. For a complete and thorough discussion, the reader should consult the original references. However, because of its importance and potential application to a wide range of problems involving ocean structures, a brief review of the method is given herein.

The major disadvantages of the method are (1) the relatively large amounts of computer time required as compared with traditional engineering approaches, and (2) the difficulty of extending the method generally (for example, to complicated three-dimensional problems with arbitrary boundaries). However, it is felt that with time these disadvantages will be overcome and that eventually the method will be widely used for the solution of many problems involving the flow of fluids. One of the often heard criticisms of the method is that non-symmetries are introduced into the flow patterns arbitrarily (i.e., via computer simulation), but one must remember that nature itself introduces 'arbitrary' non-symmetries. Thus, in a fashion, the method simulates nature. Although the method can be applied to three-dimensional problems, we will restrict our presentation to a two-dimensional, or plane flow treatment, using an incompressible viscous fluid.

6.1 INTRODUCTION

In general, there are two methods for describing the motion of fluid particles. In the Eulerian method, we fix our attention on physical points in space and determine the motion at these points as a function of time. In this method, the coordinate system can be arbitrarily chosen either moving or stationary and then the motion of the flow field is referred to the selected origin.

The other method, known as the Lagrangian method, describes the motion of the fluid particles by literally following them and describing their properties as they move with time. In essence, each particle element is tagged and then changes in the properties of the particle are expressed as functions of time.

Each of the methods has advantages and disadvantages and it is interesting to note that the original references illustrating the PIC method (Harlow (1955) and Evans and Harlow (1957)) employed the Lagrangian notation.

Langley (1959) developed an Eulerian approach in which the properties of the fluid were determined at mesh points of a fixed grid system. In the most recent development, Welch et. al. (1966) developed a mixed Eulerian-Lagrangian scheme in which the advantages of each system of notation are retained. In this scheme, certain properties (i.e., velocity and pressure) are treated in the Eulerian fashion whereas other properties (density, in this case) are treated in the Lagrangian sense.

Subsequent sections of this chapter are devoted to (i) a description of the governing equations and their finite difference equivalents, (ii) a classification of the boundary conditions, (iii) a review of the iteration scheme, and finally (iv) a general step-by-step calculation sequence. Again, the reader is urged to consult the appropriate references for a more complete discussion of the method.

6.2 GOVERNING EQUATIONS OF MOTION AND FINITE DIFFERENCE EQUIVALENTS

Most problems involving fluid motion require the solution of at least two types of equations. These are the basic or fundamental equations of continuity and momentum and one of the subsidiary equations (in this case, the equation relating shear stress and rate of shear deformation). The continuity equation is:

$$\nabla \cdot (\rho \vec{v}) + \frac{\partial p}{\partial t} = 0$$

which for an incompressible fluid becomes

$$\nabla \cdot \vec{v} = 0 \tag{6.2.1}$$

The momentum equation (or Newton's second law of motion applied to a control volume) is:

$$\vec{F} = \int_A \rho \vec{v} \, (\vec{v} \cdot \vec{n}) \, dA + \int_{\forall} \frac{\partial}{\partial t} (\rho \vec{v}) \, d\forall \tag{6.2.2}$$

The equation relating shearing stress and rate of deformation for a linear viscous fluid (Newtonian) is:

$$\tau = \mu \frac{du}{dy} \tag{6.2.3}$$

where μ is the viscosity and is a constant.

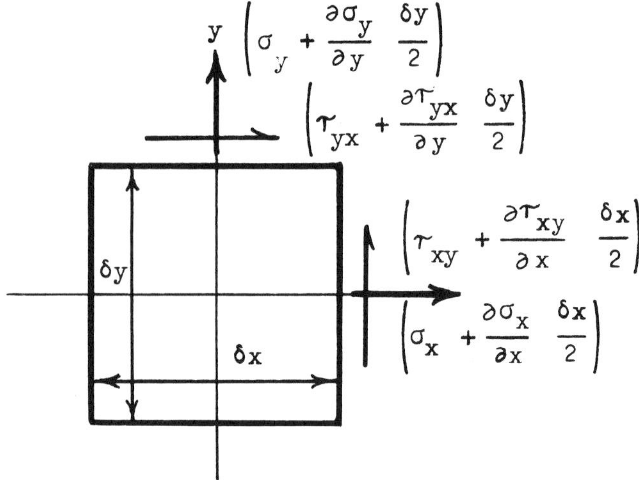

Figure 6.2.1 Definition sketch for fluid element.

By equating all of the external forces acting on the fluid element as shown in Figure 6.2.1, by assuming that the fluid is isotropic and by invoking the Stokes' hypothesis, the Navier-Stokes equation can be obtained. This equation governs the motion of a linear viscous fluid and for an incompressible fluid, it appears in vector form as

$$\rho\frac{\partial \vec{v}}{\partial t} + \rho(\vec{v}\cdot\nabla)\vec{v} = \rho\vec{g} - \nabla p - \mu\nabla^2\vec{v} \qquad (6.2.4)$$

The expressions for the components in plane flow appear as

$$\rho\frac{\partial u}{\partial t} + \rho u\frac{\partial u}{\partial x} + \rho u\frac{\partial v}{\partial y} = \rho g_x - \frac{\partial p}{\partial x} + \mu\left(\frac{\partial^2 u}{\partial x^2} + \frac{\partial^2 u}{\partial y^2}\right) \qquad (6.2.5)$$

$$\rho\frac{\partial v}{\partial t} + \rho u\frac{\partial v}{\partial x} + \rho v\frac{\partial v}{\partial y} = \rho g_y - \frac{\partial p}{\partial y} + \mu\left(\frac{\partial^2 v}{\partial x^2} + \frac{\partial^2 v}{\partial y^2}\right) \qquad (6.2.6)$$

The complete derivation of these equations is given by Schlichting (1960). By making use of the continuity relation for incompressible fluids, Equations (6.2.5) and (6.2.6) may be written as

$$\frac{\partial(\rho u)}{\partial t} + \frac{\partial}{\partial x}(\rho u^2) + \frac{\partial}{\partial y}(\rho uv) = \rho g_x - \frac{\partial p}{\partial x} + 2\left[\frac{\partial}{\partial x}\left(\mu\frac{\partial u}{\partial x}\right) + \frac{\partial}{\partial y}\left(\mu\frac{\partial u}{\partial y}\right)\right] + \frac{\partial}{\partial y}\left[\mu\left(\frac{\partial v}{\partial y} - \frac{\partial u}{\partial y}\right)\right] \qquad (6.2.7)$$

$$\frac{\partial(\rho v)}{\partial t} + \frac{\partial}{\partial x}(\rho uv) + \frac{\partial}{\partial y}(\rho v^2) = \rho g_y - \frac{\partial p}{\partial y} + 2\left[\frac{\partial}{\partial x}\left(\mu\frac{\partial v}{\partial x}\right) + \frac{\partial}{\partial y}\left(\mu\frac{\partial v}{\partial y}\right)\right] + \frac{\partial}{\partial x}\left[\mu\left(\frac{\partial v}{\partial x} - \frac{\partial u}{\partial y}\right)\right] \qquad (6.2.8)$$

It should be noted that for the calculation of the flow field for an incompressible fluid, the equation of state is superfluous. However, many problems of interest involve simultaneous flows of fluids of different densities and the numerical approach employed requires that the location of the interface (if it exists) be determined. Thus, we include the equation of state, which appears as

$$\frac{\partial \rho}{\partial t} + (\vec{v} \cdot \nabla)\rho = 0 \qquad (6.2.9)$$

It is also desirable to scale Equations (6.2.9) (6.2.1) and (6.2.4) in order to maintain some control (and limit) over the magnitudes of the variables. This is easily done by introducing a transformation of variables such that ρ, $\vec{v}$, ∇, g, p and μ appear as the starred quantities ρ^*, $\vec{v}^*$, ∇^*, g^*, p^* and μ^*, respectively. The transformations are $x = \lambda x^*$, $y = \lambda y^*$, $t = (\lambda/Q)t^*$, $\vec{v} = Q\vec{v}^*$, $\rho = R\rho^*$, $p = RQ^2p^*$, and $g^* = \lambda g/Q^2$ and $\mu^* = \mu/\lambda RQ$. The corresponding "del" operator is defined as

$$\nabla^* = \frac{\partial}{\partial x^*} + \frac{\partial}{\partial y^*} .$$

In the material to follow, it will be assumed that the equations have been non-dimensionalized and the starred notation will be dropped from further use.

Before proceeding to place the partial differential equations in finite difference form, it is convenient at this time to introduce the double grid system and notation which is employed to assist in carrying out the numerical procedure. This system, illustrated in Figure 6.2.2, employs two grids.

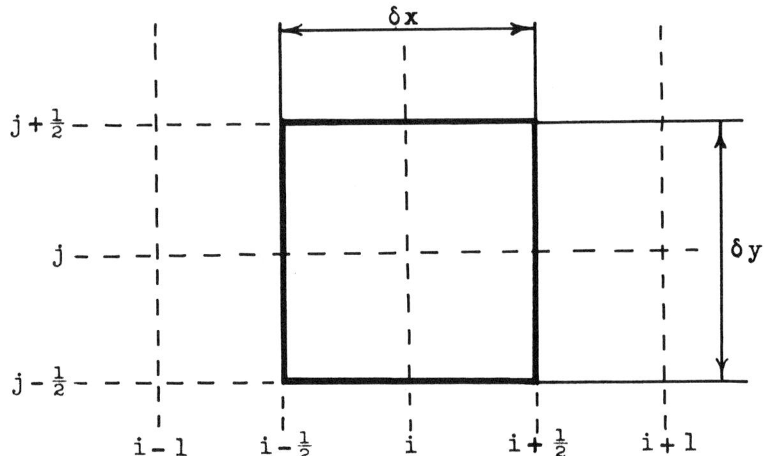

Figure 6.2.2 Illustration of double grid system for Marker and Cell calculations

The grid indicated by the solid lines outlines the cells for which the calculations apply and the dashed grid system is used to keep track of the variables. The cells have the dimensions of δx and δy as indicated; thus, any particular cell can be identified by $i\,\delta x$ and $j\,\delta y$ and properties of the fluid occupying that cell can be further identified by the time parameter $n\,\delta t$ where δt is the elemental time parameter. Although the fluid properties can be associated with any point location relative to the cell, the only one developed is that shown in Figure 6.2.2, where the horizontal velocity u is placed at the midpoints of the sides of the cell, the vertical velocity is placed at the midpoints of the top and bottom of the cell and the pressure, density, and viscosity are placed at the center of the cell.

With this notation, the finite difference equivalents of Equations (6.2.9), (6.2.1), (6.2.7) and (6.2.8) are, respectively,

$$\frac{\rho_{ij}^{n+1}-\rho_{ij}}{\delta t}+\frac{(\rho u)_{i+\frac{1}{2}j}-(\rho u)_{i-\frac{1}{2}j}}{\delta x}+\frac{(\rho v)_{ij+\frac{1}{2}}-(\rho v)_{ij-\frac{1}{2}}}{\delta y}=0,$$

$$\frac{u_{i+\frac{1}{2}j}-u_{i-\frac{1}{2}j}}{\delta x}+\frac{v_{ij+\frac{1}{2}}-v_{ij-\frac{1}{2}}}{\delta y}=0,$$

$$(\rho u)_{i+\frac{1}{2}j}^{n+1}=\xi_{i+\frac{1}{2}j}+\frac{\delta t}{\delta x}\left(p_{ij}-p_{i+1j}\right),$$

where

$$\xi_{i+\frac{1}{2}j}=(\rho u)_{i+\frac{1}{2}j}+\delta t\,\frac{(\rho u^2)_{ij}-(\rho u^2)_{i+1j}}{\delta x}+\frac{(\rho uv)_{i+\frac{1}{2}j-\frac{1}{2}}-(\rho uv)_{i+\frac{1}{2}j+\frac{1}{2}}}{\delta y}$$

$$+\frac{2}{(\delta x)^2}\left[\mu_{i+1j}\left(u_{i+\frac{3}{2}j}-u_{i+\frac{1}{2}j}\right)-\mu_{ij}\left(u_{i+\frac{1}{2}j}-u_{i-\frac{1}{2}j}\right)\right]$$

$$+\frac{1}{\delta y}\mu_{i+\frac{1}{2}j+\frac{1}{2}}\;\frac{u_{i+\frac{1}{2}j+1}-u_{i+\frac{1}{2}j}}{\delta y}+\frac{v_{i+1j+\frac{1}{2}}-v_{ij+\frac{1}{2}}}{\delta x}$$

$$-\mu_{i+\frac{1}{2}j-\frac{1}{2}}\;\frac{u_{i+\frac{1}{2}j}-u_{i+\frac{1}{2}j-1}}{\delta y}+\frac{{}_{i+1j-\frac{1}{2}}\quad{}_{ij-\frac{1}{2}}}{\delta x}+\rho_{i+\frac{1}{2}j}g_x,$$

$$(\rho v)^{n+1}_{ij+\frac{1}{2}}=\zeta_{ij+\frac{1}{2}}+\frac{\delta t}{\delta y}\,(p_{ij}-p_{ij+1}),$$

where

$$\zeta_{ij+\frac{1}{2}}=(\rho v)_{ij+\frac{1}{2}}+\delta t\;\frac{(\rho uv)_{i-\frac{1}{2}j+\frac{1}{2}}-(\rho uv)_{i+\frac{1}{2}j+\frac{1}{2}}}{\delta x}+\frac{(\rho v^2)_{ij}-(\rho v^2)_{ij+1}}{\delta y}$$

$$+\frac{2}{(\delta y)^2}\left[\mu_{ij+1}(v_{ij+\frac{3}{2}}-v_{ij+\frac{1}{2}})-\mu_{ij}(v_{ij+\frac{1}{2}}-v_{ij-\frac{1}{2}})\right]$$

$$+\frac{1}{\delta x}\mu_{i+\frac{1}{2}j+\frac{1}{2}}\;\frac{u_{i+\frac{1}{2}j+1}-u_{i+\frac{1}{2}}}{\delta y}+\frac{v_{i+1j+\frac{1}{2}}-v_{ij+\frac{1}{2}}}{\delta x}$$

$$-\mu_{i-\frac{1}{2}j+\frac{1}{2}}\;\frac{u_{i-\frac{1}{2}j+1}-u_{i-\frac{1}{2}j}}{\delta y}+\frac{v_{ij+\frac{1}{2}}-v_{i-1j+\frac{1}{2}}}{\delta x}+\rho_{ij+\frac{1}{2}}g_y.$$

It is emphasized that the terms involving density, pressure and viscosity are defined at the center of the cell, whereas their use in the above equations require that they be evaluated at the sides of the cell. In general, a single average is employed. For example,

$$\rho_{i+\frac{1}{2}j} = \tfrac{1}{2}(\rho_{i+1j} + \rho_{ij})$$

There are some exceptions to this averaging rule and the reader is again urged to consult the references. For computational purposes, the equations given above are placed in a slightly different format but the equations as given illustrate the principles quite well.

6.3 BOUNDARY EQUATIONS

A very important aspect of the Marker and Cell method is the specification of the boundaries. In general, there are always free surface and solid boundaries. Corresponding to the limits of a physical model, there are limits to the computer model. These limits are called inflow and outflow boundaries.

The equations expressing the conditions of the various boundaries are obtained via consideration of the physical model. In the case of a solid boundary, the normal velocity must vanish. In vector form, this is

$$\vec{v} \cdot \vec{n} = 0 \qquad (6.3.1)$$

where $\vec{n}$ is defined as the unit vector normal to the fluid surface (in this case, also normal to the solid boundary).

The condition existing at a free surface is that the normal and tangential components of the stress must vanish. These conditions are given by:

$$2n_x^2 \mu \frac{\partial u}{\partial x} + 2n_x n_y \mu \left(\frac{\partial u}{\partial y} - \frac{\partial v}{\partial x} \right) + 2n_y^2 \mu \frac{\partial v}{\partial y} = p \qquad (6.3.2)$$

$$2n_x n_y \left(\frac{\partial v}{\partial y} - \frac{\partial u}{\partial x} \right) + (n_x^2 - n_y^2) \left(\frac{\partial u}{\partial y} + \frac{\partial v}{\partial x} \right) = 0 \qquad (6.3.3)$$

where n_x and n_y are the components of the unit normal vector in the ith and jth directions, respectively.

As for the inlet and outlet boundaries, interest centers on the need to match the velocity profile and other fluid properties at these boundaries. For example, at an inlet the velocity may be held constant. At an outlet, however, the density must be held constant in a direction normal to the fluid surface as delineated by the cell boundaries (i.e., what amounts to the control volume). This condition is given by

$$\frac{\partial \rho}{\partial n} = 0 \qquad (6.3.4)$$

In addition, the initial values for the density, velocity, and pressure may be arbitrarily specified.

Once the region of fluid motion has been adequately covered by a grid system, it is necessary to approximate the boundary of the fluid surface in order to identify the boundary cells and thus carry out the numerical calculations. The usual calculation procedure requires the use of adjacent cells in order to evaluate the fluid properties at the cell of interest. This is accomplished by the introduction of image cells reflected about the boundary. Because of the existence of many cell types of either the interior or boundary variety, care must be taken in the computation scheme to select the proper means for evaluating the fluid properties at each step in the calculation procedure. Again, the reader is urged to consult the appropriate references. As an example, corner cells in a rectangular grid require special treatment.

6.4 CALCULATION PROCEDURE

The calculation procedure is a crucial step in application of the Marker and Cell approach. It consists roughly of the following steps:

1. Data is read in and constants are stored.

2. Cells are identified and fluid particles are tagged.

3. Initial particle configurations are calculated and density and viscosity fields are determined.

4. Velocities at the boundary cells are calculated.

5. Pressure fields are calculated via an iterative.

6. Velocity field for the interior cells are calculated and then adjusted as in step 4.

7. Particles are moved to a new position as a result of step 6.

8. New densities and viscosities are calculated as a result of step 7 and return to step 5.

9. Determine final pressures and velocities.

10. Move the particles, go to the next time increment and return to step 4.

The iterative procedure which has been most successful in terms of reduced computer time is the overrelaxation technique. It is to be pointed out that calculation of the velocities (step 6) and pressure (step 5) proceeds in the Eulerian fashion while calculation of the displacements (step 10), densities and viscosities (step 8) proceeds in the Lagrangian sense. Thus, the use of a mixed Eulerian-Lagrangian notation.

With respect to the particle displacement calculations, the velocities used to calculate the new positions is a weighted average of the nearby velocities.

A number of problems have been simulated using this approach and dramatically illustrated using time sequence photographs of the fluid particles as they are displaced on a CRT. Hopefully, advances in computer technology will reach the point where the method will become less expensive and hence more widely used. It seems to lend itself naturally to fluid-structure interaction problems.

6.5 REFERENCES

1. Evans, Martha W. and Harlow, F. H. (1957). "The Particle-in-Cell Method for Hydrodynamic Calculations," Los Alamos Scientific Laboratory, LA-2139, 73p, New Mexico.

2. Harlow, F. H. (1955). "A Machine Calculation Method for Hydrodynamic Problems," Los Alamos Scientific Laboratory, LAMS-1956, 94p, New Mexico.

3. Langley, H. J. (1959). "Methods of Differencing in Eulerian Hydrodynamics," Los Alamos Scientific Laboratory, LAMS-2379, 83p, New Mexico.

4. Schlichting, H. (1960). Boundary Layer Theory, McGraw-Hill Book Company, New York, New York.

5. Welch, J. E., Harlow, F. H., Shannon, J. P., and Daly, B. J. (1966). "The MAC Method," Los Alamos Scientific Laboratory LA-3425, 146 p., New Mexico.

Chapter VII

FLUID-INDUCED FORCES

The traditional method of determining the forces induced on structural elements has been to consider the force to be due to linear combination of an inertial contribution and a drag contribution, or

$$F_T = F_I + F_D$$

Each of these time-varying components has then been formulated in terms of (i) geometrical properties of the structure, (ii) fluid properties describing the flow field and (iii) some "variable constants" which have been determined empirically. A great deal of effort has been devoted to an improved description of the flow field as discussed in Part I of this text and perhaps even more effort has been expended on correlations of the empirically determined "variable constants". The purpose of this chapter is to describe and illustrate how these constants may be estimated and to indicate to what extent these predictions may be in error. To a large degree, these empirically determined constants exhibit a good deal of scatter in the data as published in the pertinent literature but usually (there are some notable exceptions) the scatter is not accounted for except in terms of experimental error in the acquisition of data.

One of the problems in using experimental data to determine the coefficients appearing in the inertial and drag components of the total force is that the relative contribution from each of these components is not known for a particular experiment. Thus, since the coefficients are estimated simultaneously, small errors in one of the terms whose contribution is large may introduce large errors in that component whose contribution is small. Therefore, it is absolutely essential to have some idea of the relative contribution of each of these force components. Dean and Harleman (1966) have prepared such a diagram which is based on a ratio of the maximum inertial and drag forces, but which is presented in terms of the height/diameter and water depth/wave length ratios. For deep water, a convenient rule of thumb criteria is the following expression:

$$\frac{1}{2\pi}\,\frac{C_D}{C_M}\left(\frac{H}{D}\right) = 1$$

Thus, it is immediately seen that as an obstruction becomes large relative to the wave length, the inertia force predominates and vice-versa.

It is instructive at this time to review the fluid phenomenological behavior during unidirectional flow around a circular cylinder, under non-accelerated conditions. We can do this by placing the cylinder in a fluid stream and then letting the free stream velocity increase slowly from zero well up into the completely turbulent region. In this case, the acceleration is negligible and if the Reynolds number is used as the characterizing parameter to delineate regions of flow behavior, the following zones, as widely documented in the literature, can be distinguished.

(a) For $\mathbb{R} < 1$, flow corresponds to Stokian (or creeping motion) flows around spheres and is of little practical interest.

(b) For $1 < \mathbb{R} < 50$, creeping motion flow gradually terminates throughout this range. However, two sub-regimes are noted. At a Reynolds number of about 5, a separation of the laminar boundary layer (LBL) occurs and a stable eddy appears at the rear of the cylinder. The region downstream (or wake) of the cylinder remains laminar.

(c) For $50 < \mathbb{R} < 5000$, the disturbances downstream of the cylinder increase in intensity and well developed vortices appear. The eddy downstream of the cylinder begins to be shed and reappear periodically. The periodic disturbances in the wake are the well-known

Kármán vortex street. The flow pattern in the wake is no longer completely laminar. The non-symmetrical vortex formation gives rise to lateral forces having a frequency defined by the Strouhal number.

(d) For $5000 < \mathbb{R} < 2x10^5$, the wake is entirely turbulent although the boundary layer on the forward portion of the cylinder is laminar. The point of separation moves to the plane of the half-cylinder (or slightly forward of it) and the drag force remains essentially constant.

(e) For $2x10^5 < \mathbb{R} < 5x10^5$, the boundary layer on the forward portion of the cylinder becomes turbulent and the pressure drag decreases so that the net effect is a sharp drop in the total drag force.

(f) For $\mathbb{R} < 5x10^5$, the flow may be considered to be fully turbulent.

For accelerated flow of a circular cylinder, the most noticeable occurrence would be the dramatic increase in the induced force, due to the acceleration of the fluid particles surrounding the cylinder. This aspect will be discussed more fully later.

The phenomenological occurrences which are present in the case of unidirectional flow are also present in the case of oscillatory flow. However, there is one distinction between the two types of flow which introduces many complications to oscillatory flow problems. Briefly, the distinction is that in unidirectional flows the upstream fluid flow pattern is usually well defined and the level of turbulence is low, whereas, in oscillatory flows the "upstream" flow pattern is contaminated by the presence of the preceeding flows and is usually agitated so that the turbulence level is relatively high. This disturbance of the upstream flow pattern is a direct result of the fluid separation and subsequent shedding of eddies before the reversal of the flow.

This increase in the turbulence level upstream of the obstruction is of primary importance in the case of multiple members such as groups of piles. It is well-known, for example, that an increase in the turbulence level of the oncoming flow shifts (lowers) the Reynolds number at which the transition from a laminar to a turbulent boundary layer occurs. At this critical Reynolds number, the point of separation is shifted in the downstream direction toward the rear of the cylinder and decreases the wake size thereby leading to a sharp drop in the drag coefficient. It is this difference between unidirectional and oscillatory flow which is most pronounced in the presence of multiple obstructions that also probably accounts for a good deal of the scatter reported in the empirical studies referred to previously.

Thus, in addition to the phenomenological occurrences of interest to single members, the following effects are most important where groups of members are involved: (1) sheltering and orientation, (2) solidification and (3) synchronization.

Returning for a moment to the formulation of the induced force on a single member, some studies have considered that the total induced force may be correlated with a single resistance coefficient but we will use the relation given previously (i.e., the linear combination of inertial-like and drag-related forces) as the main focus of this chapter. In the following sections we will be concerned with a more complete discussion of inertial and drag coefficients. Chapter VIII treats the ancillary topics of group effects including laterally induced forces and impact type forces.

7.1 INERTIAL COEFFICIENTS

The inertial force component of the total wave force may be visualized most easily by considering an obstruction that is being accelerated in otherwise still fluid. The total force required to accelerate the obstruction depends not only upon the mass of the obstruction but also upon the mass of the fluid particles in the vicinity of the obstruction. Thus, the obstruction behaves as though an added mass of fluid were moving with it. As a matter of fact, it is the fluid-obstruction system that is being accelerated. This relation is usually expressed as:

$$F_I = (M + M') a$$

where M is the mass of the obstruction, M' is the added mass of the fluid and a is the acceleration. Most authors refer to the sum of $M + M'$ as the virtual mass. In terms of a fixed obstruction being excited by oscillatory water waves, the acceleration a is replaced by the temporal water particle acceleration $\dot{u}$ that would exist in the absence of the obstruction. (The convective acceleration terms are conveniently ignored). For a circular cylindrical pile, the inertial force term is expressed as

$$F_I = C_M \frac{\rho}{g} \frac{\pi}{4} D^2 \dot{u}$$

where F_I is the force/unit length, D is the cylinder diameter, ρ is the density of sea water, and $\dot{u}$ is the time-varying particle acceleration and C_M is the added mass coefficient.

Since it is difficult to measure the acceleration-dependent and velocity-dependent terms separately, the effects of viscous drag and separation pressure drag on the added mass is not precisely known although, for most fixed ocean structures in the range of wave particle velocities encountered, the effect is not significant. This is not the case, however, for nearby boundaries such as might occur for a pipeline in proximity to the sea floor, or a structural member near the free surface. The latter problem has been studied by Murtha (1964) who found that the effects of the free surface in decreasing the added mass of a long cylinder were confined to a rather narrow region (two diameters) near the surface. In the case of ship hulls, the effect of the free surface on the added mass of the vessel has been intensively studied and documented in numerous references. Both viscous and boundary effects on the virtual mass have been studied by Ackermann and Arbhabhirama (1964).

Although the added mass concept is useful in visualizing the fluid dynamic behavior, it is not rewarding unless the coefficients for the particular body shape and fluid motion are known or at least can be evaluated. These coefficients have been evaluated for a number of body shapes. The method of evaluation has usually proceeded via ideal fluid considerations—thus, effects of viscosity have been neglected. It is to be emphasized that for a general body having six (6) degrees of freedom, there appear twenty-one (21) inertia (or added mass) coefficients. For a circular cylinder, we are interested in most cases in the relative motion of the cylinder normal to its longitudinal axis. Thus, the added mass coefficients due to rotational (which is zero) and longitudinal motion are of little interest.

Added mass coefficients for a number of bodies in unidirectional flow have been determined theoretically by Lamb (1932). Added mass coefficients for two-dimensional rectangular bodies have been evaluated theoretically by Riabouchinsky (1920). These results are shown in Figure 7.1.1.

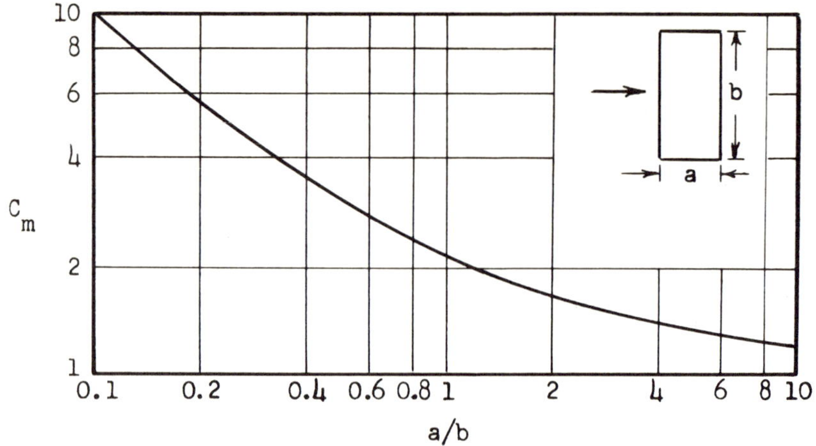

Figure 7.1.1 Added mass coefficient, C_m, for two-dimensional flow past rectangular bodies. (After Riabouchinsky, 1920)

Experimental determinations of added mass coefficients have been reported by Stelson and Mavis (1957) and by Zienkiewicz and Nath (1964). The latter study made use of an electrical analogy method, the results of which agreed very well with theory. Limited experimental results are also given by Brater, McNown and Stair (1958), for submerged barge-like structures subjected to oscillatory water waves. They found that the value of the added mass coefficient was not only a function of body geometry, but also a function of the Strouhal number (fL/U) or perhaps more correctly the Keulegan-Carpenter parameter (UT/D).

In a more recent investigation, Jen (1968) has studied the variation of inertial forces on a circular cylinder in an irregular wave environment. The scatter in the data indicates not only the dependence of the inertial coefficient on parameters other than body geometry, but what is probably more important is the influence of the level of turbulence in the "upstream" oncoming fluid. For periodic waves, the value reported was $C_M = 2.04$ which is very close to that obtained from ideal fluid flow. For irregular waves, a least squares fit yielded values of 2.20 and 2.08 for two test runs. The discrepancy between the highest value obtained and that obtained from periodic waves is thus only slightly less than 8%. Utilizing available ocean data, Brown and Borgman (1967) give various methods of estimating C_M and the statistical distribution of C_M.

In summary, the virtual mass coefficient depends on the geometry of the body and surrounding boundaries (both solid and free surface) as well as the nature of the motion (including its history) and the properties of the fluid medium. There are a number of references indicating various methods of estimating inertial coefficients from theory, reduced scale model studies, analogs and some limited field studies. Results of some of these studies will be indicated in the next section. The variation in inertial coefficients due to group effects (solidification) will also be discussed later although in essence, it is a special case of boundary or proximity influences.

7.2 DRAG COEFFICIENTS

The drag contribution to the total force consists essentially of two parts, skin friction (or viscous) drag and pressure (or form) drag. The latter also includes what is known as separation drag. For unidirectional flow, the dependence of the total drag force (induced on a completely submerged obstruction by a fluid stream) on the Reynolds number and the relative roughness of the boundary surface is well-known. The results for a large number of body geometries appear in a number of important references (for example, see Hoerner (1965)). Unfortunately, such is not the case for oscillatory flow and the limited number of experimental investigations certainly indicate a large amount of scatter in the data.

Just as in the case for the inertial force, the presence of a boundary influences the magnitude of the pressure drag (even for unidirectional flow) and thus the relative contribution of the pressure drag to the total drag. This is well illustrated in the case of a large structure (ship) near a free surface. Here the pressure (or form) drag is manifested in the form of waves induced by and radiating from the vessel. This is the well-known wave drag due to wave generation by the body and in the case of ships, the wave drag is assumed to account by far for the major portion of the pressure drag. This fact is an important aspect in the prediction of the drag of large obstructions (such as ships) from model tests since the wave drag is proportional to the Froude number.

In this context, the Froude number is defined as

$$F = \frac{V}{\sqrt{gL}}$$

where L is a characteristic length of the obstruction and V is the relative velocity between the obstruction and the free stream of the fluid. It is known, for example, that the wave drag reaches a maximum value when the Froude number approaches a value of 1/2. On the other hand, when the obstruction is submerged below a given depth relative to its length, the wave drag becomes a negligible part of the pressure drag. Even so, the pressure drag which is due to the presence of the wake (or any zones of differential pressures around the obstruction) may still dominate the total drag picture. This is certainly the case for small obstructions such as the circular cylinders used in offshore platform design. Here the area exposed to skin friction drag is small and the body shape is such that the skin friction drag is small relative to the pressure drag. On the other hand, in the case of large streamlined obstructions such as a submarine where the pressure recovery downstream is almost complete, the skin friction drag is a significant contribution to the total drag picture. The extreme case is that of a long flat plate in a zero pressure gradient environment where all of the drag is skin friction drag.

It is to be emphasized again that (i) the skin friction drag is dependent upon the Reynolds number as well as the relative roughness, and (ii) the pressure drag due to the presence of the wake is also dependent upon the Reynolds number but the pressure drag due to wave generation is dependent upon the Froude number.

For a circular cylinder, the drag force/unit length is usually formulated as

$$F_D = \frac{1}{2} C_D \rho D u |u|$$

where F_D is the force/unit length, ρ is the mass density of the fluid, D is the cylinder diameter, u is the fluid particle velocity (both wave and current) that would be present in the fluid in the absence of the cylinder, and C_D is the non-dimensional drag coefficient. Interest thus centers on evaluating the drag coefficient C_D. In the case of unidirectional flow, this would be a relatively simple task to accomplish. However, for oscillating flow, it is necessary to evaluate the drag coefficient C_D and C_M simultaneously. This, in itself, would not be too difficult except for the fact that the flow field is not well defined. Here again, the presence of eddies and other manifestations of turbulence (which, in most cases, is non-homogeneous and non-isotropic) prevent any clear analytic solution on a deterministic basis. Thus, recourse has been to statistical techniques and descriptions to account for the scatter which has been observed in the limited number of experimental studies. A number of comparisons of values of inertial coefficients, C_M, and drag coefficients, C_D, as obtained from both model and prototype measurements, have been made and appear in the published literature in both graphical and tabulated forms. These comparisons, however, have only limited value since they cover a variety of pile sizes and furthermore, the tests were conducted under a wide range of Reynolds numbers. Although it is usually mentioned that these coefficients are the average or mean values obtained from correlation of the test data with the most suitable wave theory known at the time, the range of the values so obtained from each study is usually not indicated. In other words, nothing equivalent to a standard deviation is available in every case. Nevertheless, such comparisons serve to indicate the wide range of values which have been obtained but one is cautioned about the danger inherent in using the values without also at the same time considering how the hydrodynamic flow field was determined.

There are at least the following reasons to explain the variability in the results obtained by the various investigators in evaluating these coefficients.

(1) Different wave theories were used to determine the flow field in the data analysis. Thus, estimates of particle acceleration and particle velocity were not uniform from one study to another.

(2) Each group of tests covered a limited bandwidth of Reynolds numbers. These bandwidths were not coincident but covered a fairly large range.

(3) The flow conditions and manner of conducting the tests were simply dissimilar.

In summary, the best procedure at the present time for estimating the drag and mass coefficients for a circular cylinder subject to hurricane-type waves seems to be to obtain the records of internal force distributions as measured during Wave Project I and Wave Project II, and to make a correlation of these measured forces with a suitable wave theory (e.g., Dean's stream function or Chappelear's numerical wave theory). The recorded data indicates that this approach is reliable in the prediction of the forces and moments induced by the largest waves, under similar conditions.

7.3 BEHAVIOR OF MULTIPLE MEMBERS

The discussion up to now has been concerned, for the most part, with the behavior of single structural members in an oscillating fluid environment. Frequently, however, it is necessary to determine how a group of members will react when excited similarly. It should be clear that the interaction of a group of members in close proximity will modify the flow pattern as compared with a single member. The available measurements are too limited to permit a complete quantitative description of these group effects but it is known that the most important of these effects are (i) sheltering, (ii) solidification and (iii) synchronization.

Sheltering Effects: Although our interest is general, let us confine the following discussion to a group of circular cylinders (or vertical circular piles) in close proximity with each other. As indicated in the preceeding section, an increase in the turbulence intensity in a fluid stream decreases the critical Reynolds number at which the transition from a laminar to a turbulent boundary layer takes place. At this critical Reynolds number, the point of separation is moved backwards toward the rear of the body (in this case, a circular cylinder) and decreases the wake region, thereby leading to a steep drop in the drag coefficient. Suppose then, that we have two cylinders with their centers lined up in a direction parallel to the oncoming flow, as shown in the sketch below.

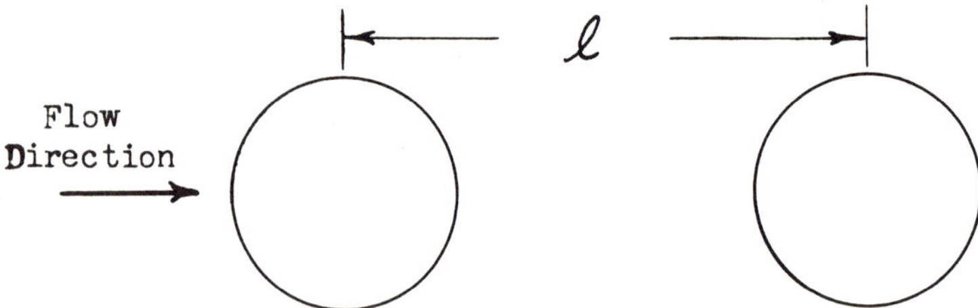

Figure 7.3.1 Illustration of **sheltering effects.**

Eddies which are shed from the upstream cylinder may so increase the turbulence intensity of the fluid flow pattern surrounding the second cylinder that the drag coefficient for the second cylinder is considerably reduced. This occurrence is known as sheltering, and one can see immediately that the degree of sheltering is dependent upon the direction of the oncoming flow. Landweber (1942), Chappelear (1959) and Laird (1961) have studied this phenomenon theoretically but they all treat the fluid as an ideal (inviscid) fluid and approach the problem via potential theory. Thus, the actual drag is not considered. However, the information developed is useful and pertinent to this discussion. Chappelear's results well illustrate the type of problem considered and some of his results appear in Figure 7.3.2. His analysis assumes that:

(a) The velocity distribution in the neighborhood of the cylinders is given by potential flow theory.

		Correction Factor
2 cylinders of radii R_1 and R_2 with separation ℓ. α is the angle between the wave direction and the axis.	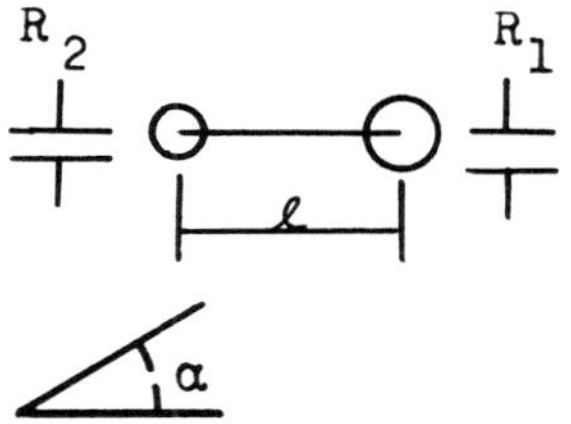 R_2, R_1, ℓ, α	On R_1 $1 - \left(\frac{R_2}{\ell}\right)^2 \cos 2\alpha$ On R_2 $1 - \left(\frac{R_1}{\ell}\right)^2 \cos 2\alpha$
3 cylinders of radius R with separation ℓ. α is the angle between the wave direction and the line of centers.	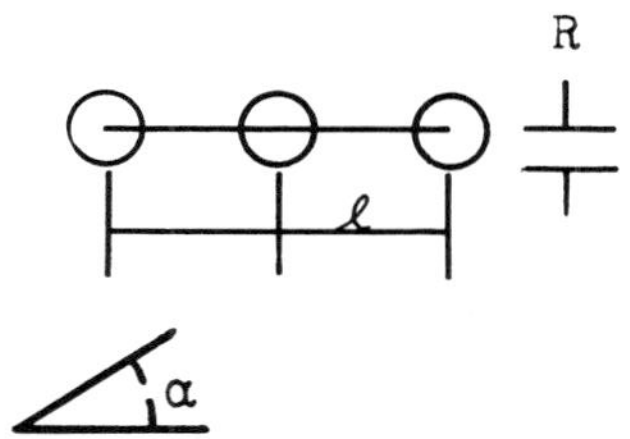 R, ℓ, α	On left, $1 - \frac{5}{4}\left(\frac{R}{\ell}\right)^2 \cos 2\alpha$ On center, $1 - 2\left(\frac{R}{\ell}\right)^2 \cos 2\alpha$
4 cylinders of radius r whose centers are on a square of diagonal 2ℓ and one cylinder of R at the center of the square. α is the angle between the wave direction and the diagonal.	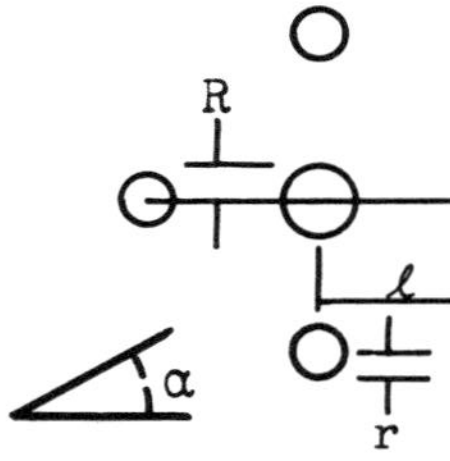 R, ℓ, r, α	On center, 1 On right, $1 - \left[\left(\frac{R}{\ell}\right)^2 + \left(\frac{r}{2\ell}\right)^2\right] \cos 2\alpha$

Figure 7.3.2a Correction factors for various arrangements of groups of cylinders. (After Chappelear, 1959).

6 cylinders of radius R on a rectangular lattice $2\ell_1$ long and ℓ_2 wide. α is the angle between the wave direction and a line through 3 cylinders.

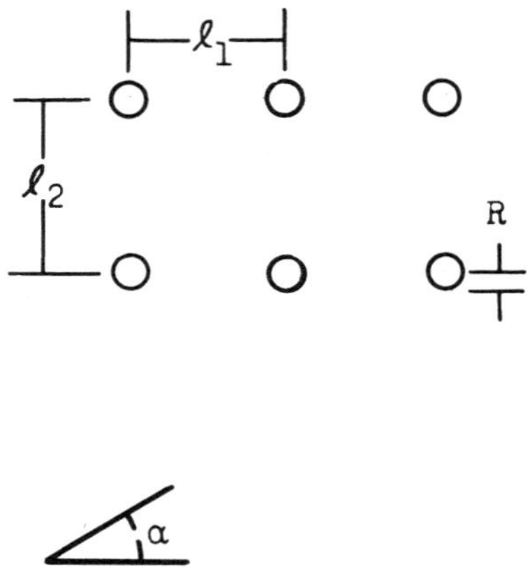

Correction Factor

On lower left,

$$1 - \left[\frac{5}{4}\left(\frac{R}{\ell_1}\right)^2 - \left(\frac{R}{\ell_2}\right)^2 + \frac{R^2(\ell_1 - \ell_2)}{(\ell_1^2 + \ell_2^2)^2} + \frac{R^2(4\ell_1^2 - \ell_2^2)}{(4\ell_1^2 + \ell_2^2)}\right] \cos 2\alpha$$

$$+ \left[\frac{2\ell_1 \ell_2 R^2}{(\ell_1^2 + \ell_2^2)^2} + \frac{4\ell_1 \ell_2 R^2}{(4\ell_1^2 + \ell_2^2)^2}\right] \cos 2\alpha$$

On lower center,

$$1 - \left[\frac{2R^2}{\ell_1^2} - \frac{R^2}{\ell_2^2} + \frac{2R^2(\ell_1^2 - \ell_2^2)^2}{(\ell_1^2 + \ell_2^2)^2}\right] \cos 2\alpha$$

Figure 7.3.2b Correction factors for various arrangements of groups of cylinders. (After Chappelear, 1959).

(b) The characteristic dimension of the group is small compared to the wave length. This is interpreted to mean that the center to center distances of the extreme cylinders comprising the group is small relative to the wave length. Thus, linear wave theory is employed.

(c) The radii of the individual cylinders are small compared with the center to center distances.

(d) The force on a single cylinder can be estimated.

The results appear in the form of correction factors applied to wave force estimations for a single cylinder. The formulation for the force on a group of cylinders thus appears as

$$F_g = N \cdot C_r \cdot F_s$$

where C_r is the correction factor (as given by expressions such as those appearing in Figure 7.3.2), N is the number of cylinders and F_s is the empirically determined force on a single cylinder.

For cylinders whose centers are in a line parallel to the direction of fluid motion, Biermann and Herrnstein (1933), Laird, Johnson and Walker (1962) and Wilson and Caldwell (1970) have conducted experimental (laboratory) tests to determine the influence of the forward cylinder on the drag of the after cylinder. The Biermann and Herrnstein tests were conducted in a wind tunnel (uniform steady flow) with the cylinders held fixed whereas the Laird, Johnson and Walker tests used the inverted model technique of moving circular cylindrical models through water. Although the models were oscillated in the latter tests, only data from the initial swing were analyzed. Dean and Harleman (1966) have pointed out differences in the tests; however, in addition to the presence of surface wave generation due to motion of the cylinders in the Laird et. al. tests, which was not present in the Biermann and Herrnstein tests, there is the fact that envelopment of the wake around the cylinder of interest takes time (which depends upon the distance separating the cylinders). To be more precise, the extent to which the vortices develop depend upon the duration of flow in one direction. It is the ratio of the duration of flow (if less than critical) to that required for vortex formation which seems to be of fundamental importance here. This time factor was not a consideration in the Biermann and Herrnstein tests whereas it definitely influenced the results of the Laird et. al. tests. The test by

Wilson and Caldwell (1970) were also conducted in a wind tunnel with stationary cylinders but included the additional complication of a nearby boundary.

In the case of cylinders whose centers are perpendicular to the direction of fluid motion, the large scale wave tank tests conducted by Ross (1959) and by Laird, Johnson and Walker (1962) may be cited. The essential distinction between the two tests is that the Reynolds number range for the two tests were considerably different. The Laird et. al. tests were in the subcritical range whereas the Ross tests were near the critical Reynolds number. Putting all of these facts together, Dean and Harleman (1966) suggest that for a pair of identical cylinders (one directly behind the other) the sheltered cylinder experiences a reduction in drag for a distance as large as nine (9) times their diameters.

On the other hand, a parallel pair of identical cylinders may be considered isolated if the separation distance is larger than twice their diameters. Thus, the sheltering effect is considerably stronger in the direction of motion than in a transverse direction.

Finally, to complete our discussion of sheltering effects, the question of large lift (or transverse) forces on individual cylinders in a group has been investigated by Laird et. al. (1962), Laird and Warren (1963), and Laird (1966). These tests were for different cylinder configurations and spacings and although large lift forces and moments on individual cylinders were recorded, the application of any of the data obtained from these tests to prototype situations seems doubtful. It is clear, however, that large lift and moment forces may occur and this is very pronounced where the cylinders are flexibly supported. This will be discussed under synchronization effects.

Solidification Effects: In contrast to sheltering effects which are drag-related, solidification is inertial-related. To illustrate, for a structure composed of a group of large cylinders, the major portion of the total wave force is inertial and is thus proportional to the total displaced volume. Thus, for n cylinders, the inertial force is

$$F_I \propto \sum_{i=1}^{n} V_i$$

where V_i is the volume of fluid displaced by each cylinder. Now, when the space between the cylinders decreases, there is a volume of fluid literally trapped between the cylinders.

This entrapment or blocking effect, which depends upon a certain critical spacing, gives rise to a sharp rise in the added mass coefficient. Wang (1969) has sketched how this effect applies to a group of cylinders in a circular orientation (shown below).

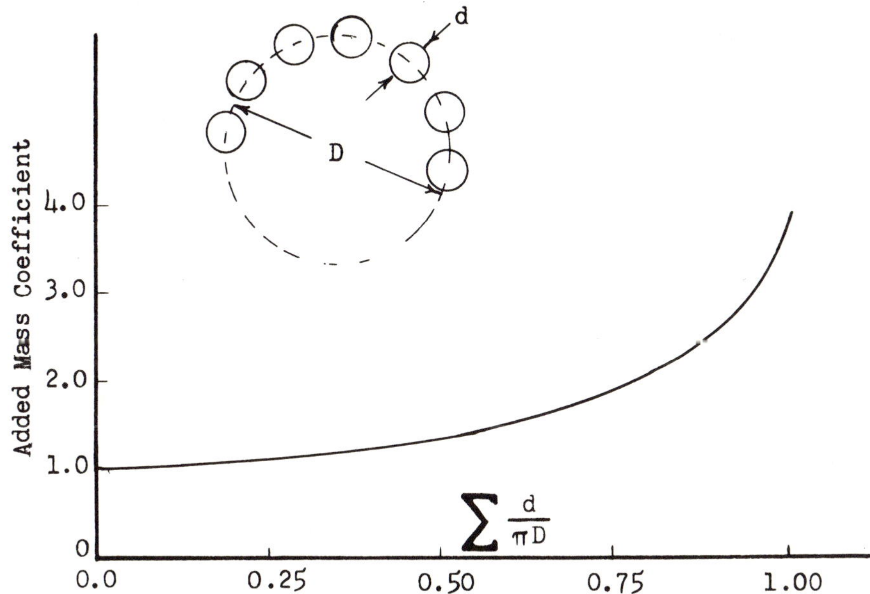

Apparently, little information is available on this occurrence. However, the added mass coefficient can be obtained very simply by vibrating the structure in the fluid.

<u>Synchronization Effects:</u> Synchronization effects are most pronounced when the structural members (cylinders in our case) are flexible. Before citing the pertinent references and significant conclusions obtained therefrom, it is illuminating to review the nature of vortex shedding and its relation to the Strouhal and Reynolds numbers. The flow pattern around any large body placed in a fluid stream will experience separation when the stream becomes sufficiently strong. The manifestation of this separation is the appearance of vortices which grow in size with increasing Reynolds numbers until they are carried into the wake, and reappear again on the downstream side of the cylinder. In the case of a circular cylinder, which is symmetric, these eddies form a vortex system which is unstable at a high enough Reynolds number and the eddies (or vortices) begin to alternate in formation and departure, first on one side and then the other. The Strouhal number is defined as

$$S_n = \frac{fD}{V}$$

where f is the frequency of vortex shedding in cycles per second, D is the characteristic length dimension of the member and V is the free stream velocity. For circular members, the average value S_n has been found to rise from 0.2 at low Reynolds numbers ($2x10^5$) to 0.4 at high Reynolds numbers. For a wide variety of structural shapes, the Strouhal numbers are extensively tabulated. The Strouhal numbers as a function of the Reynolds number for a limited number of bodies are given in numerous references.

For unidirectional flow in the subcritical Reynolds number range, the frequency of eddy shedding has been found to be given by the following:

$$f = S_n\left(1 - \frac{19.7}{R_n}\right)\frac{V}{D} \qquad (7.3.1)$$

For high Reynolds numbers, which is usually the case of ocean structural members, the corresponding relation is

$$f = S_n \frac{V}{D} \qquad (7.3.2)$$

Farquharson (1952) studied the aerodynamic stability of suspension bridges and concluded that at least three conditions (or regions of behavior) were possible. They are:

(1) When the structure is at rest, the vortex shedding frequency is controlled by the oncoming flow.

(2) At certain critical flow velocities, self-excited oscillations are induced whenever the vortex shedding frequency coincides with the fundamental frequency (or one of the harmonics of the natural frequency) of the full structure.

(3) Above these critical velocities, the oscillating structure controls the vortex shedding frequency rather than the flow velocity. This situation exists over very definite ranges of flow velocity. The lower limit is a critical velocity but the upper limit is not so well defined. Within this range, the structure is practically at rest. The situation is well depicted by Figure 7.3.3.

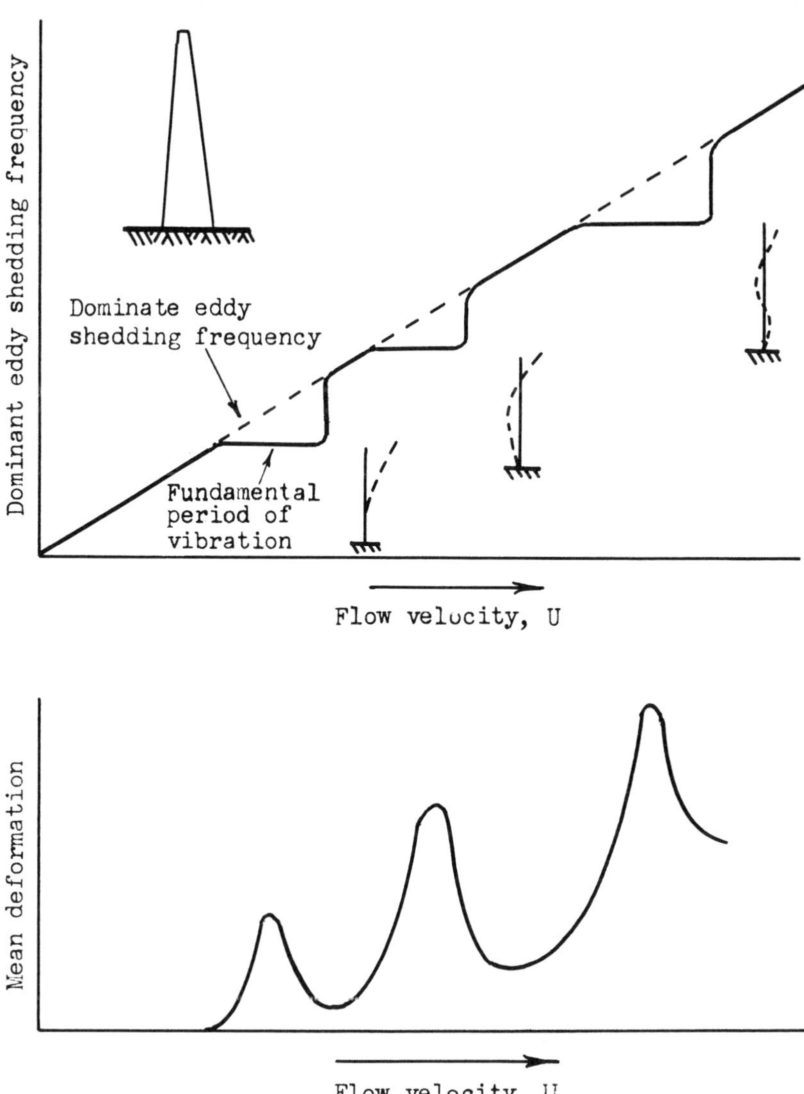

Figure 7.3.3 Example of fluid dynamic and fluid elastic resonance as illustrated by Toebes (1965).

Among the experimental studies conducted, those by Landweber (1942), Laird and Warren (1963), and Toebes (1965) are particularly informative. However, only qualitative statements concerning behavior of groups of cylinders can be made at the present time. These are:

(1) The probability of synchronization decreases as the number of members increase.

(2) Increasing the flexibility of individual members does not necessarily increase the degree of synchronization.

(3) Synchronization usually results in only a slightly higher drag but a significantly higher lift (transverse) force.

(4) Three frequencies are important and come into play. These are the (i) natural frequency of vibration, f_n, of the group; (ii) frequency of eddy shedding, f_e; (iii) excitation frequency, ω. Partial synchronization in the exciting force (drag) direction occurs when the excitation frequency is near the natural frequency of vibration of the group whereas synchronization in the transverse direction occurs when the eddy shedding frequency is near the natural frequency of the group.

To expand on the last conclusion, Laird's measurements show that the peak value in the lift and drag coefficients always occur when the ratio of the eddy shedding frequency to natural frequency is slightly less than 1. Laird further notes that lift forces can amount to as much as $4\frac{1}{2}$ times the drag force. It is obvious that the designer of an ocean structure should avoid having the structure frequency coinciding with the eddy shedding frequency. If it is unavoidable, spoilers may have to be provided in order to prevent the occurrence of large vortex induced forces.

7.4 STATISTICAL APPROACH TO THE EVALUATION OF INERTIAL AND DRAG COEFFICIENTS FOR SINGLE MEMBERS

Mention has been made of the requirement to evaluate inertial and drag coefficients separately since, in oscillatory flow, both related forces are present simultaneously. Usually the assumption is made that drag force is zero when the inertial force attains its maximum value. This may be a reasonable approach for periodic waves but hardly seems valid for non-periodic waves. There is a method, however, for evaluating these coefficients even in non-periodic waves and while the method is not yet capable of handling those portions of the member which are alternately wetted and dried, comparisons between force distributions predicted and measured for fully submerged members are in substantial agreement. The method is discussed in detail by Pierson and Holmes (1965) and has been greatly extended by Brown and Borgman (1967). The method assumes that the sea surface is a stationary, Gaussian stochastic process and that the instantaneous force can be represented by an empirical type equation such as that formulated by Morison (1950). Because of the importance of the method, details of derivation are included herein. It is assumed that the force $f_z(t)$ at some elevation z is given by an equation of the type:

$$f_z(t) = C_D \frac{\rho}{2} Du(t) \left| u(t) \right| + C_M \frac{\rho}{4} \pi D^2 \dot{u}(t) \tag{7.4.1}$$

The normal bivariate distribution of velocity, u, and acceleration $\dot{u}$, is given as follows where u and $\dot{u}$ are assumed to be independently distributed with zero mean.

$$P(u,\dot{u})\, du\, d\dot{u} = \frac{1}{2\pi\sqrt{\Psi_1\ \Psi_2}} \exp \left[\frac{-u^2}{2\Psi_1} + \frac{-\dot{u}^2}{2\Psi_2} \right] du\, d\dot{u} \tag{7.4.2}$$

Ψ_1 and Ψ_2 are the variances of the spectra of velocity and acceleration, respectively. The distribution of $f_z(t)$ is derived by performing a transform on Equation (7.4.2) which preserves measure and still remains a probability density function. The random variable $f_z(t)$ is defined separately for $0 < u < \infty$ and $-\infty < u < 0$.

For $u > 0$, $|u| = u$ and, therefore,

$$f_z(t) = k_1 u^2 + k_2 \dot{u} \tag{7.4.3}$$

Transforming variables according to

$$\dot{u}^* = \dot{u} \qquad (7.4.4)$$

and taking the inverse of Equation (7.4.3) gives

$$u = +\sqrt{\frac{f}{k_1} - \frac{k_2}{k_1}\,\dot{u}^*} \qquad (7.4.5)$$

and

$$\dot{u} = \dot{u}^* \qquad (7.4.6)$$

subject to the condition that $f > k_2\dot{u}^*$.

The Jacobian of u and $\dot{u}$ is given by

$$\frac{\partial(u,\ \dot{u})}{\partial(f,\dot{u}^*)} = \frac{1}{2k_1}\,\frac{1}{\sqrt{\dfrac{f}{k_1} - \dfrac{k_2}{k_1}\,\dot{u}^*}} \qquad (7.4.7)$$

From Equations (7.4.5), (7.4.6) and (7.4.7), Equation (7.4.2) becomes

$$P_1(f)df = \frac{1}{2\pi\sqrt{\ 1\ \ 2}}\left\{\int_{-\infty}^{f/k_2} \frac{\exp\left[-\dfrac{1}{2\psi_1}\left(\dfrac{f}{k_1} - \dfrac{k_2}{k_1}\,\dot{u}^*\right) - \dfrac{1}{2\psi_2}\,\dot{u}^{*2}\right]}{k_1\sqrt{\dfrac{f}{k_1} - \dfrac{k_2}{k_1}\,\dot{u}^*}}\,d\dot{u}\right\} df$$

(7.4.8)

An additional transformation of variable,

$$\alpha = \frac{f}{k_1} - \frac{k_2}{k_1}\dot{u}^* \tag{7.4.9}$$

and hence,

$$d\alpha = \frac{k_2}{k_1} d\dot{u}^* \tag{7.4.10}$$

yields

$$P_1(f)df = \frac{1}{2\pi\sqrt{\psi_1\psi_2}}\left\{\int_0^\infty \frac{\exp\left[\frac{-\alpha}{2\psi_1} - \frac{1}{2\psi_2}\left(\frac{f}{k_2} - \frac{k_1\alpha}{k_2}\right)^2\right]}{k_2\sqrt{\alpha}} d\alpha\right\} df \tag{7.4.11}$$

For the condition $-\infty < u < 0$,

$$f = -k_1 u^2 + k_2\dot{u} \tag{7.4.12}$$

A similar sequence of operations yields the result that

$$P_2(f)df = \frac{1}{2\pi\sqrt{\psi_1\psi_2}}\left\{\int_0^\infty \exp\left[-\frac{\alpha}{2\psi_1} - \frac{1}{2\psi_1} - \left(\frac{k_1\alpha}{k_2} + \frac{f}{k_2}\right)^2\right] d\alpha\right\} df \tag{7.4.13}$$

The sum of Equations (7.4.11) and (7.4.13) yields Equation (7.4.14), the probability distribution of the force.

$$P(f)df \quad \frac{1}{2\pi\sqrt{\psi_1 \psi_2}} \left\{ \int_0^\infty \left[\frac{\exp \frac{-\alpha}{2\psi_1} \exp \frac{-\left(\frac{k_1\alpha}{k_2} \quad \frac{f}{k_2}\right)^2}{2\psi_2}}{2k_2\sqrt{\alpha}} \right] d\alpha + \int_0^\infty \left[\frac{\exp \frac{-\alpha}{2\psi_1} \exp \frac{-\left(\frac{k_1\alpha}{k_2} \quad \frac{f}{k_2}\right)^2}{2\psi_2}}{2k_2\sqrt{\alpha}} \right] d\alpha \right\} df \qquad (7.4.14)$$

An analytic expression for this equation can be found, as noted by Pierson and Holmes (1965), but additional difficulties in evaluating the Bessel function so obtained led to an evaluation of Equation (7.4.14) by machine integration. This is accomplished as follows:

$$k_1\psi_1 = A; \quad k_2\sqrt{\psi_2} = B; \quad \gamma^2 = \frac{\alpha}{\psi_1} \qquad (7.4.15)$$

The function then reduces to

$$P(f)df = \frac{1}{2\pi\sqrt{B}} \left\{ \int_0^\infty \exp -\frac{1}{2}\left[\gamma^2 + \frac{\left(\gamma^2 + \frac{f}{A}\right)^2}{\frac{B}{A^2}}\right] d\gamma \right.$$

$$\left. + \int_0^\infty \exp -\frac{1}{2}\left[\gamma^2 + \frac{\left(\gamma^2 - \frac{f}{A}\right)^2}{\frac{B}{A^2}}\right] d\gamma \right\} df \qquad (7.4.16)$$

Let $df = \Delta f$; $f = q\Delta f$, $q = 0, 1 \ldots N$; $d\gamma = \Delta\gamma$

$$\gamma = (j + 1/2)\Delta\gamma, \quad j = 0, 1 \ldots M \qquad (7.4.17)$$

The function then further reduces to the evaluation of

$$P_q = P(q\Delta f)\,\Delta f = \frac{\Delta\gamma\Delta f}{2\pi B} \sum_{j=0}^{M} \exp - (R_j + S_j) + \exp(R_j + T_j) \qquad (7.4.18)$$

for $q = 0, 1, 2 \ldots N$

where

$$R_j = \frac{\left[(j + \frac{1}{2})\Delta\gamma\right]^2}{2} \qquad (7.4.19)$$

$$S_j = \frac{\left\{\left[(j + \frac{1}{2})\Delta\gamma\right]^2 - \frac{q(\Delta f)}{A}\right\}^2}{2\frac{B}{A^2}} \qquad (7.4.20)$$

$$T_j = \frac{\left\{\left[(j + \frac{1}{2})\Delta\gamma\right]^2 + \frac{q(\Delta f)}{A}\right\}^2}{2\frac{B}{A^2}} \qquad (7.4.21)$$

For the purposes of computation, if $\sqrt{B}/A \geq 1$, $\Delta\gamma = 0.05$ and if $\sqrt{B}/A \leq 1$, $\Delta\gamma = 0.05\sqrt{B}/A$.

If $(R_j + T_j) \geq 9$, $\exp - (R_j + T_j)$ is set equal to zero, or if $(R_j + S_j) \geq 9$, $\exp - (R_j + S_j)$ is set equal to zero. The function is integrated over the force range at each step in the calculation as $q = 0, 1, 2 \ldots N$, and the computation is terminated if

$$P_0 + 2\sum_{q=1}^{N} P_q \geq 1 - e \qquad (7.4.22)$$

where e is a constant of the order 0.001.

Since P(f) as given by Equation (7.4.14) is an even function, it has only even moments. Eaton (in Holmes (1964)) has derived these moments in the following manner. Noting the redefinition of variables according to Equation (7.4.22), for example,

$$Z = f; \quad W = k_2\dot{u}; \quad V = k_1u^2, \; u>0; \quad V = k_1u^2, \; u<0 \tag{7.4.23}$$

a solution for k_1 and k_2 is required by a derivation of the moments of the probability density function as defined by Equation (7.4.14).

Since u and $\dot{u}$ are independent and V is a function of u only and W is a function of $\dot{u}$ only, the random variables V and W are independent. Because of symmetry about the origin

$$E(Z) = E(V) = E(W) = 0$$

where E is the expected value operator, and also

$$E(VW) = 0$$

since V and W are independent.

From Equation (7.4.3),

$$Z = V + W \tag{7.4.24}$$

Therefore,

$$\begin{aligned} E(Z^2) &= E(V^2 + 2VW + W^2) \\ &= E(V^2) + E(W^2) \end{aligned} \tag{7.4.25}$$

$$E(Z^2) = E(V^4 + 4V^3W + 6V^2W^2 + 4VW^3 + W^4$$

$$= E(V^4) + 6E(V^2)E(W^2) + E(W^4) \qquad (7.4.26)$$

The random variable W has a normal distribution with standard deviation equal to $k_2\sqrt{\Psi_2}$. Therefore, its moment-generating function is

$$\frac{(k_2\sqrt{\Psi_2}\ t)^2}{2} = 1 + \frac{(k_2\sqrt{\Psi_2}\ t)^2}{2} + \frac{(k_2\sqrt{\Psi_2}\ t)^4}{4.2!} + \frac{(k_2\sqrt{\Psi_2}\ t)^6}{8.3!} + \ldots.$$

By differentiating n times and setting $t = 0$, where $u = 1, 2, 3, 4 \ldots.$, it is seen that the moments, M_n, are

$$M_0 = 1; \quad M_1 = 0; \quad M_2 = k_2^2 \Psi_2; \quad M_3 = 0$$

$$M_4 = \frac{4.3.2.}{4.2!}, \quad k_2^4 \Psi_2^2 = 3k_2^4 \Psi_2^2 \qquad (7.4.27)$$

where M_2 is the variance of W, (i.e., $E(W^2)$) and M_4 is the fourth moment of W, (i.e., $E(W^4)$). The random variable V has all its odd moments equal to zero because of symmetry. The even moments of $(V/k_1 \Psi_1)$ are the same as those about the origin of a chi-squared distribution with one degree of freedom, which has the moment-generating function

$$M_t = (1 - 2t)^{-\frac{1}{2}}$$

Successive differentiation gives

$$M_t^{I} = (1 - 2t)^{-3/2}$$

$$M_t^{II} = 3(1 - 2t)^{-5/2}$$

$$M_t^{III} = 15(1 - 2t)^{7/2}$$

$$M_t^{IV} = 105(1 - 2t)^{-9/2}$$

Setting $t = 0$, and noting that the odd moments are zero,

$$M_0 = 1; \quad M_1 = 0; \quad M_2 = 3; \quad M_3 = 0; \quad M_4 = 105 \qquad (7.4.28)$$

Therefore, the moments of the random variable V are given by

$$M_0 = 1; \quad M_1 = 0; \quad M_2 = 3k_1^2\psi_1^2; \quad M_3 = 0; \quad M_4 = 105k_1^4\psi_1^4 \qquad (7.4.29)$$

where M_2 is the variance of V, (i.e., $E(V^2)$), and M_4 is the fourth moment of W, (i.e., $E(W^4)$).

By definition,

$$E(Z^2) = \frac{1}{N}\sum_{j=1}^{N} f_j^2 = F_2 \qquad (7.4.30)$$

and

$$E(Z^4) = \frac{1}{N}\sum_{j=1}^{N} f_j^4 = F_4 \qquad (7.4.31)$$

and two simultaneous equations for k_1 and k_2, ψ_1 and ψ_2 being known, are

$$F_2 = 3k_1^2\psi_1^2 + k_2^2\psi_2 \qquad (7.4.32)$$

and

$$F_4 = 105k_1^4\psi_1^4 + 18k_1^2\psi_1^2k_2^2\psi_2 + 3k_2^4\psi_2^2 \qquad (7.4.33)$$

whence

$$k_1 = \left[\frac{(F_4 - 3F_2^2)^{\frac{1}{2}}}{\sqrt{78}\;\psi_1^2}\right]^{\frac{1}{2}} \qquad (7.4.34)$$

$$k_2 = \left[\frac{F_2 - 3k_1^2 \Psi_1^2}{\Psi_2} \right]^{\frac{1}{2}} \tag{7.4.35}$$

From Equations (7.4.30), (7.4.31), (7.4.32), (7.4.33), (7.4.34) and (7.4.35),

$$E(Z^2) = 3k_1^2 \Psi_1^2 + k_2^2 \Psi_2 \tag{7.4.36}$$

$$E(Z^4) = 105k_1^4 \psi_1^4 + 18k_1^2 \psi_1^2 k_2^2 \psi_2 + 3k_2^4 \psi_2^2 \tag{7.4.37}$$

If $f_1, f_2, f_3, \ldots. f_n$ is a large random sample from the force population, the moments of these forces are defined as

$$F_1 = \frac{1}{N} \sum_{j=1}^{N} f_j \tag{7.4.38}$$

$$F_2 = \frac{1}{N} \sum_{j=1}^{N} (f_j)^2 \tag{7.4.39}$$

$$F_3 = \frac{1}{N} \sum_{j=1}^{N} (f_j)^3 \qquad (7.4.40)$$

$$F_4 = \frac{1}{N} \sum_{j=1}^{N} (f_j)^4 \qquad (7.4.41)$$

where F_2 is an estimator of $E(Z^2)$, F_4 is an estimator of $(E(Z^4)$ and $F_1 = F_3 = 0$.

Hence, from Equations (7.4.36) and (7.4.37) two simultaneous equations for the estimation of k_1 and k_2 are

$$F_2 = 3k_1^2 \psi_1^2 + k_2^2 \psi_2 \qquad (7.4.42)$$

and

$$F_4 = 105k_1^4 \psi_1^4 + 18k_1^2 \psi_1^2 k_2^2 \psi_2 + 3k_2^4 \psi_2^2 \qquad (7.4.43)$$

The distributions obtained in the foregoing manner have been compared with measured distributions as recorded during hurricane waves. An example of the comparison is shown in Figure 7.4.1. The predicted measured and normal distribution (with the same variance as the actual data) are shown. The theory predicts that the probability of an extreme value for f is many times greater for large values of f than if the data were assumed to fit a normal curve. It is noted, however, that the theory predicts

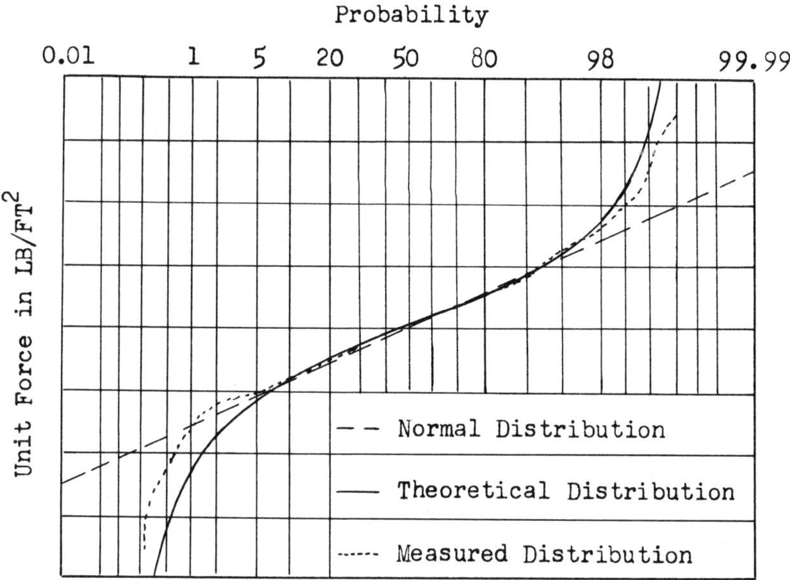

Figure 7.4.1 Comparison of cumulative probability distribution of force. (After Holmes, 1964)

forces out to $+\infty$ and $-\infty$ which is unrealistic physically. Therefore, at extremes there will inherently be differences between measured and predicted values.

7.5 REFERENCES

1. Ackermann, N. L. and Arbhabhirama, A. (1964). "Viscous and Boundary Effects on Virtual Mass," Journal of Engineering Mechanics Division, American Society of Civil Engineers, Separate Paper No. 4014, August.

2. Biermann, D., and Herrnstein, W. H. (1933). "The Interference Between Struts in Various Combinations," NACA Report No. 468.

3. Brater, E. F., McNown, J. S., and Stair, Leslie D. (1958). "Wave Forces on Submerged Structures," Journal of Hydraulics Division, American Society of Civil Engineers, Separate Paper No. 1833, November.

4. Brown, L. J., and Borgman, L. E. (1967). "Tables of the Statistical Distribution of Ocean Wave Forces and Methods of Estimating Drag and Mass Coefficients," Technical Memorandum No. 24, Coastal Engineering Research Center, October.

5. Chappelear, J. E. (1959). "Wave Forces on Groups of Vertical Piles," Journal of Geophysical Research, AGU, Vol. 64.

6. Dean, R. G., and Harleman, D.R.F., (1966). Chapter 8, "Interaction of Structures and Waves," in Estuary and Coastline Hydrodynamics, edited by Ippen, A. T. (1966), McGraw-Hill Book Company.

7. Farquharson, F. B. (1952). "Aerodynamic Stability of Suspension Bridges," University of Washington Engineering Experiment Station, Bulletin 116, Series 3.

8. Hoerner, S. F. (1965). Fluid Dynamic Drag, Published by the author, New York, New York.

9. Holmes, P. (1964). "Forces Induced by Ocean Waves on Piles," U. S. Naval Civil Engineering Laboratory, Technical Report R-328, Port Hueneme, California, October.

10. Jen, Yuan (1968). "Laboratory Study of Inertia Forces on a Pile," Journal of the Waterways and Harbors Division, Proceedings, American Society of Civil Engineers, Paper No. 5806, WW1, February.

11. Laird, A. D. K. (1961). "Eddy Forces on Rigid Cylinders," Journal of Waterways and Harbors Division, American Society of Civil Engineers, Vol. 87, November.

12. Laird, A. D. K. (1966). "Flexibility in Cylinder Groups Oscillating in Water," Journal of Waterways and Harbors Division, American Society of Civil Engineers, Vol. 92, WW3.

13. Laird, A. D. K., Johnson, C. A., and Walker, R. W. (1962). "Water Eddy Forces on Oscillating Cylinders," Transactions, American Society of Civil Engineers, Vol. 127, Part 1.

14. Laird, A. D. K., Warren, R. P. (1963). "Groups of Vertical Cylinders Oscillating in Water," Journal of the Engineering Mechanics Division, American Society of Civil Engineers, Vol. 89, February.

15. Lamb, Sir Horace (1932). Hydrodynamics, 6th Edition, Dover Publications, New York.

16. Landweber, L. (1942). "Flow About a Pair of Adjacent, Parallel Cylinders Normal to a Stream: Theoretical Analysis," David Taylor Model Basin Report 485, Washington, D. C.

17. Morison, J. R., O'Brien, M. P., Johnson, J. W., and Schaaf, S. A. (1950). "The Force Exerted by Surface Waves on Piles," Petroleum Transactions, Vol. 189, pp. 2848.

18. Murtha, J. P. (1954). "Virtual Mass of Partially Submerged Bodies," M. S. Thesis, Carnegie Institute of Technology, Department of Civil Engineering.

19. Pierson, W. J., Jr., and Holmes, P. (1965). "Irregular Wave Forces on a Pile," Journal of Waterways and Harbors Division, American Society of Civil Engineers, Vol. 91, WW4.

20. Riabouchinsky, D. (1920). "Sur la Resistance des Fluids," Comptes Rendus, Institute de France, Paris, France.

21. Ross, C. W. (1959). "Large Scale Tests of Wave Forces on Piling," Beach Erosion Board, Technical Memorandum No. 111.

22. Stelson, E., and Mavis, F. T. (1957). "Virtual Mass and Acceleration in Fluids," Transactions, American Society of Civil Engineers, Vol. 122.

23. Toebes, G. H. (1965). "Flow Induced Structural Vibrations," Journal of Engineering Mechanics, American Society of Civil Engineers, EM6.

24. Wang, H. (1969). Unpublished Lecture Notes, UCLA Short Course, "Dynamic Analysis of Ocean Structures, Los Angeles, California, August.

25. Wilson, J. F., and Caldwell, H. (1970). "Force and Stability Measurements on Models of Submerged Pipelines," Preprints, Offshore Technology Conference, Paper No. 1224, Houston, Texas.

26. Zienkiewicz, O. C., and Nath, B. (1964). "Analog Procedure for Determination of Virtual Mass," Journal of the Hydraulics Division, American Society of Civil Engineers, Separate Paper No. 4042, September.

Chapter VIII

FORCES INDUCED BY BREAKING WAVES

Chapter VII dealt principally with fluid-induced forces and where the fluid motion appeared in the form of surface gravity waves, only non-breaking waves were considered. In this chapter, our concern is with the forces induced by waves breaking on structures. Generally, information is meager; however, some specific information is available in the case of plane barriers. Two cases of interest have been studied. First, we consider barriers oriented in the vertical plane and then those in the horizontal plane.

8.1 VERTICAL PLANE BARRIERS

Forces induced by waves breaking on vertical barriers have been studied by a number of investigators. At least two mechanisms responsible for hydrodynamic impact have been suggested. One, due to von Karmán, is based on the impulse momentum principle while the other is based on a process of air entrapment and collapse of the resulting pocket. The studies of Bagnold (1939), Denny (1954), Carr (1954) and Leendertse (1961) tend to support the latter view. The principle findings of their studies are that

(1) The occurrence of impact depends upon the entrapment of a thin lens of air at the instant the wave impinges on the wall.

(2) The lower the peak load, the longer the duration of the impact load.

(3) The shock impulse seems to be a constant value of the initial wave momentum per foot of wave crest (this constant value was found to be 0.07 by Denny and up to a maximum of 0.11 by Carr.

(4) The area under the pressure-time curve (i.e., the impulse) tended to approach but never exceeded a definite maximum.

The difficulty with the air entrapment and collapse mechanism is that no scaling law is applicable so that the results cannot, in general, be applied to prototype design. This is not a problem with the impulse-momentum analysis as will be discussed when we take up the case of horizontal members. Carr (1954) also found that the impulse loads induced by breaking waves on inclined members were substantially lower than those induced on vertical members and that they occurred much less frequently. It is noted that the special case of breakwaters, which are usually permeable, is extensively dealt with in the literature and that there are adequate procedures governing their design. There have been some preliminary tests of waves breaking on circular cylinders and other curved surfaces but, at the present time, the results are inconclusive.

8.2 HORIZONTAL MEMBERS

The forces induced by waves on horizontal surfaces have been studied analytically and experimentally by Wang (1967). He assumed that the total force was the sum of the impact forces and the slowly varying uplift, or

$$F = F_i + F_h \tag{8.2.1}$$

The impact force, F_i is obtained from the impulse momentum relations, or

$$F_i = p_i \cdot A_r = \frac{d}{dt}(MV_e) \tag{8.2.2}$$

where p_i = mean impact pressure, acting over the area A_r, M is the mass of the amount of water causing impact, and

V_e is the effective velocity of the mass M in the vertical direction.

The slowly varying uplift force F_h is obtained from the Eulerian equation of motion, or

$$F_h = p_s A_r = - A_r \left[+ \gamma z + \frac{\gamma}{g} \int \left(\frac{\partial w}{\partial t} + u\frac{\partial w}{\partial x} + v\frac{\partial w}{\partial y} + w\frac{\partial w}{\partial z} \right) dz + c \right] \quad (8.2.3)$$

where γ is the specific weight of water. With these fundamental relations, various estimates are made of the pressures p_i and p_s for standing wave, progressive wave and dispersive progressive wave systems. One important feature in the analysis concerns the estimate of M and V_e at the instant of contact for a standing wave. It is reasoned that the mass, M, is equal to the water mass contained in a semi-cylinder with length b, the length of the pier and of the diameter 2s.

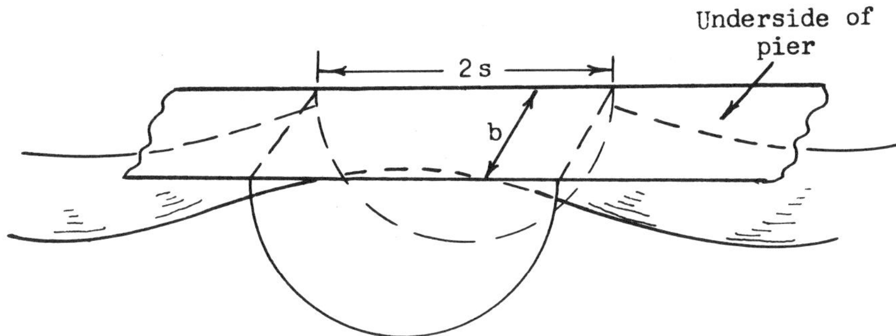

The wetted surface then is equal to 2s (i.e., the wetted length or

$$M = \frac{\rho\pi}{2} s^2 b$$

The impact equation then becomes

$$\frac{p_i}{\gamma} = \frac{\pi}{4gs} \frac{d}{dt} (s^2 V_e) = \frac{\pi}{2g} V_e \frac{ds}{dt} + \frac{\pi}{4g} s \frac{dV_e}{dt} \quad (8.2.4)$$

At the instant of contact, the wetted length s approaches zero and the effective velocity approaches w which is the vertical velocity of the water particles at the free surface. Thus, the right-hand side vanishes and the equation becomes

$$\frac{p_i}{\gamma} = \frac{\pi}{2g} V_e \frac{ds}{dt} \tag{8.2.5}$$

which states that the impact is proportional to the product of the vertical component of the water velocity at the point of contact and the rate of change of wetted length at the instant of impact. This basic concept and corresponding relation is then evaluated for the various conditions of interest. It is shown that for standing waves, ds/dt is equal to $V \cdot \delta$ where δ is a form factor depending on the shape and asymmetry of the incident wave.

The integral in the expression for the slowly varying uplift pressure is evaluated by substituting solutions of wave characteristics for various waves of interest.

For the case of standing waves, this integral becomes

$$\frac{1}{g}\int\left(\frac{\partial w}{\partial t} + u\frac{\partial w}{\partial x} + v\frac{\partial w}{\partial y} + w\frac{\partial w}{\partial z}\right) dz = \eta\left[\frac{\cosh \frac{2\pi(z+d)}{\lambda}}{\cosh \frac{2\pi d}{\lambda}}\right] + \text{higher order terms} \tag{8.2.6}$$

where d is the depth of water. The bracketed term is simply a pressure response factor β so that finally

$$\frac{p_s}{\gamma} = -z + \eta\beta \tag{8.2.7}$$

Thus, for a pier deck at an elevation h above mean water level

$$\frac{p_s}{\gamma} = -h + \eta\beta \tag{8.2.8}$$

For a regular progressive wave, the expression for the slow rise component is exactly that given for standing waves (i.e., Equation (8.2.6)). On the other hand, the expression for the fast rise (or impulse) component corresponding to Equation (8.2.5) is given by Wang (1967) as

$$\frac{p_i}{\gamma} = \frac{\pi H_1}{2} \tanh \frac{2\pi h}{\lambda} \left\{ \frac{1 - \dfrac{4d^2\left(1 + \left[\dfrac{\pi B}{\lambda}\left(1 - \dfrac{2d}{H_1}\right)\right]^2\right)}{H_1^2}}{1 - \left[\dfrac{\pi B}{\lambda}\left(1 - \dfrac{2d}{H_1}\right)\right]^2} \right\} \tag{8.2.9}$$

where h is the water depth, d is the clearance and H_i is the incident wave height, and B is the "effective" length of the pier.

Similar expressions are given for dispersive progressive waves and the analysis procedure seems to agree well with the limited number of laboratory type measurements. The effects of protrusions, such as firewalls, structural panels, etc., were not investigated but it would appear that the effect of such irregularities is pronounced.

In structural design, not only the magnitude but the impact duration is important. Thus, Wang also suggested the following empirical relation for estimating the duration of impact

$$t_d = (15 \times 40) \times \sqrt{H} \tag{8.2.10}$$

where t_d is the impact duration in milliseconds and H is the wave height in feet where impact occurs. The reader is cautioned that this equation is based on a very limited amount of data. As to the shape of the impulse, it may either be single peaked triangular or dual-peaked rectangular.

The important conclusions are:

(1) The impact pressure component is very important for both large and small structural components not only because the peak is enormous, but also

because the magnitude of the corresponding impulse is significant.

(2) The magnitude of the slowly rising "hydrostatic-type" pressure component is not excessive being from one to two times the hydrostatic pressure. The duration of this lead depends on how long the wave is in contact with the structure.

8.3 REFERENCES

1. Bagnold, R. A. (1939). "Interim Report on Wave-Pressure Research," Journal, Institute of Civil Engineers, June.

2. Carr, John H. (1954). "Breaking Wave Forces on Plane Barriers," California Institute of Technology, Hydrodynamics Laboratory Contract Noy-12561. Report No. E-11.3, November.

3. Denny, D. F. (1951). "Further Experiments on Wave Pressures," Journal, Institute of Civil Engineers, February.

4. Leendertse, J. (1961). "Forces Induced by Breaking Waves on a Vertical Wall," U. S. Naval Civil Engineering Laboratory, Technical Report R-092, Port Hueneme, California, March.

5. Wang, H. (1967). "Estimating Wave Pressures on a Horizontal Pier," U. S. Naval Civil Engineering Laboratory, Report R-546, Port Hueneme, California.

PART III
DYNAMIC BEHAVIOR WITH PARTICULAR REFERENCE TO OCEAN PROCESSES

In classical structural design, the approach is generally deterministic; that is, the responses of structures to given loads, either static or dynamic are calculated. These responses (deflection, strain, stress, etc.) are then checked so that they are kept within desired limits of specified safety factors. In the design of ocean structures, the statistical approach has much to offer. This is because the forces caused by ocean currents and waves impinging on platforms, piles, or fixed offshore structures are non-periodic and of varying amplitude or random in nature. It is the purpose of Part III to find and interpret statistically the responses of some simplified models of both structures and structural materials subjected to certain random-type forces associated with ocean waves.

With the statistical measures of the dynamic responses, one attempts to predict: (1) the probability that the response will exceed a certain value; (2) the statistical information on the time from the beginning of response to the time when the response exceeds a specified value; (3) the statistics on the amount of time the response exceeds a specified value; or (4) the statistics on material failure. In other words, when a mode of failure is assumed, the probability of failure can be predicted. For instance, if the analysis showed that, for given random forces on an offshore structure, the probability of exceeding a failure stress was 2%, that out of 100 such identical structures under identical random loading, two of the structures should be expected to fail.

An ocean structure is only as reliable as its material. Unfortunately, the reasons for material failure are not fully understood. For this reason, the design engineer must rely on test data for his structural materials, and must interpret such data in a rational way for use in structural design. For instance, one type of failure would be when the material simply cracks apart because large forces on the structure cause tensile stresses to exceed a maximum value. If the engineer knows the type of test which can give him a measure of the ultimate strength of his material, then hopefully he can design the structure to keep below the breaking stress. This assumes that he knows the forces involved, and that the test data are obtained under environmental conditions similar to that of his structure. He has to consider time as a factor too, since corrosion and cyclic forces may eventually cause structural failure at stresses far below the breaking stress observed in a short-time simple tension test.

The elementary statistical ideas needed by structural engineers to describe both the impinging forces and the resulting structural and material responses are first discussed in Chapter IX. In Chapter X, stationary responses are evaluated for a one degree of freedom and a continuous beam structural model. In Chapter XI, the dynamic behavior of structural materials with and without corrosion protection is discussed. In addition, a statistical theory for damage accumulation in metals under random type stresses is presented. A numerical example is used to illustrate this theory to estimate the probable lifetime of a fixed offshore structure.

Chapter IX

A REVIEW OF SOME STATISTICAL CONCEPTS

The forces measured on existing ocean structures have rarely had simple time histories. As a result, the structural responses such as deflection, strain, and acceleration to these "random" type forces do not have simple time histories either. Typical data for either the time behavior of a force or the response to that force at a point on an ocean structure might look like the trace shown in Fig. 9.1, where, for generality, the ordinate is designated as $z = z(t)$. At first glance, such a trace appears to have no definite

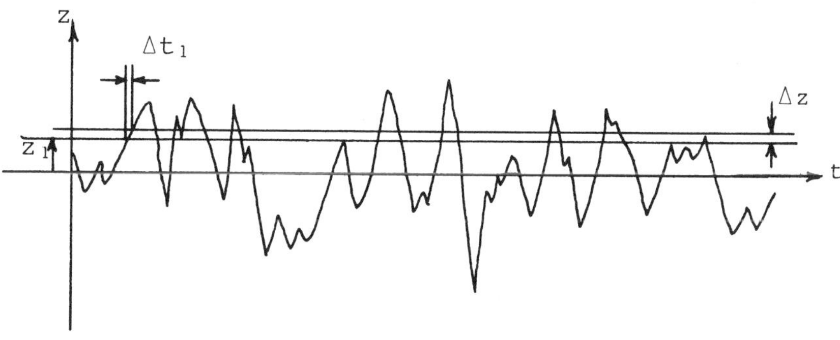

Figure 9.1 A Typical Force or Response History

frequency of fluctuation, although there appears to be an upper bound for z. If z were identified as a force, one might be tempted to simply use the upper bound for z and design a structure according to classical static methods. One might ask, then, what if an approximate forcing frequency could be associated with z over a short time period and what if this frequency were tuned to the natural frequency of the structure? Could not the structural deflections then exceed the static design conditions? It is the purpose of this chapter to define some of the statistical tools with which an engineer can obtain some answers to these questions. Crandall and Mark (1964) elaborate on these ideas.

9.1. THE STATIONARY AND ERGODIC HYPOTHESES.

It is first necessary to characterize the variable z in some mathematical sense. To do this, several assumptions are made about this random variable. The first simplifying assumption about z is that it is stationary. In principle (it is rarely done in practice) this assumption can be checked in the following way. Cut the sample trace of z in Fig. 9.1 into equal J parts. Label each separate trace as $z^{(1)}$, $z^{(2)}$, $z^{(3)}$, ... $z^{(j)}$, and take the time base as zero at the left end of each trace. For example, Fig. 9.2 shows such a

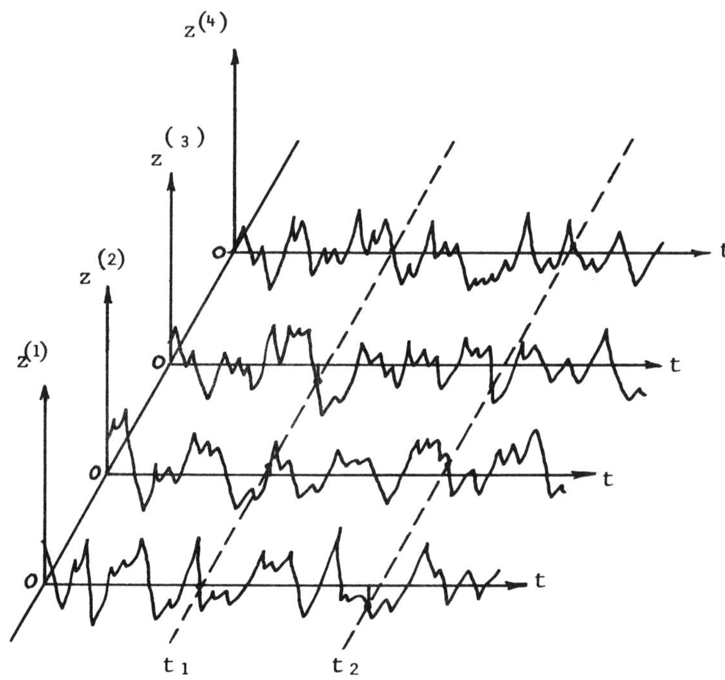

Figure 9.2 Sample Trace of Figure 9.1 Cut into Four Equal Parts

trace which has been divided into four equal parts, $J = 4$, where the times t_1 and t_2 are drawn across the ensemble of traces. The stationary hypothesis is warranted if both of the following criteria are substantially met:

(1) Consider a fixed time t_1. Measure the value of the force at t_2 on each chart and average them, giving

$$\frac{1}{J}\sum_{j=1}^{J} z^{(j)}(t_1) \tag{9.1}$$

The numerical values of expression (9.1) should be about the same for any value of t_1 which one chooses, where the value of J is not four, but a very large number.

(2) Pick a constant time interval τ. For a time t_1 and another time t_2, where $(t_2 - t_1) = \tau$, measure the values of $z^{(j)}(t_1)$ and $z^{(j)}(t_2)$ on each chart and form the sum

$$\frac{1}{J}\sum_{j=1}^{J} z^{(j)}(t_1) \quad z^{(j)}(t_2) \tag{9.2}$$

The numerical values of the expression (9.2) should be about the same for <u>any</u> value of t_1 and t_2, as long as $(t_2 - t_1) = \tau$, and J is very large.

It can be shown that for linear systems the response will also be stationary if the input force is. It should be noted that, for any time history z to be strictly stationary, it can have no beginning and no end. It is assumed here that the non-stationary effects of starting and stopping are negligible. Thus, the "essential nature" of z does not change.

The second simplifying assumption is that z obeys the <u>Ergodic Hypothesis</u>. The first part of this hypothesis states: the average value of z as given by (9.1), based on a constant t_1 over an ensemble of charts, is equal to the <u>time</u> average of z over just one typical chart. This must be true for all values of t_1. Suppose that the typical chart which is representative of the ensemble of charts, is called $z^{(k)}(t)$. Suppose further that the chosen time interval over which $z^{(k)}(t)$ is considered is long enough so that it represents "reasonably well" the behavior of z. The time average of $z^{(k)}(t)$ is given approximately by

$$\frac{1}{N}\sum_{n=1}^{N} z^{(k)}(t_n) \tag{9.3}$$

where $z^{(k)}(t_n)$ is the value of the trace at time t_n in the time interval. The values of t_n are equally spaced, and the total number of measurements is N. If expressions (9.1) and (9.3) are nearly equal, the first part of the ergodic hypothesis is approximately satisfied, or

$$\frac{1}{J}\sum_{j=1}^{J} z^{(j)}(t_1) \simeq \frac{1}{N}\sum_{n=1}^{N} z^{(k)}(t_n) \tag{9.4}$$

where J and N are both large numbers. In other words:

$$\begin{pmatrix}\text{ensemble average of z}\\ \text{across many charts}\end{pmatrix} = \begin{pmatrix}\text{time average of z}\\ \text{from a typical chart}\end{pmatrix}$$

For the second part of the ergodic hypothesis, another type of average is defined by

$$\frac{1}{N}\sum_{n=1}^{N} z^{(k)}(t_n) \quad z^{(k)}(t_n + \tau) \tag{9.5}$$

where $z^{(k)}(t)$ is measured at N discrete times $t = t_n$ and $t = t_n + \tau$ along the one typical chart. Here τ = constant. If expressions (9.2) and (9.5) are nearly equal, the second part of the ergodic hypothesis is approximately satisfied, or

$$\frac{1}{J}\sum_{j=1}^{J} z^{(j)}(t_1) \quad z^{(j)}(t_2) \simeq \frac{1}{N}\sum_{n=1}^{N} z^{(k)}(t_n) \; z^{(k)}(t_n + \tau) \tag{9.6}$$

To summarize these results, based on an "experimental" viewpoint, we say that both expressions (9.1) and (9.2) must remain nearly constant for z to be stationary. For z to obey the ergodic hypothesis, (9.4) and (9.6) must hold approximately. It should be noted that an ergodic trace is always a stationary one, whereas the converse is not necessarily true. In the statistical analysis of the structural models in the following chapters, it will be assumed that z possesses both the stationary and ergodic properties. The sample function $z^{(k)}(t)$ is designated simply as z and may represent either force input or a response, depending on the context.

9.2. MEASURES OF A RANDOM VARIABLE.

The mean value of z, over a time interval of length 2T, is given by

$$\overline{z(t)} = \lim_{T\to\infty} \frac{1}{2T}\int_{-T}^{T} z(t)dt \simeq \frac{1}{N}\sum_{n=1}^{N} z(t_n) \tag{9.7}$$

and will be assumed to be zero. If z is a force this is equivalent to saying that z excludes a static, or steady force component. For instance, forces on the structure due to steady currents and dead weight have been subtracted from z. Structural deflections from static forces can be calculated separately by usual methods and later superimposed on the statistical dynamic representation of the deflection.

The mean-square value of z is defined by

$$\overline{z^2(t)} = \lim_{T\to\infty} \frac{1}{2T} \int_{-T}^{T} z^2(t)\, dt \simeq \frac{1}{N}\sum_{n=1}^{N} z^2(t_n) \tag{9.8}$$

The autocorrelation function of z, already defined approximately by expression (9.5), is given by

$$R_z(\tau) = \lim_{T\to\infty} \frac{1}{2T} \int_{-T}^{T} z(t)\, z(t+\tau)d\tau = \int_{-\infty}^{\infty} S_z(\omega)e^{i\omega\tau}d\omega \tag{9.9a}$$

or approximately by

$$R_z(\tau) \simeq \frac{1}{N} \sum_{n=1}^{N} z(t_n)\, z(t_n + \tau) \tag{9.9b}$$

We note that $\overline{z^2(t)} = R_z(0)$, and is sometimes called the variance.

The power spectral density of z, introduced as $S_z(\omega)$ is given by

$$S_z(\omega) = \frac{1}{2\pi} \int_{-\infty}^{\infty} R_z(\tau)e^{-i\omega\tau}\, d\tau \tag{9.10a}$$

or approximately by

$$S_z(\omega) \simeq \frac{\overline{z^2(\omega)}}{2\pi\Delta\omega} \tag{9.10b}$$

If a record of z can be obtained electronically, a record of $\overline{z^2(\omega)}$ can be obtained also. A band-pass filter of pass band width $\Delta\omega$ passes z in the frequency interval ω to $(\omega + \Delta\omega)$ for a sufficiently long time. The measured value of z in this interval, called $z(\omega)$ is squared, averaged, and divided by $2\pi\Delta\omega$, as in (9.10b). Of course, time averages must be based on times T sufficiently large so that z is a representative sample, and $\Delta\omega$ must be small. The equivalence of (9.10a) and (9.10b) under these restrictions is shown by Middleton (1960).

Typical traces of $S_z(\omega)$ for two types of measured signals are shown in Figs. 9.3 and 9.4. The first is called <u>wide band</u> because the spectral density covers a wide range of frequencies.

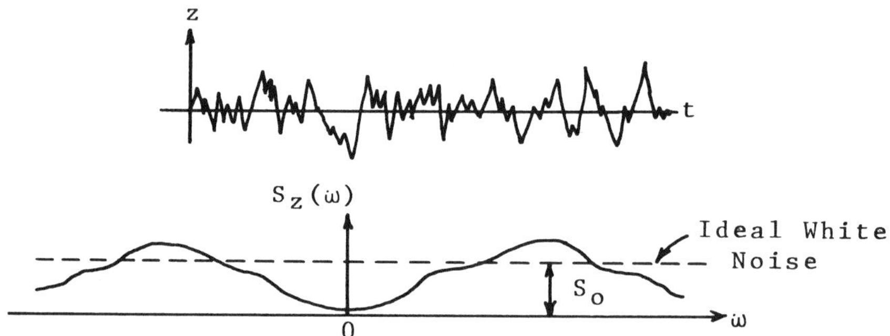

Figure 9.3 Typical Wide Band Trace and the Corresponding Power Spectral Density

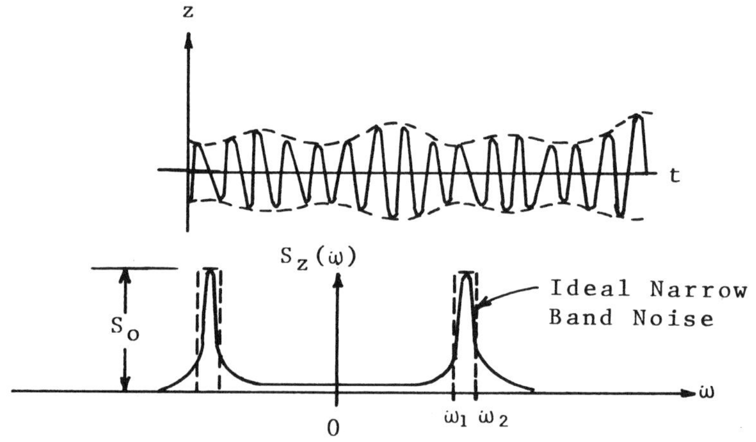

Figure 9.4 Typical Narrow Band Trace and the Corresponding Power Spectral Density

White noise is the special case of a wide band power spectral density, where $S_z(\omega) = S_o$ and S_o is a constant over all frequencies as shown by the dotted line of Fig. 9.3. The narrow-band trace is one for which all large values of $S_z(\omega)$ are concentrated near a particular frequency ω_o. In this case, ω_o is based on the instantaneous variation within the envelopes of Fig. 9.4. Force records for ocean structures indicate that $S_z(\omega)$ often has a narrow band. Just as white noise is often idealized as a constant value over all frequencies, a narrow band is often idealized as a constant value over just a narrow frequency range and is zero elsewhere, as shown in Fig. 9.4.

It should be noted that physical values of frequency are always positive, although (9.9a) implies integration over all negative values of ω as well. Since (9.10b) implies that $S_z(\omega)$ is always non-negative in an experimental sense, one can consider it symmetric about $\omega = 0$. Thus, the concept of negative ω's can be ignored by writing (9.9a) in the form

$$R_z(\tau) = 2\int_0^{\infty} S_z(\omega)e^{i\omega\tau}d\omega$$

It follows that $R_z(\omega)$ is symmetric about $\omega = 0$ also. The two important statistical parameters, which occur as transformed pairs, are thus

$$R_z(\tau) = \int_{-\infty}^{\infty} S_z(\omega)e^{i\omega\tau}d\omega = 2\int_0^{\infty} S_z(\omega)e^{i\omega\tau}d\omega \tag{9.11}$$

$$S_z(\omega) = \frac{1}{2\pi}\int_{-\infty}^{\infty} R_z(\tau)e^{-i\omega\tau}d\tau = \frac{1}{\pi}\int_0^{\infty} R_z(\tau)e^{-i\omega\tau}d\tau \tag{9.12}$$

If one is known, the other can be calculated, at least in principle. Usually $S_z(\omega)$ is obtained experimentally and $R_z(\tau)$ is calculated. $R_z(\tau)$ is used in probability calculations, which leads us to the final statistical concept, probability distributions.

9.3. PROBABILITY DISTRIBUTIONS.

For a random trace z, one might ask, what is the probability at any instant of z lying between z_1 and an increment Δz? To answer this question, one draws two horizontal lines at z_1 and $z_1 + \Delta z$, and sums the time intervals Δt_1, Δt_2, ... during which z occupies this interval. See Fig. 9.1. This sum divided by the total time T then represents the fraction of the total time that z remains in the amplitude interval z_1 to $z_1 + \Delta z$. This fraction is denoted by

$$P = \frac{\Delta t_1 + \Delta t_2 + \ldots}{T} \tag{9.13}$$

$$= P[\, z_1 \leq z \leq (z_1 + \Delta z)]$$

<u>The probability density function</u>, $p(z)$, is defined in terms to the probability P by

$$p(z) = \frac{1}{\Delta z} P[z_1 \leq z \leq (z_1 + \Delta z)] = \frac{\Delta t_1 + \Delta t_2 + \ldots}{T\ \Delta z} \tag{9.14}$$

Strictly speaking, these definitions hold only in the limit as Δt and Δz become vanishingly small. It is assumed that probabilities are mutually additive, so the probability that z is between z_1 and z_2 is given by the limiting sum

$$P\,[z_1 \leq z \leq z_2] = \int_{z_1}^{z_2} p(z)dz \tag{9.15}$$

Furthermore, the probability that z lies between $-\infty$ and $+\infty$ is one, since this is the whole range for z. That is

$$P\,[-\infty \leq z \leq \infty] = \int_{-\infty}^{\infty} p(z)dz = 1$$

It is often possible, by using Fig. 9.1 and (9.13), to find a plot of P <u>vs</u> z such as shown in Fig. 9.5. Here, the total time T is long enough so that z is truly a representative sample trace. If this P <u>vs</u> z plot is "smoothed out", the probability density function can be found graphically from the slope or,

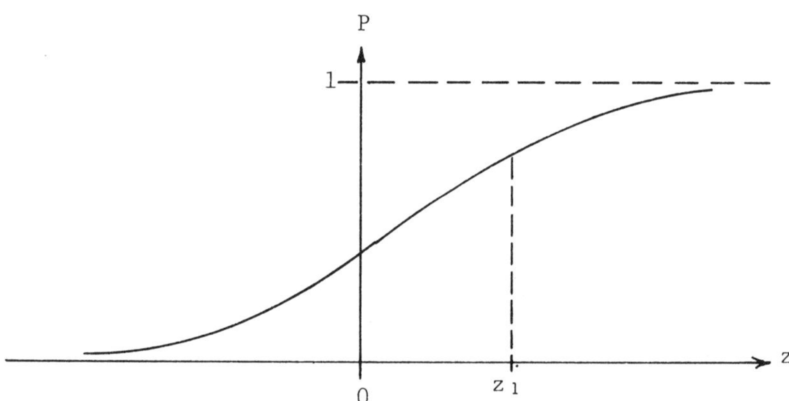

Figure 9.5 Cumulative Probability Trace, or the Probability that $z < z_1$

$$p(z) = \frac{dP}{dz}$$

which is consistent with the strict definitions of P and p.

Data often shows that z is approximately normal or stationary Gaussian process. For such a process, the probability density function for z is

$$p(z) = \frac{1}{\sqrt{2\pi R_z(0)}} e^{\frac{-z^2(t)}{2R_z(0)}} \tag{9.16}$$

where z has a zero mean value. This probability density function is shown in Fig. 9.6. A very important property of a Gaussian process is its behavior with respect to linear systems. For instance,

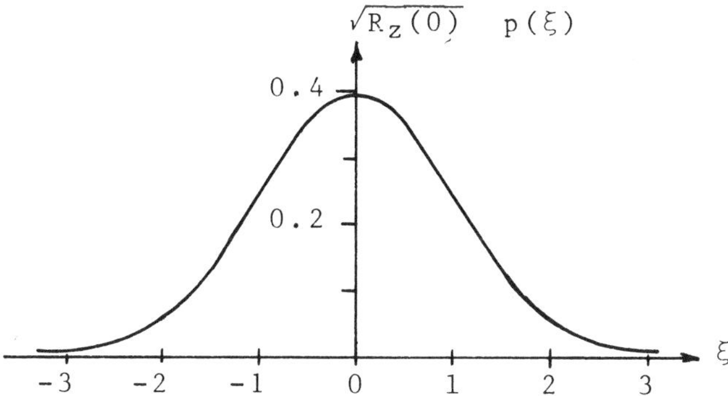

Figure 9.6 Gaussian Probability Density Function Where $\xi = z/\sqrt{R_z(0)}$

when the excitation force of a single degree of freedom of a linear system is Gaussian, then the response of this system is also Gaussian. The use of Gaussian probability in design is illustrated in the next chapter.

Data often shows that the response of a single degree of freedom, linear system is narrow band as shown in Fig. 9.4. In ocean systems which can be characterized by this description, it may be necessary to know the probability distribution of the peak amplitudes of z. If the probability density function for z is Gaussian, then the probability density function for the peaks of z at level a is

$$p(a) = \frac{a}{R_z(0)} e^{-\frac{a^2}{2R_z(0)}} \tag{9.17}$$

This is discussed by Powell (1958) in some detail. The function p(a), called the Rayleigh distribution, is shown in Fig. 9.7, where $R_z(0)$ is the variance of the response. Use will be made of the Rayleigh distribution in the study of material fatigue damage in Chapter XI. A clear discussion of probability distributions, with numerical examples, is given by Thomson (1965).

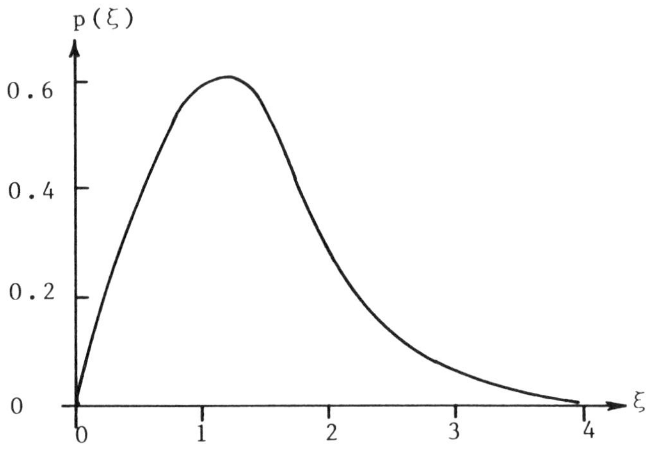

Figure 9.7 Rayleigh Probability Density Function Where $\xi = a/\sqrt{R_z(0)}$

9.4. REFERENCES

1. Crandall, S. H., and Mark, W. D. (1964), Random Vibration in Mechanical Systems, Academic Press, New York, Chapter 1.

2. Middleton, D. (1960), An Introduction to Statistical Communication Theory, McGraw-Hill Book Company, New York, p. 263.

3. Powell, A. (1958), "On the Fatigue Failure of Structures due to Vibrations Excited by Random Pressure Fields", Journal of the Acoustical Society of America, Vol. 30, 1958, pp. 1130-1135.

4. Thomson, W. T. (1965), Vibration Theory and Application, Prentice-Hall, Inc., Englewood Cliffs, N. J., Chapter 10.

Chapter X

THE ONE DEGREE OF FREEDOM AND CONTINUOUS BEAM MODELS

This chapter deals with the statistics for the vibration responses of a one degree of freedom model and also a simple beam model. In the first case, a tall offshore structure fixed to the ocean floor is assumed to vibrate in its first mode only when subjected to random wave forces. In the second case, a uniform, horizontal cross beam on a fixed ocean structure is assumed to vibrate in all modes under random forces. In each case, the statistical vibration analysis follows a definite pattern. Since the steps are rather lengthy, it is often easy to lose sight of the design goals of such an analysis. These steps to be followed are summarized:

1. Set up a reasonable mathematical model of the structure and formulate the governing equations of motion for this model.
2. Calculate the complex frequency response function where the applied force is harmonic. Deduce the response to a nonharmonic force.
3. Calculate the unit impulse response function where the applied force is step pulse. From this, deduce the response to a general type of force input.
4. Assume that the random input force is stationary and ergodic, and use the results of steps 2 and 3 to calculate the system's responses.
5. With the calculated or measured distribution for the power spectral density $S_F(\omega)$ of the input force, estimate S_o, the maximum value of $S_F(\omega)$. Calculate the mean square responses of the system for white noise, assuming S_o is constant for all frequencies.
6. Calculate the mean square responses for band-limited white noise, assuming that S_o covers a limited frequency range only, and is zero for all other frequencies.

7. From an assumed probability density function for the force, such as Gaussian, calculate the probability that a particular response will not exceed a certain specified value. Use these results to establish confidence limits for a particular design and modify the design accordingly.

10.1 - MODEL OF A FIXED OFFSHORE STRUCTURE

An actual offshore structure may consist of a platform supported by several vertical cylinders attached to the ocean bottom. The mass of the structure is distributed in the supports and in the cross-bracing between supports. A sketch of a four-legged structure with a rectangular platform, a jack-up rig used for exploration drilling, is shown in Fig. 10.1. Cross-bracing is kept to a minimum in order to minimize the wave and current forces on the structure. If the ocean floor is mainly mud, the legs may be driven more than 100 feet into the floor in order to achieve a stiff, stable structure.

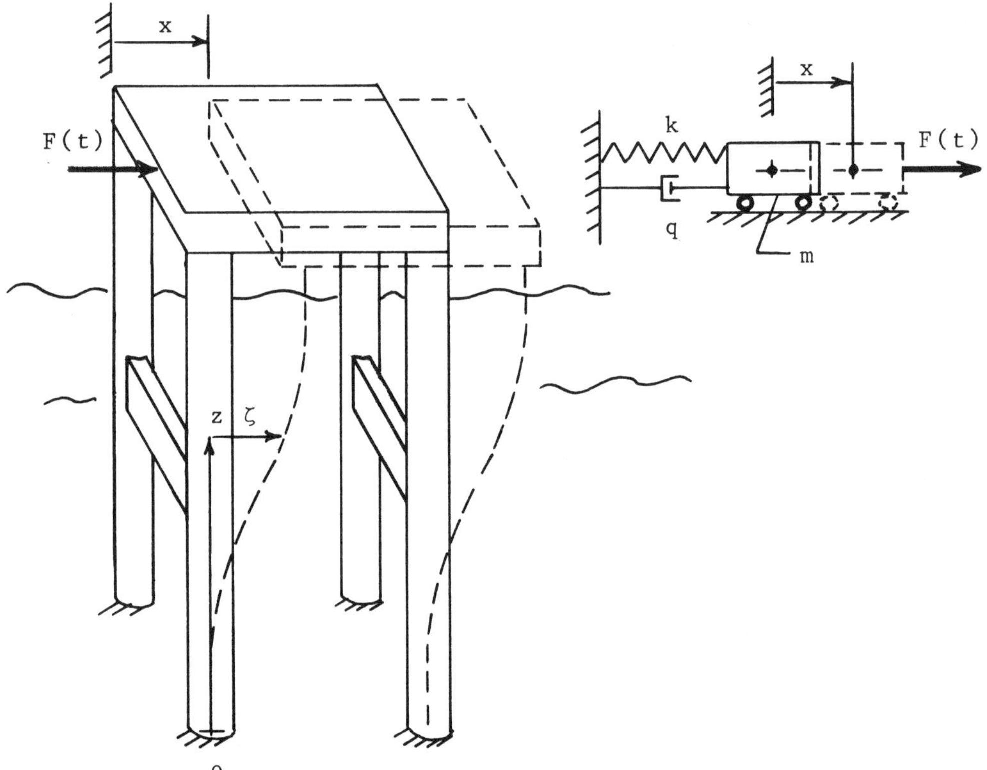

Figure 10.1 Schematic Diagram of an Offshore Structure and its Mathematical Model.

It can be imagined that waves in a random sea can cause such structures to vibrate in many modes. Nath and Harleman (1967) report that the frequency range of forces caused by waves is from a minimum of about 0.04 cps to a maximum of 0.70 cps. This corresponds to wave lengths from 3200 feet to 10 feet. Structures in 200 feet of water with natural frequencies from 0.25 to 0.125 cps are now being designed. These structures as well as ones which may be built in 400-500 feet of water, fall within the frequency spectrum of the ocean waves. The larger the number of waves with frequencies spanning the bending frequencies of the structure, the higher the probability is that the structure will undergo extreme vibrations, with subsequent failure.

The first mode shape, corresponding to the lowest or natural bending frequency of vibration, is often the most important vibration mode in these structures. Although higher modes exist and may be excited, the amplitudes of such oscillations often diminish rapidly due to damping of both the structural material and the sea environment. For this reason, the mathematical model of a fixed offshore structure will be taken as a one degree of freedom system. This mode shape is shown in Fig. 10.1. The displacement x of the platform, assumed to remain horizontal during motion, is the same as the displacement for the equivalent spring-mass system shown in this figure. The undamped natural frequency of the structure and of the equivalent system in water is

$$\omega_n = \sqrt{\frac{k}{m}} \tag{a}$$

where k is the spring constant determined from the structural stiffness. The mass m is a composite of the platform mass, the leg mass, and a portion of the water mass which is moved by the legs during free vibrations. The value of m can be calculated from (a) if k and ω_n are known.

The value of k can be found by employing the approximate method discussed by Timoshenko (1956), a method also used by Nath and Harleman (1967). The elastic deflection curve for one leg, in terms of the symbols in Fig. 10.1, can be approximated by

$$\zeta = \frac{F\ell^3}{24\ EI}\ \frac{1}{1 - \beta}\left(1 - \cos\frac{\pi z}{\ell}\right) \tag{b}$$

where

$$\beta = \frac{W\ell^2}{\pi^2\ EI}$$

E = Young's modulus

I = cross sectional moment of inertia of one leg

$4F$ = total horizontal static force on the platform, along x

$4W$ = total platform weight

The value of k is determined from (b) using the relationship $k = F/x$, where x is the value of ζ at $z = \ell$. It follows that

$$k = \frac{12\ EI}{\ell^3} \left(1 - \frac{W\ell^2}{\pi^2\ EI}\right) \qquad (c)$$

which is the spring constant for the whole structure as well as for one leg, as long as the stiffening effects of any cross bracings below the platform are negligible. Note that k decreases as W increases, and that when $W = \pi^2\ EI/\ell^2$, the critical Euler buckling load for each leg, the stiffness goes to zero.

The next step is to calculate the undamped natural frequency ω_o which the structure would have if it vibrated freely without the external resistance of the water mass. To do this, the Rayleigh method, as illustrated by Den Hartog (1947), will be employed. Assume as an approximation that the deflection curve for the structure undergoing free, harmonic vibrations has the same shape as the elastic deflection curve given by (b), or

$$\zeta = \zeta(z,\ t) = A \left(1 - \cos\frac{\pi z}{\ell}\right) \sin\omega_o t \qquad (d)$$

where A is an arbitrary constant, the amplitude of vibration, and ω_o is the natural frequency. As shown by Langhaar (1962), the total potential energy for the structural system is given by

$$V = 2\ EI \int_o^\ell \left(\frac{\partial^2\zeta}{\partial z^2}\right)^2 dz - 2\ W \int_o^\ell \left(\frac{\partial\zeta}{\partial z}\right)^2 dz \qquad (e)$$

where the first term on the right side of (e) is the strain energy in bending for all four legs. The second term is the change in potential energy of the platform weight, or the product of the platform weight and its decrease in vertical height due to the bending of the legs. With the expression for $\zeta(z,\ t)$ given by (d), V is evaluated. The maximum value of V occurs at the most extreme deflected position of the structure where the horizontal velocity and therefore the kinetic energy of the structure is zero. In this position, $\sin\omega_o t = 1$ and

$$V_{max} = \left[EI\ell \left(\frac{\pi}{\ell}\right)^4 - W\ell \left(\frac{\pi}{\ell}\right)^2\right] A^2 \qquad (f)$$

The translational kinetic energy of the structure due to motion in the x direction is assumed to be the dominant kinetic energy of the system. This energy is given by

$$T = 2\int_0^{\ell}\left(\frac{\partial\zeta}{\partial t}\right)^2 \rho dz + \frac{2W}{g}\left(\frac{\partial\zeta}{\partial t}\right)^2_{z=\ell} \tag{g}$$

where ρ is the mass density per unit length of one of the four identical legs. The first term on the right in (g) is the kinetic energy for all four legs, and the other term is the kinetic energy of the platform. The maximum value of T occurs when the structure assumes a vertical position, and when $\cos\omega_o t = 1$. With (d) and (g), the result is

$$T_{MAX} = \left[3\rho\ell + \frac{8}{g}W\right]\omega_o^2 A^2 \tag{h}$$

Since energy is conserved in this undamped system,

$$V_{MAX} = T_{MAX} \tag{i}$$

From (f), (h) and (i), the frequency of free vibration, unencumbered by water mass effects, is given by

$$\omega_o^2 = \frac{\pi^2 g\left(\frac{\pi^2 EI}{\ell^2} - W\right)}{8\ell\,(0.375\ m_1 g + W)} \tag{j}$$

where $m_1 = \rho\ell$, the mass of one leg, and W is one fourth the platform weight acting on one leg. It is observed that only 37½% of the actual mass of the leg enters into the mass of the equivalent single degree of freedom system.

Recent model tests for single, flexible cylinders vibrating in water have been performed in order to find the effects of water on the bending frequencies. Measurements of McLean, et al (1964) and Caldwell (1969) indicate that ω_n is always less than ω_o. That is

$$\omega_n = C\,\omega_o \tag{k}$$

where the value of C varies between 0.5 and 0.8 and is influenced by cylinder diameter, the end constraints, and the bending stiffness.

The results just obtained are summarized by the relations (a), (c), (j) and (k), or

$$\omega_n^2 = \frac{k}{m} = C^2\pi^2 g\,\frac{\left(\frac{\pi^2 EI}{\ell^2} - W\right)}{8\ell(0.375\ m_1 g + W)}$$

$$= \frac{12\ EI}{m\ell^3}\left(1 - \frac{W\ell^2}{\pi^2\ EI}\right) \tag{10.1}$$

For the structure shown in Fig. 10.1 where W, m_1, EI, ℓ and C are known, the parameters ω_n, k and m for the equivalent one degree of freedom system can be calculated from (10.1).

To complete the mathematical model of the structural system, one must consider the nature of the wave forces on the structure and also the damping effect of the water. With regard to the time varying forces, a conservative assumption would be to take a composite measure of all of the wave forces acting on the legs and apply the result as a single horizontal force F(t) which acts on the platform as shown in Fig. 10.1. Assume that this force has a time history similar to that of Fig. 9.1, and is known. Such traces are either deduced from direct measurements on existing similar structures, or from wave measurements, as discussed in Part II. As will be discussed later, only power spectral densities of F(t) need to be known. The effect of damping on the equivalent system can be studied by the inclusion of a damping force in the equation of motion, which is given by

$$m\ddot{x} + q\dot{x} + kx = F(t) \tag{10.2}$$

where

$F(t)$ = time varying force on mass m
$q\dot{x}$ = viscous damping force (assumed linear)
kx = spring force (linear)

The viscous drag force which is proportional to velocity squared is omitted, as it is only minor compared to other forces, as long as the structure is in relatively deep water.

The steady state solution to (10.2) when $F(t) = F_1 \cos \omega t$, a harmonically varying force of amplitude F_1 and frequency ω, is

$$x = \frac{F_1}{k} \frac{\cos(\omega t + \theta)}{\left[\left(1 - \frac{\omega^2}{\omega_n^2}\right)^2 + \left(2\xi \frac{\omega}{\omega_n}\right)^2\right]^{\frac{1}{2}}} \tag{10.3}$$

where

$\omega_n = \sqrt{k/m}$ = undamped natural frequency

$\xi = q\left(2\sqrt{km}\right)^{-1}$ = damping ratio

$$\theta = \tan^{-1}\left(\frac{-2\xi\dfrac{\omega}{\omega_n}}{1-\dfrac{\omega^2}{\omega_n^2}}\right) = \text{phase angle}$$

The maximum value of (10.3), x_o, is given when the cosine term is one. Under a static load F(t) = F_1, the static deflection x_s is simply F_1/k. The ratio x_o/x_s is thus, from (10.3)

$$\frac{x_o}{x_s} = \frac{1}{\left[\left(1-\dfrac{\omega^2}{\omega_n^2}\right)^2 + \left(2\xi\dfrac{\omega}{\omega_n}\right)^2\right]^{\frac{1}{2}}} \equiv |H(\omega)| \qquad (10.4)$$

which defines the magnitude of the frequency response function, H(ω), for this linear system. $|H(\omega)|$ is plotted in Fig. 10.2 as a function of the frequency ratio ω/ω_n for various values of damping. When $\omega = \omega_n$, resonance exists, and when $\omega = 0$, the static load condition exists. If ω_n is high enough, the case for some offshore structures in shallow water, $|H(\omega)| \simeq 1$, $x_o \simeq x_s$ by (10.4) and a static design of structures can be made with some confidence. For deep water structures, however, ω_n decreases as the structure is built taller. Thus, some of the wave frequencies ω may be close to ω_n, and those waves may cause $x_o >> x_s$. The magnitude of $|H(\omega)|$ in (10.4) indicates which values of ω are "dangerous".

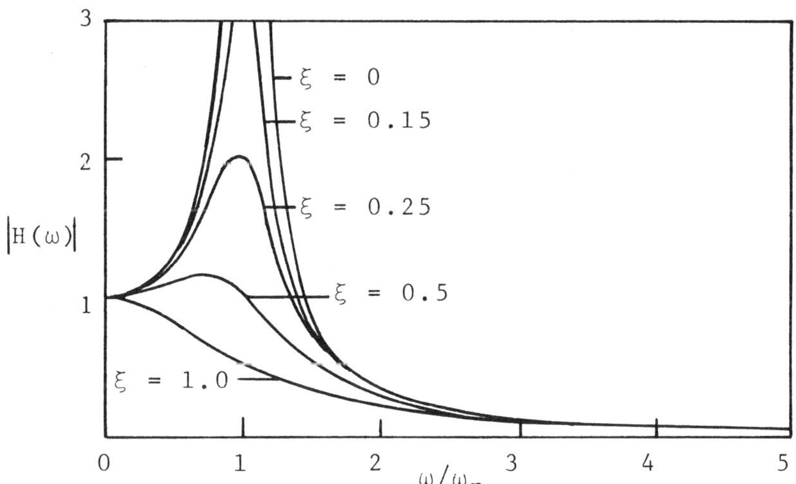

Figure 10.2 Frequency Response Function for the Linear System with Harmonic Excitation.

In this case, response of the system is the deflection x due to one harmonically varying force at one frequency. However, the wave forces acting on a real ocean structure have time histories as shown in Fig. 9.1 in which no well-defined frequencies or amplitudes are apparent. After studying such traces, one might ask "How can one predict x for design purposes?" The answer is that we can not predict x exactly. However, if F(t) is stationary-ergodic which often appears to be a reasonable assumption, the statistics defined in Chapter IX will allow us to answer the question: "For an arbitrary value of x, what is the probability that this value will be exceeded?" Such answers can be extremely valuable to a designer in predicting confidence limits for his design. Two quantities needed to find the response to a random force are now discussed. They are the complex frequency response, $H(\omega)$, and the response to a unit impulse.

10.2 - THE COMPLEX FREQUENCY RESPONSE

Since (10.2) is a linear differential equation with constant coefficients, a simple harmonic input F(t) of frequency ω led to a simple harmonic steady state response. The solution (10.3) shows however that both the amplitude and phase of x depends on ω. A concise way to describe the frequency dependence of both amplitude and phase is to derive the complex frequency response, $H(\omega)$. To do this, we describe the harmonic force excitation by

$$F(t) = F_1\, e^{i\omega t} \tag{10.5}$$

where the input force is defined by the real part of (10.5). In this case, since F_1 is real and $e^{i\omega t} = \cos \omega t + i \sin \omega t$ by definition, the force is $F_1 \cos \omega t$.

Assume that a solution x is of this same form. Thus

$$x = \frac{F_1}{k} H(\omega)\, e^{i\omega t} \tag{10.6}$$

When (10.5) and (10.6) are substituted into (10.2) and $H(\omega)$ is solved for algebraically, the result is

$$H(\omega) = \frac{1}{1 - \left(\frac{\omega}{\omega_n}\right)^2 + i2\zeta\frac{\omega}{\omega_n}} \tag{10.7}$$

In this problem, $H(\omega)$ has no physical units. This is consistent with (10.6) where the dimensions of both (F_1/k) and x are length. The reader can verify that (10.6) with (10.7) can be written in the form

$$x = A + iB$$

where A and B are real, time-dependent numbers. From this, it follows that amplitude of the response is

$$\frac{F_1}{k} \left(A^2 + B^2\right)^{\frac{1}{2}} = |H(\omega)|$$

and the phase is that of (10.3), or

$$\tan^{-1}\left(\frac{B}{A}\right) = \theta$$

All that this demonstrates is that the complex forms given by (10.5 - 10.7) are consistent with (10.3) and (10.4). This has also been demonstrated by Chen (1966). In the present development, the complex forms are easier to use.

The solutions to (10.2) will now be found for the case where F(t) is composed of N separate harmonically varying forces, each with a distinct amplitude, each out of phase with the other, where each frequency is an integer multiple n of a base frequency ω_0. As demonstrated by Crandall (1958), this force can then be represented by the exponential form of the Fourier series, given by

$$F(t) = \sum_{n=1}^{N} A_n e^{in\omega_0 t} \tag{10.8}$$

It is assumed that F(t) is periodic over the time interval $T = \omega_0/2\pi$. The coefficients A_n are given by

$$A_n = \frac{1}{T} \int_t^{t+T} F(t)\, e^{-in\omega_0 t}\, dt \tag{10.9}$$

For each force of frequency $n\omega_0$, there will be a corresponding response function $H(n\omega_0)$ given by (10.7), where ω is replaced by $n\omega_0$. The steady state response to all of these periodic forces is then given by the superposition of the responses of each component. This result is found by summing the solutions (10.6), where $H(n\omega_0)$ replaces $H(\omega)$ and the series form of F(t) replaces $F_1 e^{i\omega t}$. The result is

$$x(t) = \sum_{n=1}^{N} \frac{1}{k} A_n H(n\omega_o) e^{in\omega_o t}$$

$$= \sum_{n=1}^{N} \frac{H(n\omega_o)}{kT} e^{in\omega_o t} \int_{\alpha=t}^{\alpha=t+T} F(\alpha) e^{-in\omega_o\alpha} d\alpha \qquad (10.10)$$

For a random-type force such as shown in Fig. 9.1 there are no discrete frequencies $n\omega_o$, and no periodic time intervals T. However, assume that the force can be represented by a Fourier series analogous to (10.8). That is

$$F(t) = \sum_{n=-\infty}^{\infty} C_n e^{in\omega_o t} \qquad (10.11)$$

where

$$C_n = \frac{1}{2T} \int_{-T}^{T} F(\alpha) e^{-in\omega_o\alpha} d\alpha \qquad (10.12)$$

The response is then given approximately by an equation analogous to (10.10) or

$$x(t) = \sum_{n=-\infty}^{\infty} C_n H(n\omega_o) e^{in\omega_o t}$$

$$= \sum_{n=-\infty}^{\infty} \frac{H(n\omega_o)}{k(2T)} e^{in\omega_o t} \int_{-T}^{T} F(\alpha) e^{-in\omega_o\alpha} d\alpha$$

$$(10.13)$$

Here, the time interval has been redefined, where $T = \pi/\omega_o$ instead of $2\pi/\omega_o$. The idea of discrete frequencies only is now overcome. To do this, the incremental frequency is used

$$\Delta\omega = (n + 1)\,\omega_o - n\omega_o = \omega_o = \frac{2\pi}{2T}$$

and $1/T$ is replaced by $\frac{\Delta\omega}{\pi}$ in (10.13). Thus

$$x(t) \approx \sum_{n=-\infty}^{\infty} \frac{H(n\omega_o)\ \Delta\omega}{2k\pi} \int_{-T}^{T} F(\alpha) e^{-in\omega_o\alpha}\ e^{in\omega_o t}\ d\alpha$$

In the absence of a time period 2T over which a random force repeats itself, the time period must be extended to cover all time. Also, in the absence of discrete frequencies $n\omega_o$, all values of frequency should be included. These extensions are done by letting $T \to \infty$, $n\omega_o \to \omega$ and $\Delta\omega \to d\omega$, and by replacing the sum in the last equation by an integral. The response to a non-periodic force thus becomes

$$x(t) = \frac{1}{2\pi} \int_{-\infty}^{\infty} \frac{H(\omega)}{k} e^{i\omega t}\ d\omega \int_{-\infty}^{\infty} F(\alpha) e^{-i\omega\alpha}\ d\alpha \tag{10.14}$$

which is a result derived in another way by Crandall and Mark (1963). Thus, if F(t) is a given random force and $H(\omega)$ is given by (10.7), it is possible, in principle, to calculate the response. If the integrations can be performed, the result will again be a complex number of the form

$$x = A + iB$$

where A and B are real numbers which depend on time, from which the amplitude and phase can be calculated in the same manner as already discussed for (10.6).

Since it is not usually practical to carry out the integrations of (10.14), certain assumptions will be made about the nature of F(t) which allows one to calculate and interpret responses. Toward this end, the response to a unit impulsive force is now found.

10.3 - THE UNIT IMPULSE RESPONSE

The response of the spring-mass system is now found when F(t) is a unit impulsive force. That is

$$F(t) = \delta(t) \begin{cases} = 1 \text{ at } t = 0 \\ = 0 \text{ otherwise} \end{cases} \tag{10.15}$$

where $\delta(t)$ is the Dirac delta function. This force history, pictured in Fig. 10.3 represents a blow on mass m.

We will now find the solution $x = h(t)$ to (10.2) where F(t) is given by (10.15). The general solution to the governing equation

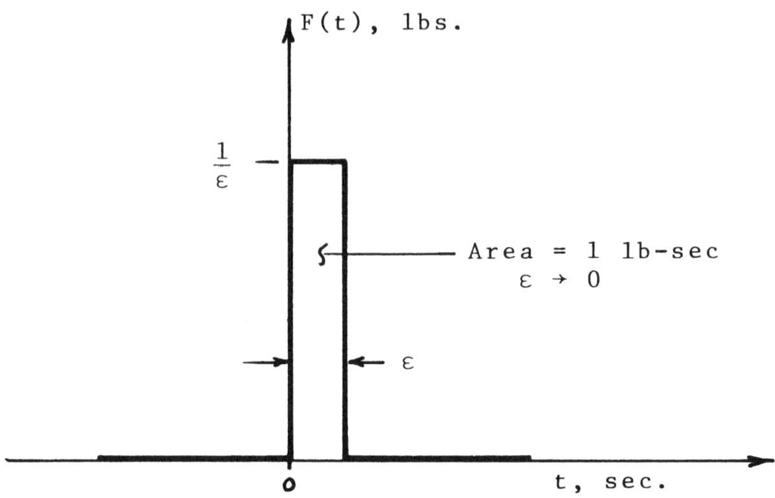

Figure 10.3 Time History of a Unit Impulsive Force.

$$m\ddot{h}(t) + q\dot{h}(t) + kh(t) = \delta(t) \tag{10.16}$$

for $t \geq 0$, or when the right hand side is zero, is given by

$$h(t) = e^{-\xi\omega_n t}\left(C_3 \sin\sqrt{1-\xi^2}\,\omega_n t + C_4 \cos\sqrt{1-\xi^2}\,\omega_n t\right) \tag{10.17}$$

All quantities in (10.17) are defined under (10.3), except the arbitrary constants C_3 and C_4. These are determined from the initial conditions, or the displacement and velocity of the mass at the instant $t = \varepsilon$, where $\varepsilon \to 0$. At this time, the mass has not yet moved, so the initial displacement $h(\varepsilon) = 0$, giving $C_4 = 0$. To find C_3 the initial value of $\dot{h}(\varepsilon)$ must be known. This can be found by integrating (10.16) over the short time interval ε. Thus

$$m\dot{h}(\varepsilon) + qh(\varepsilon) + k\int_0^{\varepsilon} h(t)\,dt = \int_0^{\varepsilon} \delta(t)\,dt$$

Since the integral of $\delta(t)$ is defined as one as $\varepsilon \to 0$ and since $h(\varepsilon) = 0$ as $\varepsilon \to 0$, then $\dot{h}(\varepsilon) = 1/m$. From (10.17), the result at time $t = \varepsilon \to 0$ is

$$\frac{1}{m} = C_3 \sqrt{1-\xi^2}\,\omega_n$$

It follows that the unit impulse-response function is given by

$$h(t) = 0 \qquad , \quad t < 0$$

$$h(t) = \frac{e^{-\xi\omega_n t}}{m\omega_n \sqrt{1 - \xi^2}} \sin\left[\sqrt{1 - \xi^2}\ \omega_n t\right] , \quad t > 0 \tag{10.18}$$

This function is pictured in Fig. 10.4.

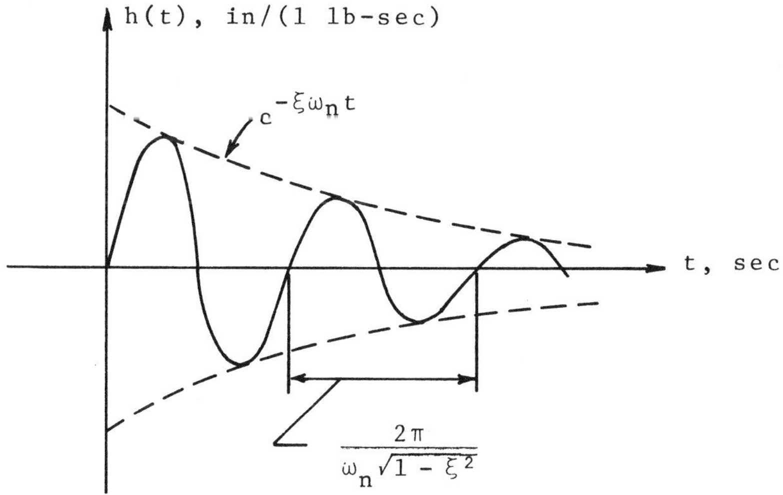

Figure 10.4 Response Function for a Unit Impulsive Force on the Linear System.

Suppose now that the imposed force is arbitrary. At a time t_i, suppose that the average value of force is $F(t_i)$ and that this force acts for time Δt_i. The response of the spring mass system at a time $t > t_i$ due to the impulse $F(t_i)\ \Delta t_i$ is then approximately

$$x(t) \simeq F(t_i)\ h(t - t_i)\ \Delta t_i$$

It is noted that the base or reference time was taken as t_i instead as zero, as it was for (10.18). Because the system is linear, superposition holds and one can approximate the response due to n such rectangular impulses, each occurring at different times t_i, as the sum

$$x(t) \simeq \sum_{i=1}^{n} F(t_i)\ h(t - t_i)\ \Delta t_i$$

$$t_i = i\ \Delta t_i\ ,\quad i = 1, 2, \ldots n.$$

In the limit $\Delta t_i \to d\tau$, and $t_i \to \tau$ where τ covers all time from $-\infty$ to t. Thus, the finite sum becomes the integral

$$x(t) = \int_{-\infty}^{t} F(\tau)\ h(t - \tau)\ d\tau$$

Furthermore, since $h(t - \tau)$ vanishes when $t < \tau$, the upper limit can be replaced by $+\infty$ without changing the integral, or

$$x(t) = \int_{-\infty}^{\infty} F(\tau)\ h(t - \tau)\ d\tau \qquad (10.19)$$

An alternate form of (10.19) can be obtained by interchanging the integration variable from τ to $\theta = t - \tau$, or

$$x(t) = \int_{0}^{\infty} F(t - \theta)\ h(\theta)\ d\theta \qquad (10.20)$$

where the lower limit was replaced by zero since $h(\theta)$ vanishes for negative θ by (10.18). From (10.18) and (10.19) the response due to an arbitrary force F(t) is

$$x(t) = \int_{-\infty}^{\infty} F(\tau)\ \frac{e^{-\xi\omega_n(t-\tau)}}{m\omega_n\sqrt{1 - \xi^2}}\ \sin\left[\sqrt{1 - \xi^2}\ \omega_n(t - \tau)\right]\ d\tau \qquad (10.21)$$

This is comparable to the result of (10.14). The connection is that $H(\omega)$ is the Fourier transform of h(t), to within a constant multiplying factor. As was the case for (10.14), the integrations of (10.21) are usually not practical to perform when F(t) is random in nature. Together, these results will now be used to obtain the mean square values for x(t) when F(t) is random, but when it obeys the stationary and ergodic assumptions.

10.4 - RESPONSE TO A STATIONARY, ERGODIC FORCE

The autocorrelation function and power spectral density for the displacement response x(t), defined by (9.11) and (9.12), are

$$R_x(\tau) = \int_{-\infty}^{\infty} S_x(\omega)e^{i\omega\tau}\, d\omega \tag{10.22}$$

and

$$S_x(\omega) = \frac{1}{2\pi}\int_{-\infty}^{\infty} R_x(\tau)e^{-i\omega\tau}\, d\tau \tag{10.23}$$

Since F(t) is stationary and ergodic and the spring-mass system is a linear one, the parameters $R_x(\tau)$ and $S_x(\omega)$ can be related to the corresponding (known) parameters for F(t). A brief sketch of the procedure used to find this connection is summarized. The reader may wish to complete the detailed proof himself, and then check his calculations with those of Crandall and Mark (1963).

One starts with the definition given by (9.9a), or

$$R_x(\tau) = \lim_{T\to\infty} \frac{1}{2T}\int_{-T}^{T} x(t)\; x(t+\tau)\; dt$$

To evaluate this, replace x(t) and x(t + τ) in the integrand by their equivalent integral forms as given by (10.20). After employing the definitions of $R_F(\tau)$, $S_F(\omega)$ and $S_x(\omega)$, and deducing that x(t) will be stationary and ergodic since F(t) is, the result is

$$S_x(\omega) = S_F(\omega)\int_{-\infty}^{\infty} h(\theta_1)e^{i\omega\theta_1}\, d\theta_1 \int_{-\infty}^{\infty} h(\theta_2)e^{-i\omega\theta_2} \tag{10.24}$$

To evaluate these integrals, the connection between H(ω) and h(t) is noted. This is seen by letting F(t) = δ(t) in (10.14) for which the response is defined as h(t). It follows from (10.14) that

$$h(t) = \frac{1}{2\pi}\int_{-\infty}^{\infty} \frac{H(\omega)}{k}\, e^{i\omega t}\, d\omega$$

since

$$\int_{-\infty}^{\infty} F(\alpha)e^{-i\omega\alpha}\, d\alpha = \int_{-\infty}^{\infty} \delta(\alpha)e^{-i\omega\alpha}\, d\alpha = 1$$

This is the Fourier integral representation of h(t) in which $H(\omega)/k$ is the Fourier transform of h(t), or

$$\frac{H(\omega)}{k} = \int_{-\infty}^{\infty} h(t)e^{-i\omega t}\, dt \qquad (10.25)$$

The integrals in (10.24) are evaluated using the result of (10.25). It follows

$$S_x(\omega) = \frac{1}{k^2}\, H(\omega) \cdot H(-\omega)\, S_F(\omega)$$

or

$$S_x(\omega) = \frac{1}{k^2}\, |H(\omega)|^2 \quad S_F(\omega) \qquad (10.26)$$

It is recalled that the mean square response is given by $\overline{x^2(t)} = R_x(0)$. It follows from (10.22) and (10.26) that

$$\overline{x^2(t)} = \int_{-\infty}^{\infty} \frac{1}{k^2}|H(\omega)|^2 \quad S_F(\omega)\, d\omega \qquad (10.27)$$

Since $|H(\omega)|$ has already been evaluated for this system in (10.4) a knowledge of the power spectral density of the force is all that is needed to calculate the mean square value of the displacement. As was mentioned, $S_F(\omega)$ can be determined from F(t) either by electronic measurements or by integrating (9.10a), where $z = F$ and where $R_F(\tau)$ is calculated from (9.9a). In the following illustrations it is assumed that $S_F(\omega)$ is known.

10.5 - RESPONSE TO IDEAL WHITE NOISE

Although measurements may show that $S_F(\omega)$ follows a trace something like the one shown in Fig. 9.3, this trace is idealized to the dotted line, where $S_F(\omega) = S_o$ for all ω. In this case, the mean square displacement is, from (10.27) and (10.4)

$$\overline{x^2(t)} = R_x(0) = \int_{-\infty}^{\infty} \frac{S_o}{k^2} |H(\omega)|^2 \, d\omega$$

$$= \frac{2s_o}{k^2} \int_o^{\infty} \frac{d\omega}{\left(1 - \frac{\omega^2}{\omega_n^2}\right)^2 + \left(2\xi \frac{\omega}{\omega_n}\right)^2} = \frac{\pi \, \omega_n \, S_o}{2 \, \xi \, k^2} \qquad (10.28)$$

Integrations such as these can be found in mathematical tables or can be performed by the method of residues. The result (10.28) is physically reasonable, as one would expect small displacements for strong springs, or high k, and for high damping, or large ξ. It should be observed that the units are consistent if frequencies are always in radians/unit time. For ease of reference, the nomenclature used, with a typical set of engineering units, is summarized in Table 10.1.

NOMENCLATURE		TYPICAL UNITS
g	gravity constant	386. in/sec^2
k	equivalent spring constant	lb/in
m	equivalent mass	lb sec^2/in.
t,τ,T	time	sec
$F(t)$	random force	lb
$\delta(t)$	impulsive force	one lb
$h(t)$	impulsive response	in/lb-sec
ω,ω_n	frequency	radians/sec
x	displacement	inches
q	damping constant	lb-sec/in.
ξ	damping ratio, (10.3)	---
$H(\omega)$	complex frequency response, (10.7)	---
$R_F(\tau)$	Autocorrelation function for F(t)	lb^2
$R_x(\tau)$	Autocorrelation function for x(t)	in^2
$S_F(\omega)$	Power spectral density for F(t)	lb^{2}sec/radian
$S_x(\omega)$	Power spectral density for x(t)	in^{2}sec/radian
$P(F)$	Probability for F(t)	---
$p(F)$	probability density function for F(t)	(lb)$^{-1}$

TABLE 10.1

The power spectral density for the displacement response from (10.26) is

$$S_x(\omega) = \frac{S_o}{k^2} \, |H(\omega)|^2$$

$$= \frac{S_o}{k^2} \cdot \frac{1}{\left(1 - \frac{\omega^2}{\omega_n^2}\right)^2 + \left(2\xi \frac{\omega}{\omega_n}\right)^2} \qquad (10.29)$$

which is sketched in Fig. 10.5 for the case of light damping ($\xi = 0.05$, for instance). It is noted that $S_x(\omega)$ resembles a typical narrow band spectrum, with the peak value at the natural frequency ω_n. Moreover, Fig. 9.4 now gives a clue to the actual behavior of x(t); its trace will look something like z(t) in this figure. It will be recalled that F(t) actually looks more like the trace in Fig. 9.3.

An infinitely wide band excitation (S_F = const.) is never realized in the physical world. Nevertheless, the conclusions based on this idealization often provide a good approximation to x(t) for a lightly damped system.

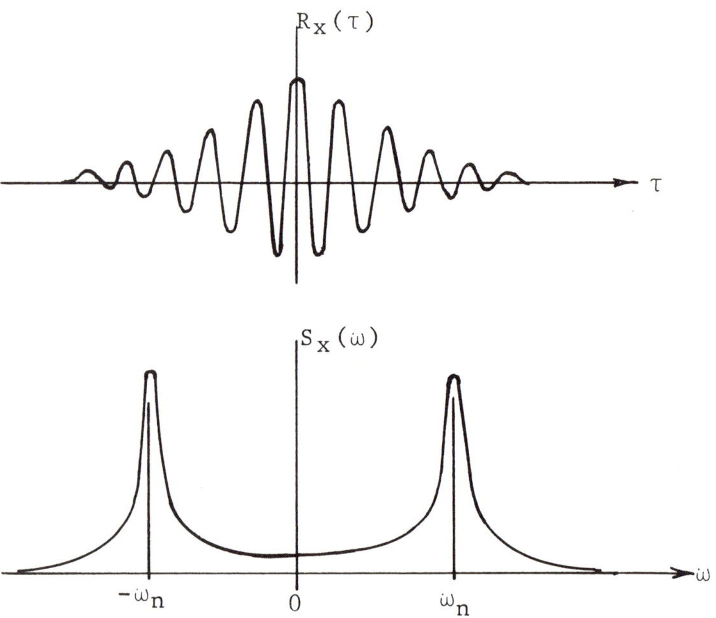

Figure 10.5 Response Statistics for the Single Degree of Freedom Model with White Noise Excitation.

10.6 - RESPONSE TO BAND-LIMITED WHITE NOISE

Some deep water ocean structures are being planned which have frequencies ω_n within the range of wave frequencies. Suppose that the wave frequencies, and therefore the frequencies of F(t), lie predominantly with a narrow band, $\omega_1 \leq \omega \leq \omega_2$, and that ω_n lies in the middle of this range. Suppose further that $S_F(\omega)$ is idealized as the dotted rectangular function shown in Fig. 9.4. The mean square displacement is found from (10.27) for $\omega_1 < |\omega| < \omega_2$ or

$$\overline{x^2(t)} = \int_{-\omega_2}^{-\omega_1} \frac{S_o}{k^2} |H(\omega)|^2 \, d\omega + \int_{\omega_1}^{\omega_2} \frac{S_o}{k^2} |H(\omega)|^2 \, d\omega$$

$$= \frac{\pi \omega_n S_o}{2 \zeta k^2} \left[I\left(\frac{\omega_2}{\omega_n}, \xi\right) - I\left(\frac{\omega_1}{\omega_n}, \xi\right) \right] \tag{10.30}$$

where the integral factor $I\left(\frac{\omega}{\omega_n}, \xi\right)$ is

$$I\left(\frac{\omega}{\omega_n}, \xi\right) = \frac{1}{\pi} \tan^{-1}\left[\frac{2\xi\left(\frac{\omega}{\omega_n}\right)}{1-\left(\frac{\omega}{\omega_n}\right)^2}\right] + \frac{\xi}{2\pi\sqrt{1-\xi}} \ln\left[\frac{1+\left(\frac{\omega}{\omega_n}\right)^2 + 2\sqrt{1-\xi^2}\left(\frac{\omega}{\omega_n}\right)}{1+\left(\frac{\omega}{\omega_n}\right)^2 - 2\sqrt{1-\xi^2}\left(\frac{\omega}{\omega_n}\right)}\right] \tag{10.31}$$

A graph of (10.31) for two values of ξ is shown in Fig. 10.6. For ideal white noise

$$I(\infty, \xi) - I(0, \xi) = 1$$

and for limited-band excitation

$$I\left(\frac{\omega_2}{\omega_n}, \xi\right) - I\left(\frac{\omega_1}{\omega_n}, \xi\right) < 1$$

It is clear from Fig. 10.6, however, that the factor in brackets in (10.30) is only slightly less than one as long as ω_1 and ω_2 span ω_n and the damping is light.

The power spectral density for the displacement is, from (10.26)

$$S_x(\omega) = \begin{cases} \dfrac{S_o}{k^2} \dfrac{1}{\left(1 - \dfrac{\omega^2}{\omega_n^2}\right)^2 + \left(2\xi \dfrac{\omega}{\omega_n}\right)^2}, & \omega_1 < \omega < \omega_2 \\ 0, & \omega < \omega_1 \text{ and } \omega > \omega_2 \end{cases} \tag{10.32}$$

which is sketched in Fig. 10.7. This result should be compared with the result for $S_x(\omega)$ of Fig. 10.5.

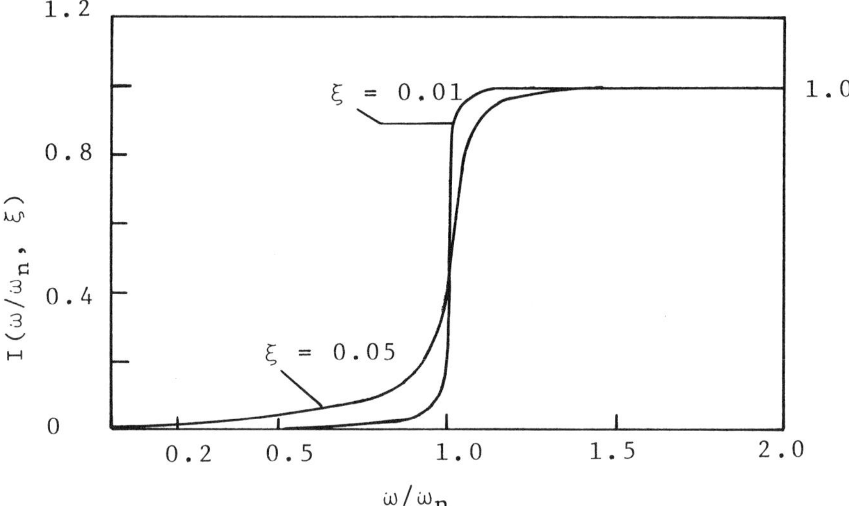

Figure 10.6 Plot of the Function Given by Equation (10.31).

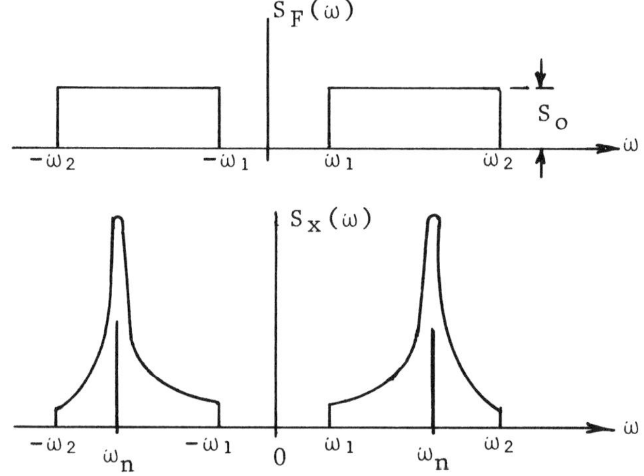

Figure 10.7 Power Spectral Densities for a Narrow Band Input $S_F(\omega)$, and the Response $S_x(\omega)$ of a Single Degree of Freedom System.

For ease of reference, the statistics for the single degree of freedom linear system are summarized in Table 10.2. Inspection of these results show that the ideal white noise assumption is justified even if the input is narrow band, provided that the following two criteria are met.

$$\begin{aligned} &(1) \quad \omega_1 < \omega_n < \omega_2 \\ &(2) \quad (\omega_2 - \omega_1) >> 2\xi\,\omega_n \end{aligned} \tag{10.33}$$

If these are met, then

$$\overline{x^2(t)} = \frac{\pi\,\omega_n\,S_o}{2\,\xi k^2} \tag{10.34}$$

where S_o is the average power spectral density of F(t) for $\omega_1 < \omega < \omega_2$.

10.7 - PROBABILITY AND DESIGN LIMITS

Assume that F(t) is Gaussian. It is stated without proof that x(t) is also Gaussian in this example, or

$$p(x) = \frac{1}{\sqrt{2\pi R_x(0)}}\, e^{-\frac{x^2(t)}{2R_x(0)}} \tag{10.35}$$

where x(t) has a zero mean value. If F(t) is stationary-ergodic, then x(t) is, and p(x) can be found when $R_x(0) = \overline{x^2(t)}$ is known. If this is the case, the probability that x(t) has a value within

$$\pm\lambda\sqrt{\overline{x^2(t)}},$$

where λ is an arbitrary constant, is given by

STATISTICS FOR THE LINEAR, SINGLE DEGREE OF FREEDOM SYSTEM

F(t) t

$$m\ddot{x} + q\dot{x} + kx = F(t)$$

F(t) INPUT
x(t) OUTPUT

INPUT QUANTITIES	OUTPUT QUANTITIES
1. $F(t) = F_1 e^{i\omega t}$	$x(t) = \frac{F_1}{k} H(\omega) e^{i\omega t}$
HARMONIC	$H(\omega) = \dfrac{1}{1 - \left(\frac{\omega}{\omega_n}\right)^2 + 2i\xi\left(\frac{\omega}{\omega_n}\right)}$
	$\omega_n = \sqrt{k/m}$; $\xi = q/(2\sqrt{km})$
2. $F(t) = \delta(t)$ UNIT IMPULSE	$h(t) = \dfrac{e^{-\xi\omega_n t}}{m\omega_n\sqrt{1-\xi^2}} \sin\left[\sqrt{1-\xi^2}\,\omega_n t\right]$
3. $F(t)$ FROM STEP 1., ALL ω'S	$x(t) = \frac{1}{2\pi}\int_{-\infty}^{\infty} \frac{H(\omega)}{k} e^{i\omega t}\, d\omega \int_{-\infty}^{\infty} F(\alpha) e^{-i\omega\alpha}\, d\alpha$
4. $F(t)$ FROM STEP 2., ALL ω'S	$x(t) = \int_{-\infty}^{\infty} F(\tau) \dfrac{e^{-\xi\omega_n(t-\tau)}}{m\omega_n\sqrt{1-\xi^2}} \sin\left[\sqrt{1-\xi^2}\,\omega_n(t-\tau)\right] d\tau$
	$= \int_{-\infty}^{\infty} F(t-\theta)\, h(\theta)\, d\theta$
5. MEAN VALUE:	MEAN VALUE:
$\overline{F(t)} = \lim_{T\to\infty} \frac{1}{2T}\int_{-T}^{T} F(t)dt$	$\overline{x(t)} = \lim_{T\to\infty} \frac{1}{2T}\int_{-T}^{T} x(t)\, dt$

TABLE 10.2

6. MEAN SQUARE VALUE:	MEAN SQUARE VALUE
$\overline{F^2(t)} = \lim_{T\to\infty} \frac{1}{2T} \int_{-T}^{T} F^2(t)\, dt$	$\overline{x^2(t)} = \lim_{T\to\infty} \frac{1}{2T} \int_{-T}^{T} x^2(t)\, dt$
$= R_F(0)$	$= R_x(0)$
7. AUTOCORRELATION FUNCTION:	AUTOCORRELATION FUNCTION:
$R_F(\tau) = \lim_{T\to\infty} \frac{1}{2T} \int_{-T}^{T} F(t)\, F(t+\tau) dt$	$R_x(\tau) = \lim_{T\to\infty} \frac{1}{2T} \int_{-T}^{T} x(t) x(t+\tau) dt$
$= \int_{-\infty}^{\infty} S_F(\omega) e^{i\omega\tau}\, d\omega$	$= \int_{-\infty}^{\infty} S_x(\omega) e^{i\omega\tau}\, d\omega$
8. POWER SPECTRAL DENSITY:	POWER SPECTRAL DENSITY:
$S_F(\omega) = \frac{1}{2\pi} \int_{-\infty}^{\infty} R_F(\tau) e^{-i\omega\tau}\, d\tau$	$S_x(\omega) = \frac{1}{2\pi} \int_{-\infty}^{\infty} R_x(\tau) e^{-i\omega\tau}\, d\tau$

INPUT-OUTPUT RELATIONS FOR STATIONARY ERGODIC SYSTEMS:

$$R_x(\tau) = \int_{-\infty}^{\infty} \int_{-\infty}^{\infty} R_F(\tau + \theta_1 - \theta_2)\, h(\theta_1)\, h(\theta_2)\, d\theta_1\, d\theta_2$$

$$S_x(\omega) = \frac{|H(\omega)|^2}{k^2}\, S_F(\omega)$$

$$\overline{x^2(t)} = R_x(0) = \int_{-\infty}^{\infty} S_x(\omega)\, d\omega = \int_{-\infty}^{\infty} \frac{|H(\omega)|^2}{k^2} S_F(\omega)\, d\omega$$

TABLE 10.2 - CONTINUED

EXAMPLES

A. IDEAL WHITE NOISE

$$\overline{x^2(t)} = 2 \int_0^\infty \frac{S_o}{k^2} |H(\omega)|^2 \, d\omega = \frac{\pi}{2} \frac{\omega_n S_o}{\xi k^2}$$

$$S_x(\omega) = \frac{S_o}{k^2} |H(\omega)|^2$$

B. BAND-LIMITED WHITE NOISE

(NARROW-BAND SPECTRAL DENSITY)

$$\overline{x^2(t)} = \frac{\pi \omega_n S_o}{2 \xi k^2} \left[I\left(\frac{\omega_2}{\omega_n}, \xi\right) - I\left(\frac{\omega_1}{\omega_n}, \xi\right) \right]$$

$$S_x(\omega) = \frac{S_o}{k^2} |H(\omega)|^2 \quad \text{for} \quad \omega_1 < \omega < \omega_2$$

$$= 0 \quad \text{ELSEWHERE}$$

where

$$I\left(\frac{\omega}{\omega_n}, \xi\right) = \text{Eq. (10.31)}$$

$$|H(\omega)|^2 = \text{Eq. (10.4)}$$

TABLE 10.2 - CONTINUED

$$P[-\lambda\bar{\sigma} \le x(t) \le \lambda\bar{\sigma}] = \int_{-\lambda\bar{\sigma}}^{+\lambda\bar{\sigma}} p(x)\,dx$$

$$= \frac{1}{\bar{\sigma}\sqrt{2\pi}} \int_{-\lambda\bar{\sigma}}^{\lambda\bar{\sigma}} e^{-\frac{x^2(t)}{2\bar{\sigma}^2}}\,dx \qquad (10.36)$$

where

$$\bar{\sigma} = \sqrt{\overline{x^2(t)}}$$

The probability that x(t) lies outside $\pm\lambda\bar{\sigma}$ is simply one minus the above value, or

$$P[|x(t)| > \lambda\bar{\sigma}] = 1 - P[-\lambda\bar{\sigma} \le x(t) \le \lambda\bar{\sigma}]$$

Table 10.3 shows numerical values of (10.36) for $\lambda = 1, 2,$ and 3. They were evaluated from "error function" tables. We see that the probability of $|x|$ exceeding its root mean square value, $\sqrt{\overline{x^2(t)}}$, is 0.317 or 31.7%. The probability of exceeding twice its rms value is 4.6%, and the probability of exceeding three times the rms value is only 0.3%.

$\bar{\sigma} = \sqrt{\overline{x^2(t)}} = \sqrt{R_x(0)}$		
λ	$P[-\lambda\bar{\sigma} \le x(t) \le \lambda\bar{\sigma}]$	$P[\|x(t)\| > \lambda\bar{\sigma}]$
1	68.3%	31.7%
2	95.4%	4.6%
3	99.7%	0.3%

TABLE 10.3

To check the "reliability" of a particular design, one would (1), assume F(t) stationary, ergodic, and Gaussian; (2), check the frequency range and band width criteria of (10.33). If these are met, then $\overline{x^2(t)}$ would be calculated from (10.34) for known values of ω_n, S_o, ξ, and k.

Consider the following numerical example. If $\overline{x^2(t)}$ were calculated to be 16.0 in., for instance, the probability that the displacement would exceed 4.0 inches would be 31.7%. However, the probability $|x|$ would exceed 12 inches is only 0.3%. If the engineer calculates that a 12 inch horizontal static displacement at the top of his structure would not produce excessive stresses and fatigue failure in any other part of his structure, then he can proceed with "reasonable confidence" with his design. It is noted, however, that the actual confidence limits are quite arbitrary.

10.8 - THE CONTINUOUS BEAM MODEL

One of the uniform cross-beams of a deep-water structure, shown in Fig. 10.1 will now be considered. The action of the currents and waves impinging normal to the beam causes forces which look something like the trace of Fig. 9.1. The responses considered will be the mean square values of deflection, slope, bending moment, and shear force along the beam. With an estimate of the bending moments M, bending stresses S can be estimated to the same degree of certainty using the classical formula $S = Mc/I$. Here, M represents the root mean square value, $\sqrt{\overline{M^2}}$, where c and I are defined in Fig. 10.8 for the beam shown.

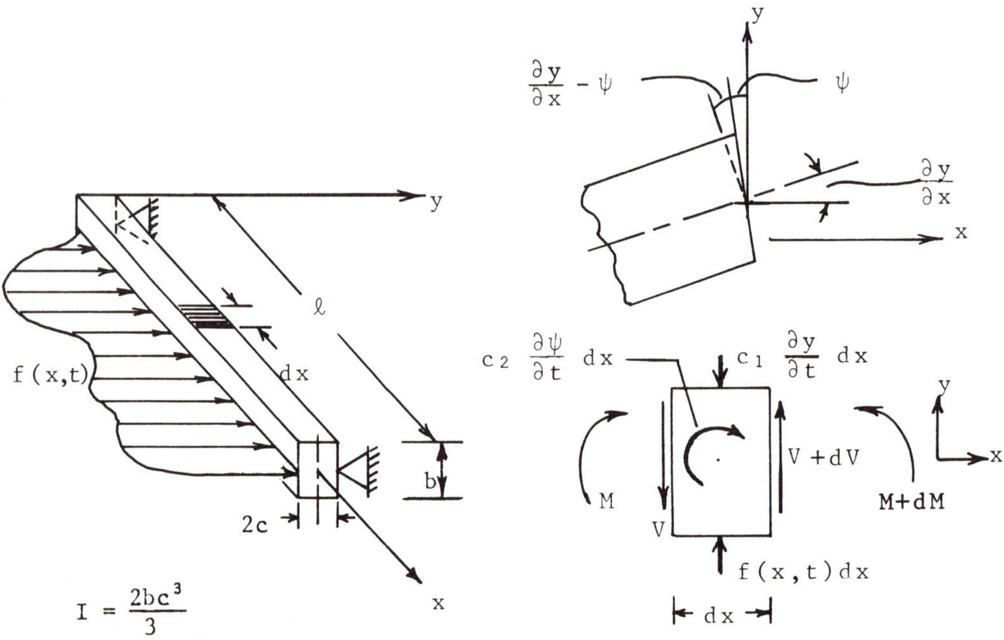

Figure 10.8 The Bernoulli-Euler Beam Model.

The structure shown in Fig. 10.1 will vibrate as a whole under the impinging forces. However, only the displacements of the cross-beam relative to the flexible uprights, to which the ends are attached, give rise to cross-beam stresses. For simplicity, it is assumed the uprights are rigid, and that each end of the cross-beam is perfectly hinged to them. When the end restraints and the motion of the uprights are neglected, one obtains an overestimate of the cross-beam's deflection, and also its stresses. The mathematical model in this case is a conservative one.

A small element of the beam in Fig. 10.8 is isolated as a free body upon which act shear forces V, bending moments M, and the external excitation force per unit length f(x,t). The x-directed forces, set up by an axial stretching due to end restraints during bending, is assumed negligible. External damping is also included, just as it was for the one-degree-of-freedom system. Here, damping is assumed to have two forms: the usual Coulomb damping force which resists y-directional motion, and is linearly proportional to the vertical velocity $\partial y/\partial t$; and a second Coulomb damping force which resists beam rotation in the x y-plane, and is proportional to the angular velocity $\partial\psi/\partial t$. These damping coefficients, transverse and rotatory, are respectively c_1 and c_2.

The dynamic force and moment equilibrium equations, obtained from applying Newton's second law to the beam element are combined. The result as shown by Crandall and Yildiz (1962) is

$$r^2 a^2 \frac{\partial^4 y}{\partial x^4} + \left(\beta_1 - \beta_2\, r^2 \frac{\partial^2}{\partial x^2}\right) \frac{\partial y}{\partial t} + \frac{\partial^2 y}{\partial t^2} = \frac{f(x,t)}{\rho A} \tag{10.37}$$

where

$$\beta_1 = \frac{c_1}{\rho A}, \qquad \beta_2 = \frac{c_2}{\rho I} = \frac{c_2}{\rho A r^2}$$

$$a^2 = \frac{E}{\rho} = \frac{EI}{\rho A r^2}$$

and where ρ is the mass density, A is the cross-sectional area, I is the cross-sectional moment of inertia, and r is the cross-sectional radius of gyration of the beam. This is the famous Bernoulli-Euler beam in which transverse deformations due to shear force V, the effects of rotatory inertia, and all internal or material damping have been neglected. Although Crandall and Yildiz (1962) have included these neglected effects in their more complex beam models, the Bernoulli-Euler beam model is physically realistic here, considering the relatively low frequency range expected for f(x,t) in the ocean environment.

10.9 - BEAM RESPONSES TO HARMONIC AND IMPULSE FORCES

Consider first the steady-state responses to simple harmonic forces. If the load per unit length has the form

$$f(x,t) = e^{i\omega t} \sin \frac{n\pi}{\ell} x \tag{10.38}$$

then a steady-state solution to (10.37) can be assumed in the form

$$y(x, t) = H_y \, e^{i\omega t} \sin \frac{n\pi}{\ell} x \tag{10.39}$$

where the quantity H_y, the complex frequency response for displacement, depends on the integer n and upon the excitation frequency ω. The term $\sin\left(\frac{n\pi x}{\ell}\right)$ was chosen since, for all times, the end conditions of zero deflections and zero bending moments on the beam model are met. These end conditions are, respectively

$$y(0, t) = y(\ell, t) = 0$$

$$\frac{\partial^2 y(0,t)}{\partial x^2} = \frac{\partial^2 y(\ell, t)}{\partial x^2} = 0$$

For n = 1, the beam vibrates in the first mode shape; for n = 2, the second mode shape, and so on. In the one degree of freedom example, only the first mode shape was considered. Actually, (10.37) is not physically accurate for large n (or high ω). Only the first few values of n are important in structural design, since damping usually "washes out" the higher mode shapes. Now, all possible mode shapes are included.

When (10.38) and (10.39) are substituted into (10.37) the result is an algebraic equation which can be solved for H_y, which is

$$H_y = \frac{\frac{1}{\rho A}}{\left(a^2 r^2 \lambda_n^4 - \omega^2\right) + i\left(\beta_1 + \beta_2 r^2 \lambda_n^2\right)} \tag{10.40}$$

where

$$\lambda_n = \frac{n\pi}{\ell} \quad , \quad a^2 = \frac{E}{\rho}$$

The complex frequency responses H_ψ, H_M, and H_V, for slope, bending moment, and shear force, respectively, are found by assuming solutions

$$\left.\begin{aligned} \psi(x,t) &= H_\psi\, e^{i\omega t} \cos \frac{n\pi}{\ell} x \\ M(x,t) &= H_M\, e^{i\omega t} \sin \frac{n\pi}{\ell} x \\ V(x,t) &= H_V\, e^{i\omega t} \cos \frac{n\pi}{\ell} x \end{aligned}\right\} \tag{10.41}$$

Since $\psi = \frac{\partial y}{\partial x}$ it follows from (10.39) and the first of (10.41) that

$$H_\psi = \lambda_n H_y \tag{10.42}$$

The moment-curvature relation for small deflections is

$$M = M(x,t) = EI \frac{\partial^2 y}{\partial x^2} = EI \frac{\partial \psi}{\partial x}$$

It follows from (10.39) and the second of (10.41) that

$$H_M = (-EI\lambda_n^2)\, H_y \tag{10.43}$$

From the dynamic equilibrium equation for the element, it follows that

$$V = -\frac{\partial M}{\partial x} + \rho A r^2 \left(\frac{\partial^2 \psi}{\partial t^2} + \beta_2 \frac{\partial \psi}{\partial t}\right)$$

Since $\psi = \partial y/\partial x$, the shear force becomes

$$V = -\frac{\partial M}{\partial x} + \rho A r^2 \left(\frac{\partial^3 y}{\partial x \partial t^2} + \beta_2 \frac{\partial^2 y}{\partial x \partial t}\right)$$

This, with (10.39) and the last of (10.41) gives

$$H_V = \rho A r^2 \lambda_n \left(a^2 \lambda_n^2 + i\omega\beta_2\right) \cdot H_y \tag{10.44}$$

Thus the responses H_ψ, H_M and H_V for harmonic excitation can all be expressed as multiples of H_y.

For a linear system of one degree of freedom, the impulse-response function was found in Section 10.4 to be in the form

$$h(t) = \frac{1}{2\pi} \int_{-\infty}^{\infty} H(\omega) e^{i\omega t}\, d\omega$$

Note that in the present analysis, $H(\omega)$, which is equivalent to H_y, H_ψ, H_M or H_v, is no longer dimensionless, and includes constants equivalent to k of the previous system. Suppose that there are now n such linear systems whose corresponding complex frequency responses are $H(n,\omega)$ as first derived. The corresponding response $h(n,t)$ of the beam due to the nth impulse is of this same form. For instance for the displacement this is given by

$$h_y(n,t) = \frac{1}{2\pi} \sum_{-\infty}^{\infty} H_y \, e^{i\omega t} \, d\omega \tag{10.45}$$

The corresponding impulse responses for moment and shear are also given by (10.45) where the subscript y is replaced by M and v, respectively.

10.10 - BEAM RESPONSES TO STATIONARY, ERGODIC FORCES

Assume that the stationary random load can be decomposed into n loadings, each with a space variation of $\sin(n\pi x/\ell)$, or

$$f(x,t) = \sum_{n=1}^{\infty} f_n(t) \sin \frac{n\pi}{\ell} x \tag{10.46}$$

where the coefficient of this Fourier sine series is

$$f_n(t) = \frac{2}{\ell} \int_0^{\ell} f(x,t) \sin \frac{n\pi}{\ell} x \, dx \tag{10.47}$$

Now, it was shown for a one-degree-of-freedom linear system that the response $x(t)$ was of the form

$$x(t) = \int_{-\infty}^{\infty} F(\theta) \, h(t - \theta) \, d\theta$$

where $h(t)$ was the response to a unit impulsive force $\delta(t)$. Again suppose that there are n such systems, each with a mode shape sin $(n\pi x/\ell)$. The response $y(x,t)$ for the beam can then be represented as the sum of n individual impulse-responses $h_y(n,t)$.

$$y(x,t) = \sum_{n=1}^{\infty} \sin \frac{n\pi}{\ell} x \int_{-\infty}^{\infty} f_n(\theta) \cdot h_y(n,t-\theta) \, d\theta \tag{10.48}$$

Here, $h_y(n,t)$ is given by (10.45) and $f_n(t)$ is given by (10.47).

If one uses the same procedures as outlined for the one-degree-of-freedom system, the mean square responses for the beam can be calculated. The proof follows the steps used to obtain (10.22) - (10.27), except that the results of n modes are superimposed using the results of (10.44) - (10.47). The mean square displacement response at location x, or the autocorrelation function evaluated at time zero, is

$$\overline{y^2(x,t)} = \sum_{n=1}^{\infty} \sin^2 \frac{n\pi}{\ell} x \int_{-\infty}^{\infty} S_f(\omega) \cdot |H_y|^2 \, d\omega \tag{10.49}$$

The ergodic hypothesis for f(x,t) was envoked to obtain Eq. (10.49).

If one replaces the symbol y in (10.45) and (10.48) by ψ, the mean square response of the slope is found to be

$$\overline{\psi^2(x,t)} = \sum_{n=1}^{\infty} \cos^2 \frac{n\pi}{\ell} x \int_{-\infty}^{\infty} S_f(\omega) \, |H_\psi|^2 \, d\omega \tag{10.50}$$

In like manner, it follows

$$\overline{M^2(x,t)} = \sum_{n=1}^{\infty} \sin^2 \frac{n\pi}{\ell} x \int_{-\infty}^{\infty} S_f(\omega) \, |H_M|^2 \, d\omega \tag{10.51}$$

and

$$\overline{V^2(x,t)} = \sum_{n=1}^{\infty} \cos^2 \frac{n\pi}{\ell} x \int_{-\infty}^{\infty} S_f(\omega) \, |H_V|^2 \, d\omega \tag{10.52}$$

Equations (10.49) - (10.52) represent <u>time</u> average responses. One can obtain <u>space</u> average responses by averaging each of these over the length of the beam. For instance, for the displacement, this type of mean square average, represented simply by $\overline{E}(y^2)$, is given by

$$\overline{E}(y^2) = \frac{1}{\ell} \int_0^{\ell} \overline{y^2(x,t)} \, dx$$

$$= \frac{1}{2} \sum_{n=1}^{\infty} \int_{-\infty}^{\infty} S_f(\omega) \, |H_y|^2 \, d\omega \tag{10.53}$$

Similar results can easily be obtained for $\overline{E}(\psi^2)$, $\overline{E}(M^2)$ and $\overline{E}(V^2)$.

10.11 - BEAM RESPONSES TO WHITE NOISE

If $S_f(\omega) = S_o$, a constant value for all ω, the integrals of the form of (10.53) can be evaluated. Such integrations are listed by Crandall and Yildiz (1962). Only one of the results will be repeated here. That is

$$\int_{-\infty}^{\infty} |H_y|^2 \, d\omega = \frac{\pi}{(\rho A a\ r)^2 \left(\frac{\pi}{\ell}\right)^4 \left[\beta_1 n^4 + \beta_2 r^2 \left(\frac{\pi}{\ell}\right)^2 n^6\right]} \tag{10.54}$$

The mean square response for displacement at location x becomes, from (10.49) and (10.54)

$$\overline{y^2(x,t)} = \frac{\pi\ S_o}{(\rho A a\ r)^2 \left(\frac{\pi}{\ell}\right)^4} \sum_{n=1}^{\infty} \frac{\sin^2 \frac{n\pi}{\ell} x}{\left[\beta_1 n^4 + \beta_2 r^2 \left(\frac{\pi}{\ell}\right)^2 n^6\right]} \tag{10.55}$$

The mean square response averaged over x, given by (10.53) and (10.54), is

$$\overline{E}(y^2) = \frac{\pi S_o}{2(\rho A a\ r)^2 \left(\frac{\pi}{\ell}\right)^4} \sum_{n=1}^{\infty} \frac{1}{\left[\beta_1 n^4 + \beta_2 r^2 \left(\frac{\pi}{\ell}\right)^2 n^6\right]} \tag{10.56}$$

With the result of (10.54), the reader can now easily find the mean square responses for slope, moment, and shear. This follows because H_ψ, H_M, and H_V are simple multiples of H_y, as shown in (10.42), (10.43), and (10.44).

Crandall and Yildiz (1962) discuss the convergence of the forms $\overline{E}(y^2)$, $\overline{E}(\psi^2)$, $\overline{E}(M^2)$, and $\overline{E}(V^2)$ in detail. The results which hold for these mean square values, also hold for $\overline{y^2(x,t)}$, $\overline{\psi^2(x,t)}$, $\overline{M^2(x,t)}$ and $\overline{V^2(x,t)}$. The responses to ideal white noise are now summarized:

1. Both $\overline{E}(y^2)$ and $\overline{E}(\psi^2)$ converge only if damping is present. At least one of the constants c_1 or c_2 must be non-zero.

2. Values of $\overline{E}(M^2)$ diverge if rotatory damping is absent ($c_2 = 0$). However $\overline{E}(M^2)$ converges if rotatory damping is present, either

alone ($c_1 = 0$) or in combination with transverse damping ($c_1 \neq 0$).

3. Values of $\overline{E}(V^2)$ diverge for all combinations of damping.

The engineer is most interested in $\overline{E}(M^2)$ because, from this he can estimate the root mean square bending stress, given by $S = c\sqrt{\overline{E}(M^2)}/I$, where c is the beam dimension of Fig. 10.8. Without at least some rotatory damping, however, such a calculation is meaningless for ideal white noise.

Crandall and Yildiz (1962) also consider the case where $S_f(\omega) = S_o$, a constant for $\omega \leq \omega_c$, where ω_c is a "high" cutoff frequency. For $\omega > \omega_c$, $S_f(\omega) = 0$. The conclusions are summarized for this case of band-limited white noise:

1. Same as conclusion 1. for ideal white noise.

2. Values of $\overline{E}(M^2)$ will be the order of $\sqrt{\omega_c}$ if $c_2 = 0$ but $c_1 \neq 0$. As before, $\overline{E}(M^2)$ converges if rotatory damping is present, either alone or in combination with transverse damping.

3. Values of $\overline{E}(V^2)$ are convergent to the order of $\sqrt{\omega_c}$ when $c_1 = 0$ and $c_2 \neq 0$.

4. Values of $\overline{E}(V^2)$ are convergent to the order of $(\omega_c)^{3/2}$ when $c_1 \neq 0$ and $c_2 = 0$.

5. Values of $\overline{E}(V^2)$ are convergent to the order of $\sqrt{\omega_c}$ if $c_1 \neq 0$ and $c_2 \neq 0$.

6. All mean square values diverge for no damping at all.

10.12 - DESIGN IMPLICATIONS FOR THE BEAM

The useful conclusions which a design engineer can draw from this problem is that mean square values of displacement and bending moment can be calculated from the analysis of the Bernoulli-Euler beam model when at least rotatory damping is included. Also, results based on the ideal white noise assumption where f(x,t) is stationary and ergodic, may be quite useful in design, even if $S_f(\omega)$ is narrow band. Finally, if f(x,t) is Gaussian then the probability that an arbitrary value of displacement or moment can exceed its rms value can be calculated here, as it was for the one-degree-of-freedom system. Again, it would be a matter of judgment as to what the design limits should be. If design limits were on stress for instance, then careful account should be taken of the type of material to be used. Too many excursions of stress beyond a certain limit for a given material could cause fatigue failure of the beam. Such design considerations will be discussed in the next chapter.

10.13 - REFERENCES

1. Caldwell, H. M. (1969), "Fluid Effects on Offshore Pipelines: A Static and Dynamic Model Study", M.S. thesis, Department of Civil Engineering, Duke University.

2. Chen, Y. (1966), Vibrations: Theoretical Methods, Addison-Wesley Publishing Company, Reading, Mass., Chapter 1.

3. Crandall, S. H., editor (1958), Random Vibrations, John Wiley and Sons, New York, Chapter 1.

4. Crandall, S. H., and Mark, W. D. (1963), Random Vibration in Mechanical Systems, Academic Press, New York, p. 58.

5. Crandall, S. H., and Yildiz, A. (1962), "Random Vibration of Beams", ASME Journal of Applied Mechanics, pp. 267-275.

6. Den Hartog, J. P. (1947), Mechanical Vibrations, Third Edition, McGraw-Hill Book Co., Inc., New York, p. 191.

7. Langhaar, H. L. (1962), Energy Methods in Applied Mechanics, John Wiley and Sons, Inc., New York, p. 45.

8. Nath, J. H., and Harleman, D. R. F. (1967), The Dynamic Response of Fixed Offshore Structures to Periodic and Random Waves, TR 102, Hydrodynamics Laboratory, M.I.T.

9. Timoshenko, S. (1956), Strength of Materials, Part II, Third edition, Van Nostrand Co., Princeton, New Jersey.

Chapter XI

DYNAMIC BEHAVIOR OF MATERIALS IN AN ENVIRONMENT

In the first part of this chapter, mechanical properties of engineering metals and the tests used to characterize them are briefly reviewed. Since standard mechanical tests are generally performed in non-corrosive laboratory environments, the structural designer can use such test results with a degree of confidence only if the structure has corrosion protection. As this protection wears away, the integrity of the exposed structure is severely threatened. In the second part of this chapter, the common mechanisms of these destructive corrosion attacks are summarized. In the third section, materials suitable for both present and future ocean structures are surveyed, and several methods of corrosion protection are discussed. In the last sections, a statistical approach is used to describe the damage accumulation in materials subjected to random type loads and to corrosive environments. Numerical examples are included which illustrate the use of the theory to estimate probable lifetimes of certain ocean structures.

11.1 BEHAVIOR OF STRUCTURAL METALS WITH CORROSION PROTECTION

Dynamic data on the mechanical properties of structural metals in non-corrosive environments are generally available from the material supplier. Meaningful interpretation of these data for use in design can be made only if the basic mechanical tests are understood. These tests are now reviewed briefly.

The Tensile Test is an important quantitative measure of the dynamic behavior of a material. It is used universally by scientists in research involving the development of new materials, and by engineers concerned with the use of materials in design and fabrication. The test specimen, a cylindrical or rectangular bar, is fixed at one end and pulled with a prescribed axial speed at the other end. From load extension measurements of the bar, two important characteristics are obtained, strength and ductility. These will be defined presently in terms of axial stress, or load per unit instantaneous cross-sectional area, and axial strain or the unit instantaneous extension.

Consider first the implications of tensile test measurements. One asks, "For a given material, what variables affect load-extension measurements?" Experiments have shown that axial stress S, depends not only on axial strain ε, but also upon the time rate of straining $\dot{\varepsilon}$. In addition, S may depend on the test temperature θ; the test environment $\mathcal{E}$ (air, sea water, hydrogen, etc.); the initial state $\mathcal{A}$ of internal or residual stress; the metal phase $\mathcal{P}$; the time rate of change of each of these variables; and so on. The tensile stress can thus be represented as a function of all of these variables, or

$$S = f(\varepsilon, \dot{\varepsilon}, \ldots; \theta, \dot{\theta}, \ldots; \mathcal{E}, \dot{\mathcal{E}}, \ldots; \mathcal{A}, \dot{\mathcal{A}}, \ldots; \mathcal{P}, \dot{\mathcal{P}}, \ldots) \tag{11.1}$$

In a properly performed tensile test, every variable in f except strain is held as constant as possible, so that (11.1) becomes

$$S = f(\varepsilon) \tag{11.2}$$

The way of reducing load-extension data to stress-strain data affects the function f. As long as the specimen is uniform in cross-section, or nearly so, the following data reduction procedures make the most sense. When the axial load P vs length extension $\Delta\ell$ over gate length ℓ_o, is linear

$$S = \frac{P}{A_o} \tag{11.3}$$

where A_o is the initial cross-sectional area of the specimen. For $\Delta\ell/\ell_o < 0.05$, the strain is given reasonably accurately by

$$\varepsilon = \frac{\Delta\ell}{\ell_o} \tag{11.4}$$

After the load-extension curve appears to become quite non-linear ($\Delta\ell/\ell_o > 0.001$ for most metals), the material is undergoing plastic flow, Poisson's ratic becomes 0.50, and material volume is conserved. In this case, assuming uniform deformation, the material volume is given by

$$\ell_o A_o = \ell A = \text{constant} \tag{11.5}$$

For each increment of length change $d\ell$, the strain increment for the specimen of instantaneous length ℓ is defined by

$$d\varepsilon = \frac{d\ell}{\ell}$$

When all strain increments are summed over the length of the specimen, the instantaneous (true) strain is

$$\varepsilon = \int_{\ell_o}^{\ell} \frac{d\ell}{\ell} = \ell n \frac{\ell}{\ell_o} \tag{11.6}$$

For plastic flow, the true stress for uniform tensile deformation is based on the instantaneous area A. Thus, from (11.5) and (11.6), the true stress is

$$S = \frac{P}{A} = \frac{P}{A_o}\frac{\ell}{\ell_o} = \frac{P}{A_o} e^{\varepsilon} \tag{11.7}$$

The designer will often find stress-strain data for materials based on (11.3) and (11.4), which are the definitions of the "engineering stress" and "engineering strain" respectively. These results will be nearly equivalent to the results based on (11.6) and (11.7), as long as $\Delta\ell/\ell_o < 0.05$. This can be seen by expanding (11.6) in a Taylor series, or

$$\varepsilon = \ell n \frac{\ell}{\ell_o} = \ell n \left(\frac{\ell_o + \Delta\ell}{\ell_o}\right) = \ell n \left(1 + \frac{\Delta\ell}{\ell_o}\right) = \frac{\Delta\ell}{\ell_o} - \frac{1}{2}\left(\frac{\Delta\ell}{\ell_o}\right)^2 + \frac{1}{3}\left(\frac{\Delta\ell}{\ell_o}\right)^3 - \dots \tag{11.8}$$

When $\Delta\ell/\ell_o << 1$, then (11.7) reduces to (11.4). In the same way, when the exponential term of (11.7) is expanded,

$$S = \frac{P}{A_o}\left(1 + \varepsilon + \frac{\varepsilon^2}{2} + \dots\right) \tag{11.9}$$

and (11.9) reduces to (11.3) for small values of ε.

When the design engineer is presented with stress-strain data, it is important to know how this data has been reduced. For design in the small strain range (elastic design), the difference between the "engineering" and "true" values of S and ε are insignificant. However, in considering plastic design, which is necessary since occasional high wave forces may overload the structure, (11.6) and (11.7) should be used for data reduction.

Tensile test data which typify the behavior of nearly all engineering metals are shown in Fig. 11.1. Comparisons among metals are made by using the following quantitative measures obtained from tensile behavior. Strength is characterized by: (1) the proportional

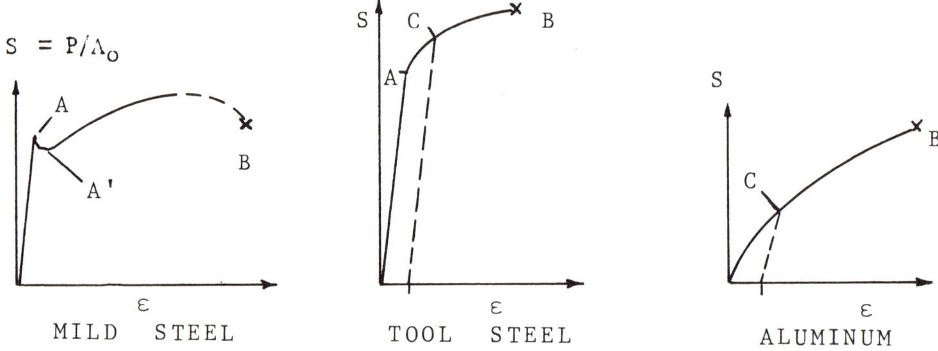

Figure 11.1 Typical Tensile Behavior of Engineering Metals

limit, or point A in Fig. 11.1, where dS/dε is no longer constant; (2) the ultimate strength at fracture or point B; and (3) the yield point, which has two values A and A' for mild steel, called the upper and lower yield point respectively. In the other two metals, no pronounced yield point exists. However, a yield strength is often defined as the stress at an off-set strain of ε = 0.002 for steel and aluminum alloys. It is determined by drawing the slope dS/dε, evaluated at ε = 0, through the point ε = 0.002 as shown by points C in Fig. 11.1. Young's Modulus E is defined as the slope of the S vs ε curve at zero strain.

The ductility of materials is characterized by two quantities, (1) the maximum value of uniform tensile strain before necking occurs; and (2), the percent reduction in cross sectional area at fracture, $f = 100\ (A_o - A_N)/A_o$ where A_N is the area of the necked portion at fracture. It is important that metals used for ocean structures have a large measure of ductility. It is this property of a metal that allows the high localized stresses which build up around sharp notches and junctions of a structure to be relieved and redistributed to lower levels by permanent deformations.

The measure of a material's ability to withstand impact whether or not the material has a sharp notch in it is called toughness. The area under the true stress-strain curve up until fracture is one measure of toughness, which is expressed in the units of energy per unit volume of the material. Unfortunately, toughness measured in this way for structural steel does not correlate with fracture resistance in service, especially at room temperature. For this reason, the Charpy test is commonly used by designers as a comparative measure of impact resistance of steels. In this test, the energy required to break beams notched in specific ways is used to correlate the relative toughness of metals. These correlations are explained more fully by Polakowski and Ripling (1966) and by Pellini (1954). A shortcoming of the Charpy tests is that these toughness correlations depend on the specimen size and notch geometry, and thus do not give fundamental knowledge of toughness. It would not be possible to use such a test, for instance, to predict the breaking strength of a large structure with a notch which is unlike a Charpy specimen. However, Charpy tests continue to be useful to designers for qualitative comparisons of the relative toughness among different engineering materials, where the test specimens are always identical.

The impact test results suggest that strain rate $\dot{\varepsilon}$ affects the fracture stress of a material, or that a material may be rate-sensitive. A rate-sensitive metal exhibits higher stress levels for higher values of $\dot{\varepsilon}$. Wilson (1964) measured this effect in hot rolled steel at room temperature for strain rates available on standard testing machines.* These results are shown in Fig. 11.2, for which

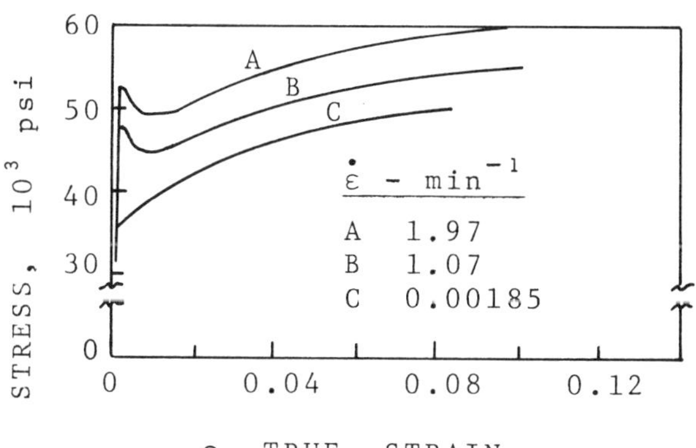

Figure 11.2 Constant Strain-Rate Behavior of Hot-Rolled Iron

*The usual range for the crosshead velocity v of a standard universal testing machine is 0.01 to 10.0 in./min. The strain rate of a specimen of total length ℓ is thus limited to the range $0.01/\ell \leq \dot{\varepsilon} \leq 10.0/\ell$, where $\dot{\varepsilon} = v/\ell$.

the general representation of stress in (11.1) can be written as

$$S = f(\varepsilon, \dot{\varepsilon}) \qquad (11.10)$$

It is important to note that although the functional relationship (11.10) may be unique for constant strain rate tests for a particular metal, where all of the other variables in (11.1) are kept constant, this relationship is not unique if the strain is held constant, where only S and $\dot{\varepsilon}$ are allowed to vary. This phenomenon, or stress relaxation, occurs at all temperature levels, especially in plastically deformed metals. In a stress relaxation test, a metal rod is prestressed to some initial value, held at constant strain, and the load it supports is measured as it decreases in time. Recent careful measurements of this phenomenon for engineering materials are reported by Wilson and Garofalo (1966). Typical data for high purity copper and aluminum are shown in Fig. 11.3.

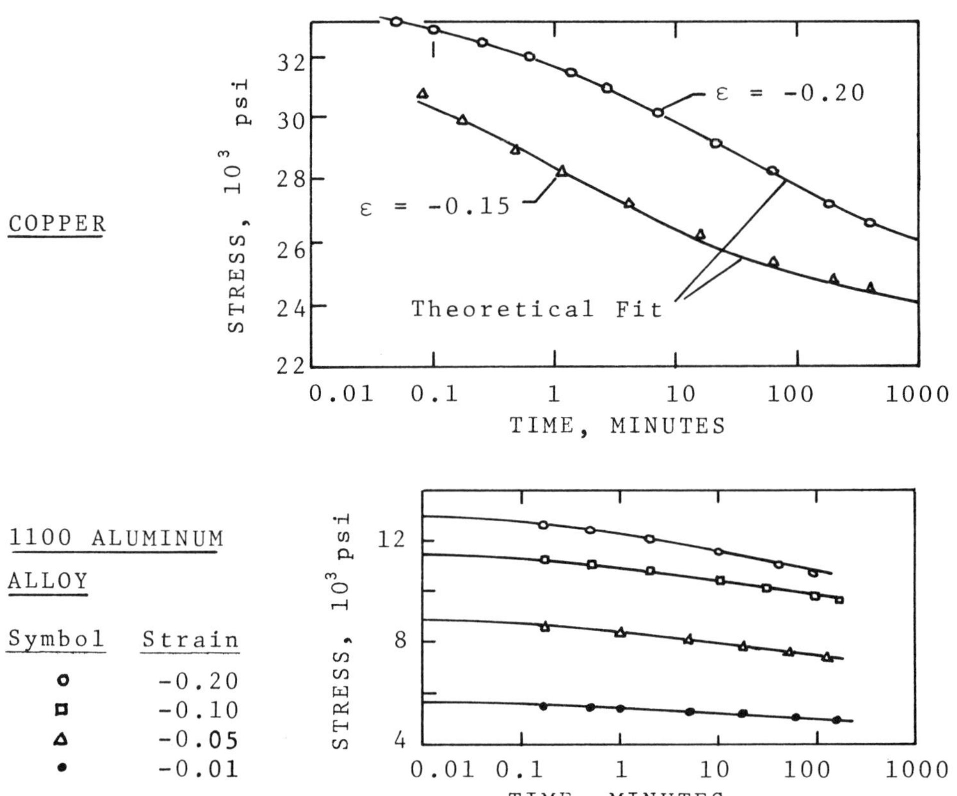

Fig. 11.3 Stress Relaxation Behavior of Copper and Aluminum Alloy 1100 at Room Temperature

Absolute reaction rate theory has been used by Wilson and Wilson (1966) to correlate stress relaxation data for a wide variety of metals over wide temperature range. A reasonably complete summary of stress relaxation data up until 1965 for engineering metals is given in this report. The experimental results are found to correlate well with the following differential equation

$$\dot{S}_i = -S_i^* k_i \left[1 - \exp\left(-2\, \alpha_i S_i/S_i^*\right) \cdot \exp\left(2\, \mu_i \alpha_i S_i/S_i^*\right)\right] \qquad (11.11)$$

where

S_i = stress for the ith type of reaction

S_i^* = initial stress of the ith type

k_i, α_i, μ_i = rate constants of the ith type.

According to Eyring and Ree (1955), μ_i = 1/2 for metals which have symmetrical energy barriers during relaxation. For this case, (11.11) can be integrated and when the total macroscopic relaxation stress is assumed to be composed of the sum of stresses from each reaction i, the result is

$$S = \sum_{i=1}^{N} \frac{2S_i^*}{\alpha_i} \tanh^{-1} \left[\left(\tanh \frac{\alpha_i}{2}\right) \exp\left(-2\, k_i \alpha_i t\right)\right] \qquad (11.12)$$

where N is the total number of reactions and t is time. For the aluminum alloy shown in Fig. 11.3, it was possible to distinguish only one type of reaction, or N = 1. The rate constants for this metal are given in Table 11.1 as a function of plastic prestrain.

Similar results are given by Wilson and Wilson (1966) for other metals, although for some cases N = 2 and μ = 0. In this reference,

Strain, ε	Initial Stress σ^*, psi	k_1, min^{-1}	α_1
.01	5,750	4.08×10^{-21}	51.8
.05	10,005	1.35×10^{-20}	47.8
.10	12,250	3.12×10^{-21}	51.0
.15	13,750	5.61×10^{-21}	54.4
.20	14,850	9.38×10^{-21}	49.7

TABLE 11.1 - Rate constants for 1100 aluminum alloy. The solid curves in Fig. 11.3 are based on these constants and Equation (11.12) for N = 1 and μ_1 = 1/2.

it is shown that (11.11) includes as special cases the exponential decay stress relaxation model proposed originally by Maxwell (1868) and the logarithmic decay model of Trouton and Rankine (1904). Maxwell's model, which correlates stress relaxation data for metals only as the stress becomes small, can be found by integrating (11.11) after assuming that $(\alpha_i S_i / S_i^*) << 1$, or

$$S_i = S_i^* \exp(-2\ \alpha_i\ k_i\ t) \qquad (11.13)$$

The logarithmic model, which correlates stress relaxation data at the beginning of the process where stresses are high, can be found by integrating (11.11) after assuming that $\exp(-2\ \alpha_i\ S_i/S_i^*) << 1$, or

$$S_i = S_i^* - \beta_i\ \ell n(1 + \gamma_i t) \qquad (11.14)$$

where

$$\beta_i = \frac{S_i^*}{2\ \mu_i\ \alpha_i}\ ;\ \mu_i \neq 0$$

and

$$\gamma_i = 2\ \mu_i\ \alpha_i\ k_i \exp(2\ \mu_i\ \alpha_i)$$

From a practical standpoint, (11.14) is the most convenient form to use to calculate the rate constants from short time stress relaxation data, although general forms such as (11.12) should be employed in engineering design for long time periods.

The design engineer should consider the stress relaxation phenomenon in all parts of ocean structures where bolts, pins, or rivets hold components together with their load carrying capacity. This load carrying capacity can decrease in time, separation of components can occur, and structural failure could result because of stress relaxation.

Another characteristic of plastic flow in metals is the Bauschinger effect. This is manifested by a lowering of the yield strength when the direction of plastic deformation is reversed. This phenomenon for a typical engineering metal is shown in Fig. 11.4. Path A B C C' is the usual tensile behavior. If the tensile test specimen is unloaded at C, the path will be C E D, and the permanent set strain will be of length A D. Now, the compression curve D E F is measured, and the compressive stress at yield E is found to be less than the original tensile yield stress at B. This effect results from anisotropy, or reorientation of the grains of the material. The mechanism is rationalized by Polakowski and Ripling (1966). The phenomenon will appear not only when compression follows extension and vice versa, but also when torsion is

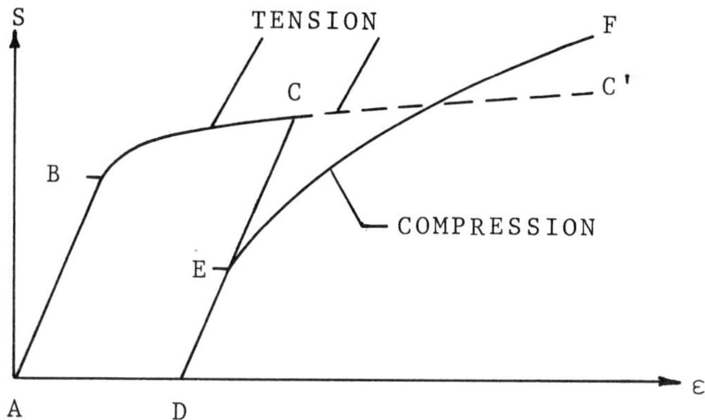

Figure 11.4 The Bauschinger Effect in Metals

followed by a reversed twist, or bending by bending in the opposite direction. The Bauschinger effect, then is one example of how the variable $\mathscr{A}$, the initial state of a metal, can affect the function f which relates stress to strain in the functional representation (11.1).

A more direct measure than the Bauschinger effect of the damage done to engineering metals under dynamic loading is the <u>fatigue test</u>. When metal rods such as shown in Fig. 11.5 are subjected to fluctuating loads, localized microscopic plastic deformation takes place. This happens even if the maximum macroscopic stress in the specimen is calculated to be well below the observed yield point stress of the tensile test. Localized cracks form in plastic regions of the rod, and these cracks grow under stress cycling. Eventually, if the stress excursions are large enough, one of the cracks grows rapidly and the material ruptures. Such a rupture is called fatigue failure.

In the usual test, the stress is made to vary harmonically as indicated in Figs. 11.5 and 11.6. The stress amplitude S is held constant and the number of cycles N until fracture is counted. The major fraction of fatigue data has been obtained from tests in which the mean stress is zero. It has been found that if the mean stress is not too large, or less than one third of the fluctuation amplitude, the mean stress level has a relatively weak effect on the number of stress cycles for failure. The following discussion is limited to the case of zero mean stress.

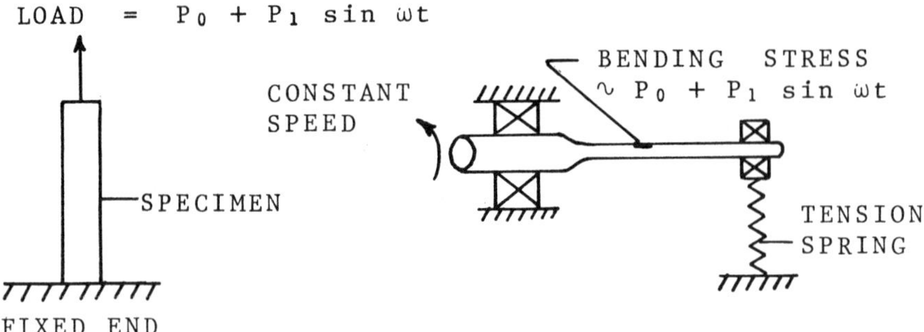

Figure 11.5 Two Fatigue Test Schemes, where P_0 and P_1 are load parameters and ω is the frequency of stress fluctuation.

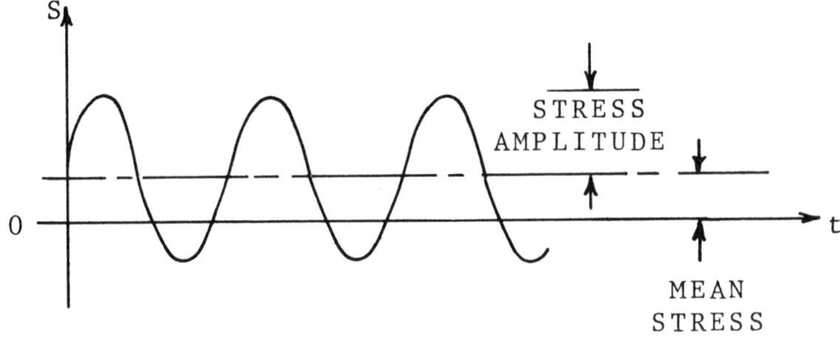

Figure 11.6 Harmonically Varying Stress in a Fatigue Test.

Suppose that a large number of identical metal specimens are tested at various stress amplitudes and the number of cycles to failure is counted. The results, when plotted as in Fig. 11.7, define the S-N or fatigue curve for that material. It is particularly important to note that this curve, a composite of many tests, depends strongly on the test environment. A corrosive environment such as sea water can greatly reduce the fatigue life of an unprotected structural steel beam, for instance. More will be said about the complications of corrosion in Section 11.2 of this chapter.

Much more could be said on the topic of metal fatigue, both on the physical mechanisms of fatigue fracture and in the analysis of various test data useful in design. Good discussions are given by Polakowski and Ripling (1966), chapter 18. Methods of statistical analysis of fatigue data are cited in a book compiled by Grover (1960) for the U. S. Navy, where it is shown how it is possible to establish probability distribution curves and hence the probabilities

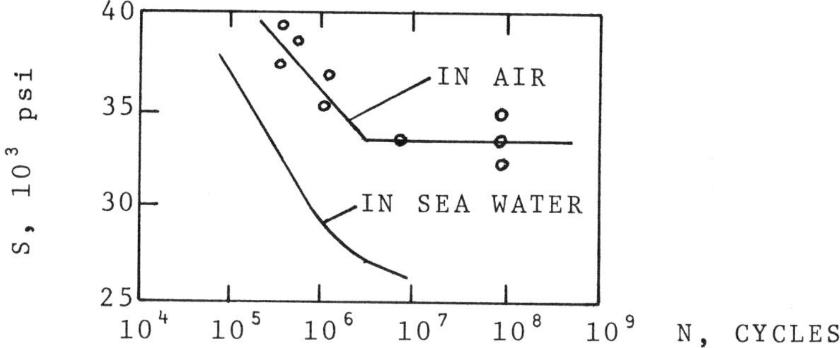

Figure 11.7 A Typical S-N or Fatigue Curve for Structural Steel.

of survival for any given N as the cyclic stress varies. One can then draw continuous S-N curves such as shown in Fig. 11.7 for any selected probability of survival. These data are then used as the starting point to calculate the material damage under random stress cycling, the topic of Section 11.3. Regardless of the method used to establish S-N curves, it is important to recognize that such curves are really lines representing probabilities of material failure under harmonic stresses whose amplitudes lie above these lines. For stress amplitudes lying below these lines, materials may still have a finite probability of failing.

The next important variable in the functional relationship (11.1) relating stress to strain is θ, or temperature. Metals at elevated temperature exhibit temperature sensitivity in which the stress level in the stress-strain curves is lowered for increasing θ and constant $\dot{\varepsilon}$. A related phenomenon is the creep deformation and subsequent rupture of metal under constant stress, such as reported by Wray and Richmond (1968). For the metals presently used for ocean structures, however, temperature effects giving rise to creep are insignificant. Significant creep in steels, for instance, is observed only at temperatures exceeding 1,000 degrees Fahrenheit.

The remaining variables in the functional form (11.1) which relate stress to strain are $\mathcal{P}$ and $\mathcal{S}$. The phase state $\mathcal{P}$ of engineering metals is quite important in terms of materials selection, and this state is reflected in the tests for mechanical behavior already discussed. With the exception of age hardening aluminum alloys, phase changes which may occur in engineering metals of ocean structures are generally insignificant at usual ocean temperatures. However, stress corrosion cracking and reduced fatigue life for unprotected metals are important considerations in this hostile environment. These environmental effects are considered next.

11.2 BEHAVIOR OF METALS IN A CORROSIVE ENVIRONMENT

Corrosion has been described as the destruction of metals by chemical or electro-chemical reactions with their environment. Rusting applies only to the corrosion of iron and steels. Corrosion mechanisms, including rust, are discussed briefly. After this, the four general types of observed corrosion damage, which are uniform attack, pitting, intergranular attack, and corrosion cracking, are reviewed. Finally, several of the environmental variables which cause these four types of corrosion on engineering materials are identified, using data from field tests and controlled laboratory experiments. Much of the source material for this short review is from the excellent book by Uhlig (1963).

Consider first some of the basic mechanisms of corrosion common to all engineering metals. Metal surfaces are composites of electrodes electrically short-circuited through the metal, as shown in Fig. 11.8. For instance, for iron and steel in the aerated water of the ocean, the positive electrodes or anodes are surface areas covered by porous rust, and negative electrodes, or cathodes, are areas exposed to oxygen. As corrosion proceeds these cell areas shift continually from place to place.

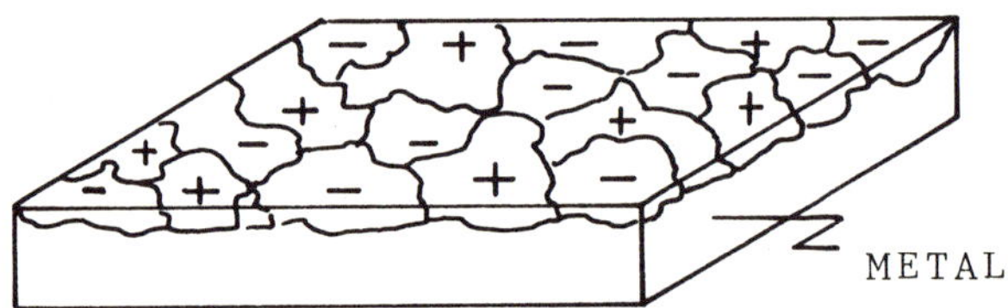

Figure 11.8 A Schematic View of Action Cells in a Metal

One of these cell areas, called a differential aeration cell, may look something like the drawing in Fig. 11.9 for iron. A similar cell area formed by water-line corrosion is also shown in this Figure. In both cases, less oxygen reaches the area covered up with rust, so that local localized currents are set up. The higher the current, the greater the amount of rust is formed in a given time. Other types of aeration cells also exist, but the corrosion mechanisms differ with the type of metal and its environment.

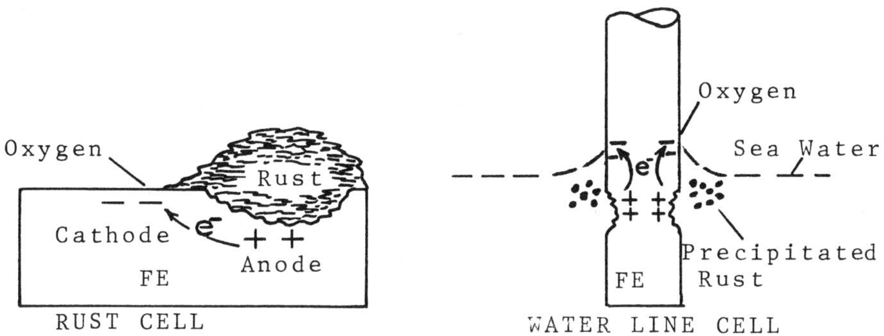

Figure 11.9 Differential Aeration Cells Formed by Rust and by Water-Line Corrosion

Two other general types of cells often take part in the corrosion of ocean structures. One is the dissimilar electrode cell which can be set up between dissimilar metals; between cold-worked or highly prestrained metal in contact with the same unstrained metal; and between grains and grain boundaries of the same metal. A third type of cell is the differential temperature cell, in which the same metal has a temperature gradient when immersed in an electrolyte such as sea water. For instance, when iron is immersed in aerated sea water, the warmer portion is anodic to the colder portion; but after a matter of hours, depending on aeration and stirring rates, the polarity may reverse.

For structures exposed to an ocean environment, it is probable that all three types of cells are responsible for corrosion although the effects of the temperature differential cell are probably the least significant. If such cells could be isolated, the corrosion which each causes would probably obey Faraday's Law, or

$$(\text{WT. OF METAL CORRODED}) = k_c I_c t$$

where I_c is the current, t is time, and k_c is a constant called the electro-chemical equivalent. This law is obeyed in ideal experiments and explains, for instance, the corrosion of electrodes in a flashlight dry cell.

Consider now the formation of rust on an unprotected steel ocean structure. Ferrous ions go into solution at the anodic areas in an amount equivalent to the reaction at the cathodic areas. At anodic areas, the reaction is

$$Fe \rightarrow Fe^{++} + 2e^- \qquad (11.15)$$

In the presence of the dissolved oxygen in sea water, the following reaction takes place at the cathodic areas

$$2H^+ + \frac{1}{2} O_2 \rightarrow H_2O - 2e^- \qquad (11.16)$$

Here, dissolved oxygen reacts with the hydrogen atoms absorbed on the iron surface. The oxidation reaction proceeds as rapidly as the oxygen can be supplied to the metal surface. The cathodic reaction is the controlling one.

When (11.15) and (11.16) are added, and use is made of the reaction

$$H_2O \rightarrow H^+ + OH^- \qquad (11.17)$$

The result is

$$Fe + H_2O + \frac{1}{2} O_2 \rightarrow Fe(OH)_2 \qquad (11.18)$$

Hydrous ferrous oxide, $FeO \cdot n\ H_2O$, and ferrous hydroxide, $Fe\ (OH)_2$, generally compose the diffusion barrier layer next to the iron surface through which O_2 must diffuse. The color of $Fe\ (OH)_2$ is normally green to greenish-black because of incipient oxidation by air.

At the outer surface of the oxide film, access to dissolved oxygen converts ferrous oxide to hydrous ferric oxide or ferric hydroxide according to the reaction

$$Fe\ (OH)_2 + \frac{1}{2}\ H_2O + \frac{1}{4} O_2 \rightarrow Fe\ (OH)_3$$

Hydrous ferric oxide which is orange to red-brown in color, comprises most of ordinary rust. It exists as non-magnetic αFe_2O_3 (Hematite) or magnetic γFe_2O_3. The α form is more stable and comprises most of the top layer of rust. Under this, one often finds a black layer of magnetic hydrous ferrous ferrite $Fe_3O_4 \cdot n\ H_2O$. The bottom layer is FeO. It is thus concluded that without the dissolved oxygen from the air, rusting in sea water would be negligible.

Corrosion consists of more than the rusting of iron and the tarnishing of other metals. Corrosion damage also results in the cracking of metals and subsequent failure due to loss of strength and ductility. It was already mentioned that this last type of corrosion damage reduces the fatigue life in steel. Even though the amount of corrosion products observed may be negligible, tiny cracks due to corrosion can cause structural failure. The four main types of corrosion which commonly occur on structural metals exposed to an ocean environment are uniform attack, pitting, intergrannular corro-

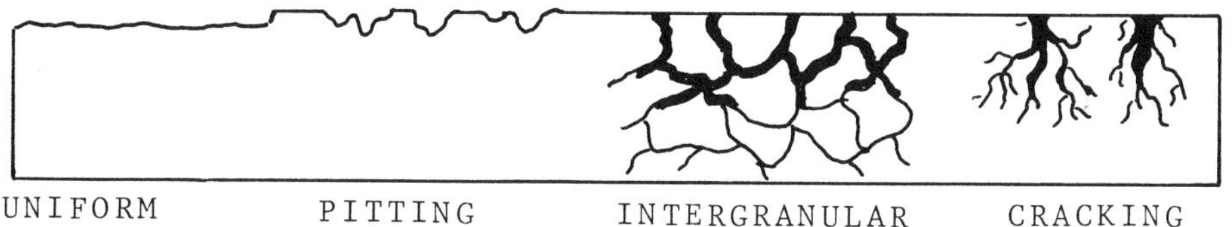

Figure 11.10 Four Main Types of Corrosion Damage in Metals

sion, and corrosion cracking. These types, sketched in Fig. 11.10, are now discussed briefly.

Uniform attack includes the rusting of iron, tarnishing of silver, "fogging" of nickel, and high temperature oxidation of metals in general. The rate of uniform attack, reported as inches of penetration per year (ipy), or as milligrams per square decimeter per day (mdd), refers to the average metal penetration or to weight loss by corrosion. Such data are reported by Uhlig (1948), for instance. When one considers these reported corrosion rates, the following facts should be kept in mind: (1) these rates represent time-averaged values; (2) the initial corrosion rates are higher than the final ones; (3) the effects of the velocity of the corrosive environment may not be included; (4) a very low corrosion rate does not preclude premature failure by cracking; (5) and, finally, the conversion of ipy to mdd or vice versa, requires a knowledge of the metal density. A given weight loss per unit area for aluminum, for instance, represents a greater actual loss of metal thickness than the same weight loss for the heavier metal, steel. To convert ipy data to mdd data, one multiplies ipy by 696 times the metal density. A convenient nomograph for this purpose is given by Fontana and Greene (1967).

Pitting is a general term to describe localized corrosion attack on a level observable by the naked eye. Depth of pitting is often expressed by the pitting factor, the ratio of deepest metal penetration to average metal penetration as determined by weight loss of a specimen. This ratio is defined in Fig. 11.11. Chromium, aluminum, and their alloys are especially susceptible to this type of attack. Thus, stainless steels are extremely susceptible to pitting, especially when subjected to the flow of liquids such as sea water. Uhlig (1948) also gives results for tests of this type. Pitting due to liquid flow is sometimes called impingement attack or corrosion-erosion. If conditions of fluid velocity around a metal are such

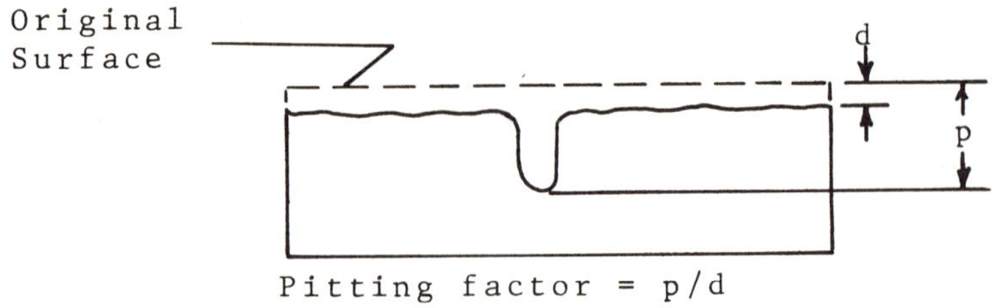

Figure 11.11 Uniform Attack Accompanied by Pitting

that repetitive low and high pressure areas are developed, bubbles form and collapse at the metal-liquid interface. In this case, the pitting is called cavitation-erosion, which can occur on unprotected steel in sea water. The deeply-pitted metal surfaces often have a spongy appearance.

Intergranular attack is the third general type of corrosion in which metal is depleted at the grain boundaries. Here, the large areas of the grains themselves act as cathodes in contact with anodic material between the grains. The attack is often rapid and deep, resulting in catastrophic failures. Improperly heat-treated 18-8 stainless steel or Duralumin-type alloys (4% Cu-Al) are subject to such attack. Pure metals are not susceptible to intergranular attack.

Corrosion cracking is the fourth general type of corrosion damage. This comes about after intergranular attack when the material is under stress. Fissures open up along the grain boundaries and across the grains themselves, causing fresh anodic material to be exposed and further corroded. Adsorption of O_2 and H_2O contribute to this crack formation. Crack-sensitive paths are formed during cooling of the melt, and by subsequent heat treatment. Corrosion fatigue, where the life-time to failure is reduced in a constant amplitude fatigue test, is one example of corrosion cracking. The lower curve in Fig. 11.7 illustrates this effect. Another example is the failure of a metal under constant tensile stress after it has been exposed for a length of time to a corrosive environment. Hydrogen cracking is another common phenomenon in carbon steels, martensitic and ferritic Cr-Fe alloys, and in Mn-Fe alloys. Hydrogen enters the metal, causing loss of metal ductility and cracking if the applied stress or residual stress is high enough. In all of these cases, failure due to corrosion cracking can occur after just a few minutes or after years, depending on the stress level, the time behavior of stress, the type of metal, and the type of environment.

Metallurgists and chemical engineers are now performing experiments on alloys, studying the interactions of controlled environments and material corrosion. Six interaction studies which may be useful to the engineer who designs the ocean structures are summarized.

1. Effects of Oxygen. Corrosion of iron and steel immersed in nearly neutral water, at room temperature without dissolved oxygen, is almost negligible. The role of oxygen in rust formation was discussed previously in this chapter. Experimental results showing the effect of oxygen concentration on the corrosion rate of mild steel were obtained by Uhlig, Triadis and Stern (1955), and are shown in Fig. 11.12.

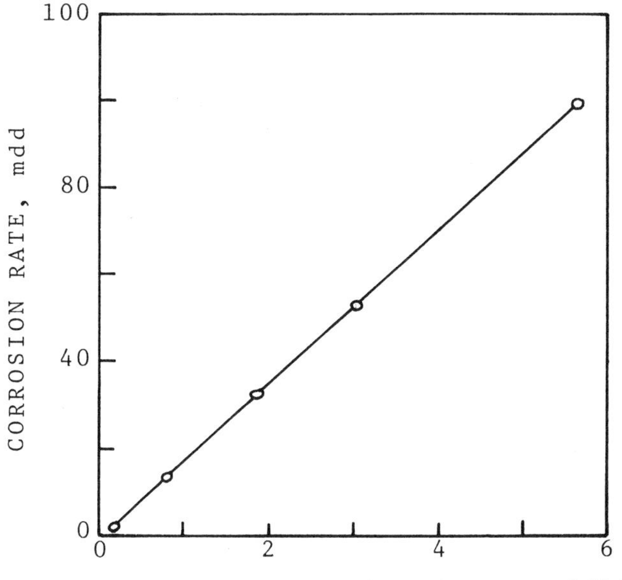

Figure 11.12 Effect of Dissolved Oxygen on the Corrosion of Mild Steel in Water at 75°F with 165 ppm $CaCl_2$.

2. Effects of Anaerobic Bacteria. In nearly neutral water without appreciable oxygen and in sea water, iron and steel have been found to corrode at high rates due to the presence of sulfate-reducing bacteria (Sporovibrio Desulfuricans). These bacteria, which live best in still water,consume hydrogen on the iron surface.

For each equivalent of hydrogen atoms that these bacteria consume, one equivalent of Fe^{++} enters solution to form rust and FeS . Severe corrosion damage of this sort has been observed in oil well casings and in pipes from deep water wells. It could also occur on deep water ocean structures.

3. Effects of Dissolved NaCl. The effect of NaCl concentration on the corrosion rate of iron in air-saturated water at room temperature, given by Uhlig (1963), is shown in Fig. 11.13. It is

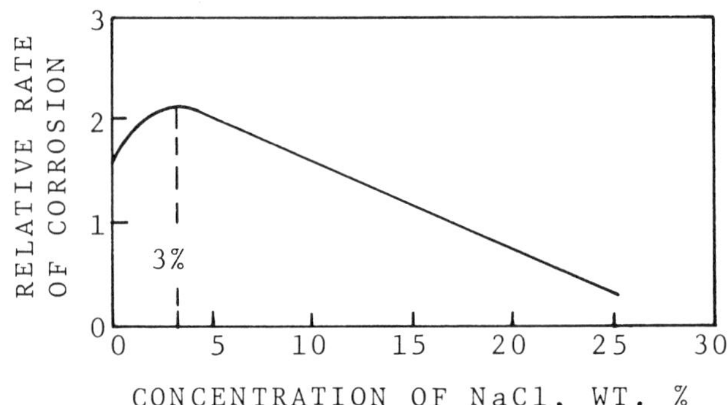

Figure 11.13 Dependency of the Corrosion Rate of Iron on NaCl Concentration

noted that the maximum corrosion rate occurs at a 3% NaCl concentration, which coincides with the concentration of NaCl in the oceans. These experiments were probably performed in still water.

4. Effects of Water Velocity. In natural fresh waters, the corrosion rates of iron are reduced after the water velocity reaches certain maximum values. Such data, due to Russell, et al (1927), are illustrated in Fig. 11.14 for both rough and polished steel. In ocean waters, however, such passivity to corrosion of steel is not established, as shown by Fig. 11.15. In undisturbed ocean waters, steel corrodes at about 0.005 ipy, or 25 mdd. At a water velocity of 25 ft./sec., however, the corrosion rate is about seven times this rate. In addition to uniform corrosion, pitting or cavitation-erosion is observed in the salt water tests.

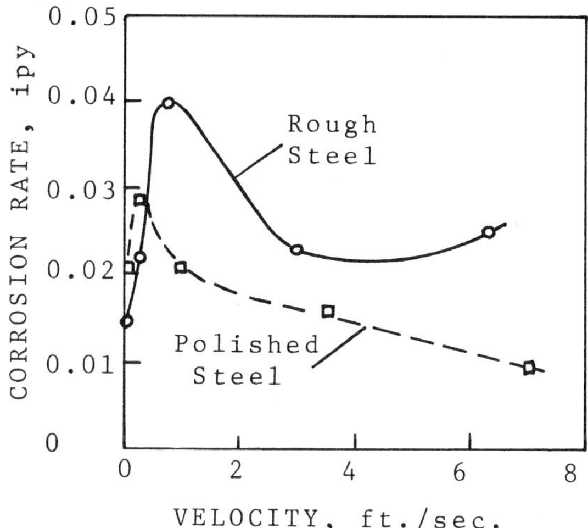

Figure 11.14 Effects of Fresh Water Velocity on the Corrosion Rate of Mild Steel at Room Temperature.

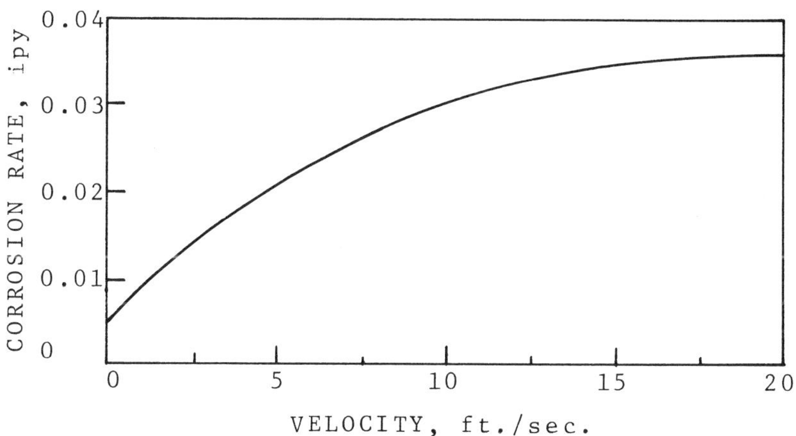

Figure 11.15 Effects of Ocean Water Velocity on the Corrosion Rate of Mild Steel, (Uhlig, 1963).

5. Effects of Composition and Environment. Corrosion rates in still ocean water and distilled water for various types of steels under no external stress are listed in Table 11.2. Whether a steel

% Carbon	Treatment	Environment	Temperature of Test	Corrosion Rate, ipy
	Effect of Heat Treatment			
0.39	Cold drawn, annealed 930°F	Distilled water	150°F	0.0036
0.39	Normalized 20 min. 1650°F	Distilled water	150°F	0.0034
0.39	Quenched 850°C (1560°F) Various specimens tempered 570°F to 1470°F	Distilled water	150°F	0.0033
	Effect of Carbon			
0.05	Not stated	3% NaCl	Room	0.0014
0.11	Not stated	3% NaCl	Room	0.0015
0.32	Not stated	3% NaCl	Room	0.0016
	Effect of Alloying			
0.13	Not stated	Sea water		0.004
0.10, 0.34% Cu	Not stated	Sea water		0.005
0.06, 2.2% Ni	Not stated	Sea water		0.005
Wrought Iron	Not stated	Sea water		0.005

TABLE 11.2 Corrosion rates for various steels when oxygen diffusion is controlling (Uhlig, 1948).

is manufactured by the Bessamer or open-hearth process, or is wrought iron or cast iron, makes little or no difference to the corrosion rate in natural waters, including ocean water. Although the carbon content of steel has no effect on the corrosion rate in fresh waters, an increase in this rate to a maximum of 20% is observed in sea water when the carbon content is raised from 0.1% to 0.8%. A 3% chromium addition to steel is found to decrease its overall weight loss in sea water. For either higher or lower percentages of chromium, localized corrosion and pitting is more pronounced in steel. Investigations of metal corrosion rates in the ocean air have also been made. In the American Society for Testing and Materials publication 435 (1967) a large amount of this data has been compiled.

6. Effects of Stress. Stresses near or above the elastic limit can cause accelerated corrosion damage to metals in an ocean environment. Since rivets always exceed the elastic limit, these parts of a structure can be quite susceptible to stress corrosion cracking. This problem is often minimized by the elimination of rivets in favor of welding. As was mentioned previously, an alternating stress can produce cracks leading to failure even when the maximum calculated stress is below the elastic limit of the material. Table 11.3 shows some typical experimental results, where the fatigue limits in air, well water, and salty river water are compared for various metals. The results of Table 11.3 may not be reliable enough to use in engineering design, since they depend upon the testing stress rate and upon the oxygen content as well as the salinity of the water. However, some general conclusions can be stated based on Table 11.3. (1) There is no relation between corrosion fatigue strength and tensile strength. (2) Medium alloy steels have only slightly higher corrosion fatigue strength than carbon steels. (3) Heat treatment does not improve corrosion fatigue strength of either carbon or medium-alloy steels; residual stresses are deleterious. (4) Corrosion resistant steels, particularly those containing chromium, have higher corrosion fatigue strength than other steels. (5) Corrosion fatigue strength for all steels is lower in salt water than in fresh water.

11.3 MATERIALS RESISTANT TO DYNAMIC LOADS AND CORROSION

There are several materials which are commonly employed in the construction of ocean structural systems. Steel is used widely for offshore drilling platforms, pipeline systems, and ship's hulls. Pilings of aluminum and lightweight underwater vehicles of titanium are becoming more common. There are several types of alloys and protective coatings which have been found suitable to the corrosive ocean environment. Structural metals with high toughness characteristics are especially important. The dynamic forces of winds, waves and currents acting on sea structures make design for high impact resistance imperative.

Metal	Fatigue Limit psi	Corrosion Fatigue Strength, psi		Damage Ratio, Corrosion Fatigue Strength ÷ Fatigue Limit	
		(10^7 to 10^8 cycles at 1450 cycles/min)			
Composition, treatment	Air	Well Water*	Salt Water#	Well Water	Salt Water
0.11% C steel, annealed	25,000	16,000		0.64	
0.16% C steel, quenched, drawn	35,000	20,000	7,000	0.57	0.20
1.09% C steel, annealed	42,000	23,000		0.55	
3.5% Ni, 0.3% C steel, annealed	49,000	29,000		0.59	
0.9% Cr, 0.1% v, 0.5% C steel, annealed	42,000	22,000		0.52	
13.8% Cr, 0.1% C steel, quenched, drawn	50,000	35,000	18,000	0.70	0.36
17% Cr, 8% Ni, 0.2% C steel, hot-rolled	50,000	50,000	25,000	1.00	0.50
Nickel, 98.96%, annealed 760°C	33,000	23,500	21,500	0.71	0.65
Monel, 67.5% Ni, 29.5% Cu, annealed 760°C	36,500	26,000	28,000	0.71	0.77
Cupro-Nickel, 21% Ni, 78% Cu, annealed 760°C	19,000	18,000	18,000	0.95	0.95
Copper, annealed 650°C	9,800	10,000	10,000	1.02	1.02
Aluminum, 99.4%, annealed	5,900		2,100		0.36
Aluminum, 98%, 1.2% Mn, hard	10,700	5,500	3,800	0.51	0.36
Duralumin, tempered	17,000	7,700	6,500	0.45	0.38
Brass, 60-40, annealed	21,000	18,000		0.86	

*2 ppm $CaSO_4$, 200 ppm $CaCO_3$, 17 ppm $MgCl_2$, 140 ppm NaCl
#Severn River Water having about 1/6 the salinity of sea water

TABLE 11.3 Fatigue limits and fatigue strength of several metals (McAdam, 1927)

Important results of research in these areas were presented at the First International Symposium, "Materials-Key to Effective Use of the Sea (1967)" and are included here. More detailed information has been reported by Lederman and Kallas (1968) and by the references listed in their article. Current information on the subject appears from time to time in The Journal of Ocean Engineering, published by Pergamon Press.

Steel

The ideal structural steel should be tough and resistant to fatigue and corrosion cracking. For underwater vehicle applications, the steel should also be able to withstand high ocean pressures, yet be light enough to provide a habitable space with maximum buoyancy and payload. From 1940 to 1958, 50 ksi tensile yield strength grade steel (50 HTS) was used for submarine hulls. The Nautilus, for instance, was constructed from this grade of steel, as reported by Shankman (1968). From 1958 to the present, 80 ksi tensile yield strength steel (HY-80) has been the basic alloy used in the construction of nuclear powered submarines in the United States. For this steel, typical impact toughness is about 100 ft-lb Charpy. Weldability and castability are excellent as are other pertinent physical and mechanical properties. Structures fabricated from HY-80 do not require heat treatment after welding.

Current research is on the development of steels with 130 ksi yield strength (HY-130) as well as on steels of 180 and 200 ksi capabilities. HY-130 has been developed by the United States Steel Corporation. The Navy has specified HY-130 for a deep sea recovery vehicle, for which Lockheed Aircraft Corporation is using a vacuum melted grade of the alloy produced by the Cameron Steel Company.

These new high strength steels can have high brittleness (or low toughness), high crack sensitivity, and high susceptibility to corrosion, unless special care is taken while they are processed. Sulfur content, for instance, greatly reduces toughness in HY-130, as shown in Fig. 11.16. Other impurities also have deleterious effects, and are very expensive to remove from the melt. Special processing, such as vacuum melting and degassing, can improve toughness by as much as 300% in HY-180 steel. For example, Republic, Inco and U. S. Steel have new alloys that require special heat-aging treatment. In 1967, U. S. Steel obtained an exceptionally strong and tough steel by aging. The composition and properties of this steel are shown in Table 11.4.

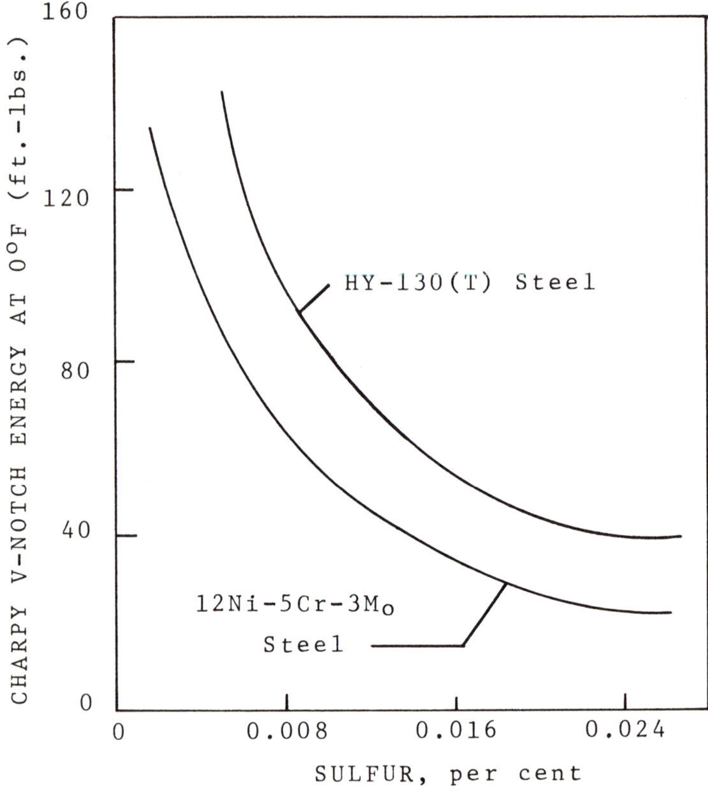

Figure 11.16. Effect of Sulfur Content on the Toughness of Two High Strength Steels (Porter, 1966).

Research engineers report that high-strength steels often suffer stress-cracking at ambient temperatures in the ocean. The Navy has developed tests to obtain more information on stress corrosion cracking. Improved welding techniques to minimize hydrogen embrittlement in welded areas have been reported by the U. S. Naval Applied Science Laboratory in Brooklyn. It is also important to note that mechanical peening of welds has produced welds with the same fatigue life as the base material. Normally, the fatigue life of a weld is much lower than that of the base material.

	Chemical Composition %										
	C	Mn	P	S	Si	Ni	Cr	Mo	Co	N	O
Base Metal	0.12	0.094	0.003	0.007	0.030	9.88	2.09	1.05	7.96	0.001	7*
Weld Metal** . .	0.098	0.014	0.002	0.005	0.095	10.20	2.09	1.04	8.06	0.002	6*

Tensile Properties

	Rolling Direction	Yield Strength (0.2% Offset), ksi	Tensile Strength, ksi	Elongation in 1 Inch, %	Reduction of Area, %	Charpy V-Notch Energy Absorption, ft.-lb.
Base Metal	Longitudinal	191	209	17.0	71	71
	Transverse	192	209	16.0	68	62
Weld Metal** . .		187	206	19.0	71	79

The weldment was aged at 950 F. for 5 hr.
*ppm.
**The weld metal was deposited in a double-Vee joint by using the inert-gas-shielded tungsten-arc welding process.

TABLE 11.4 Composition and Mechanical Properties at Room Temperature for a High Strength Steel Developed by the U. S. Steel Corporation (Lederman and Kallas, 1963).

High strength steels, prime candidates for deep-diving submersible vehicles, are commercially available through the Republic Steel Corporation. These are HP 9-4-20 and HP 9-4-25; HY-130(T); and 18% Ni maraging (190) steel.

Titanium

Titanium is 57% as dense as steel. Thus, titanium alloys have been developed to help fulfill the need for strong yet lightweight underwater vehicles. Some of these alloys are:

(A)	Ti-6Al-2Cb-1Ta-0.8 Mo	(100 ksi yield)
(B)	Ti-7Al-4V	(120 ksi yield)
(C)	Ti-8Al-2Cb-1Ta	(100 ksi yield)

Recent evidence showed that alloy (C), with 0.08% oxygen as an impurity, undergoes severe degradation in strength when prenotched specimens are tested in sea water. This is shown in Fig. 11.17. It is important to note that unnotched specimens of titanium alloy (C) showed no tendencies to crack in conventional stress corrosion tests. This indicates that the designer must exercise extreme caution in using these alloys.

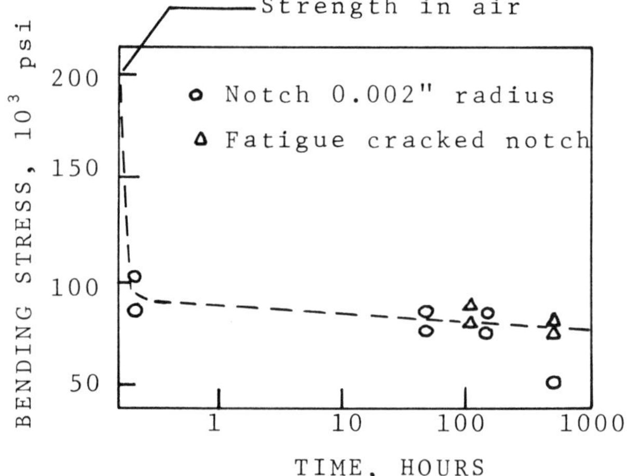

Figure 11.17 Bending Strength for Notched Specimens of Titanium Alloy Ti-8Al-2Cb-1Ta with 0.08% Oxygen, Tested in Sea Water (Lederman and Kallas, 1968).

Recent studies indicate that the addition of molybdenum to titanium, as in alloy (A), eliminates the stress corrosion problem in most circumstances. Alloy (B) also is resistant to stress corrosion because of its extremely low content of oxygen. Still, because of the problem of stress corrosion, it is advisable to design for about 85% of yield for alloys (A) and (B), where the yield strength is based on a 0.02% offset strain. Much research on titanium alloying and fabrication is presently being done at the Naval Applied Sciences Laboratory.

Aluminum

There are numerous examples of the successful use of unpainted aluminum alloys in brackish water and sea water. For instance, bare aluminum pipelines have been used without deterioration in the brackish water of Lake Maracaibo for ten years. Elsewhere, pilings of aluminum sheet have withstood the combined effects of barnacles and wave action for more than four years without deterioration. In general, these alloys have included 6061-and the various 5000-series alloys. Unprotected, higher strength 7000-series alloys have limited use because of their higher susceptibility to stress-corrosion. The 6061 alloy represents a fair compromise between required strength and corrosion resistance. Weldable marine alloys 5083, 5086 and 5456 are used extensively in small boats. Rheinhart (1965) observed that alloy 6061 specimens became severely pitted at the low oxygen zone within a few feet of the ocean bottom. Near the well-aerated ocean surface, however, he found that corrosion damage was negligible.

There are two special hazards to aluminum: low fatigue life and high galvanic corrosion. Here, aluminum becomes anodic to nearly all other structural metals when in electrical contact with them. Coatings or sacrificial anodic metals may be helpful here.

Other Materials

Lederman and Kallas (1968) report that new copper-nickel alloys and reinforced plastics are both candidates for naval surface vessels and submersibles. The cost of these alloys and the inherent structural weakness of plastics may make them impractical for fixed ocean structures. However, coating structural steel with these materials may be quite practical. The U. S. Naval Civil Engineering Laboratory, in Port Hueneme, California, is currently engaged in an extensive testing program to evaluate the performance of various materials in the deep ocean environment, particularly on, near, or just beneath the sea floor. More than 3,700 specimens of about 600 materials, including alloys, plastics and coatings, are being tested. Precise effects of exposure, such as weight loss, pit depth, stress corrosion cracking and loss of fatigue strength, are still to be evaluated. The first qualitative results have been summarized by Reinhart (1965).

Corrosion Protection

Some oil companies solve the corrosion problem of the fixed steel legs of their ocean platforms by both painting and structural encasement with other metals. The most important single factor influencing the life of a paint is the proper preparation of the metal surface, which should be clean and dry and free of any scale. Phosphate coatings on steel form an excellent bond for subsequent coats of organic-type paints. A minimum of four top coats with a total thickness of at least 0.13 mm is recommended for steel by authorities such as Uhlig (1963).

Some oil companies cover their whole structure with the commercial product, Dimetcoat, although other companies use it only on structural parts exposed to the atmosphere. On this corrosion-protection product, Tait (1964) reports:

> Anybody can get Dimetcoat and put it on. It involves strict weather and temperature criteria and sandblasting, with immediate application of the various coats. It goes on quite thick and heavy and has to be cured and treated between coats. It is expensive, but it seems to do a very good job in salt water, and the oil companies use it a great deal.

The splash zone on an ocean structure is particularly susceptible to corrosion. Sometimes monel metal or one-half inch steel plate is wrapped around the structure from approximately twenty feet above to twenty feet below the mean water line. The longitudinal seam is then welded and sealed top and bottom, giving corrosion protection to the structure at the expense of the wrapped metal. Sometimes the air space between the structural member and the surrounding tube is filled with concrete before the top is welded.

Much research is being devoted to the development of elastomer and resin coatings for use in the ocean. Lederman and Kallas (1968) list several reports in this area. Just how successfully these new coatings can be used to preserve the integrity of metals for large ocean structures is not known. It is suspected that the present high cost of many of these coatings prohibits their use except, perhaps, for submerged electrical cables.

11.4 DAMAGE ACCUMULATION IN METALS UNDER UNDER RANDOM-STATIONARY STRESS

The detailed study of damage accumulation, crack propagation and the subsequent failure of metals under external loads properly belongs to fracture mechanics. The purpose of this section is not to explore detailed damage or fracture mechanisms, but to present a simplified fatigue failure theory which can be applied to the design of ocean structures subjected to random-type forces. It will be shown how, with certain rational assumptions, the probable length of time it takes for such a structure to fail may be predicted. The present development is based on the work of Crandall and Mark (1963). It should be kept in mind that the validity of this approach still needs to be verified experimentally.

Damage Accumulation under Harmonic Stress

Assume that an S-N or fatigue curve for the structural metal has been experimentally determined in an environment which closely resembles that of the proposed ocean structure. Since the surface roughness of the metal also affects S-N data, the surface finish of the test specimens should closely resemble that of the future structural member. These data are then approximated by a straight line when log S is plotted against log N. That is

$$NS^b = c \tag{11.19}$$

where b and c are constants fitted in some way to the data. For most materials, the range of b is from 5 to 20. If these constants are chosen so that (11.19) defines a curve just under these data, Fig. 11.18, then (11.19) can be used to find a conservative estimate of the number of cycles to failure for a structural member at a given constant stress amplitude. The time to failure, T_F, under a constant frequency ν_1 cps, and a stress of constant amplitude S_1 is deduced from (11.19), or

$$T_F = \frac{N}{\nu_1} = \frac{c}{\nu_1 \; S_1{}^b}$$

When the harmonically-varying stress is not of constant amplitude, but is random in nature, the problem of fatigue failure becomes vastly more difficult.* There is not as yet enough experimental

*Throughout this analysis it is always assumed that the stress S or S(t) is the maximum macroscopic tensile stress at a point on the surface of the metal. It is calculated by the usual "strength of materials" methods of stress analysis.

evidence to build an empirical theory, nor has fracture mechanics advanced to a point where fundamental theory is possible. As an expedient, Palmgren (1924) and Miner (1954) proposed a rational extrapolation of fixed amplitude fatigue data. They suggested that if a specimen were tested first at the stress level S_1 and then at the stress level S_2, its total life could be predicted in the following manner. Let the number of cycles to failure at a constant stress amplitude S_1 be N_1, and that at S_2 be N_2, as shown in Fig. 11.18.

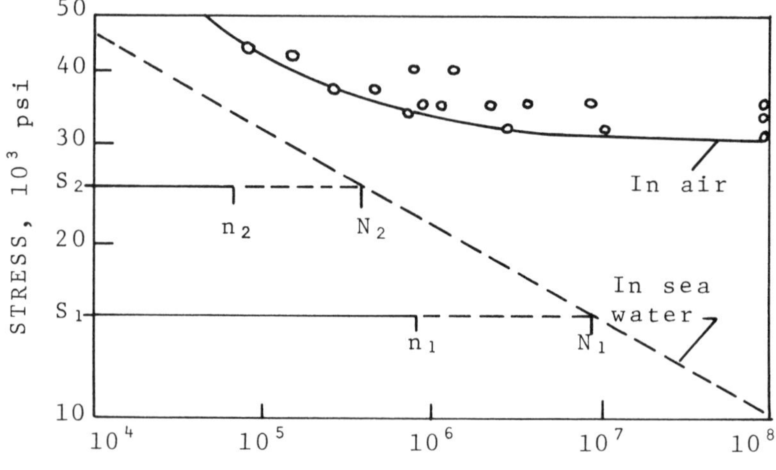

Figure 11.18 S-N Data in Air and Salt Water Environments for Annealed Steel, 0.37% Carbon. (Estimates from Timoshenko and MacCullough, 1954, and Table 11.3.)

The fraction of the material's life which is "used up" for n_1 cycles at S_1 is assumed to be n_1/N_1. The fraction of life remaining is thus $(1 - n_1/N_1)$. This remaining life is assumed to be used up at stress level S_2 in n_2 cycles, where the fraction of life remaining is n_2/N_2. Thus, failure occurs when

$$\left(1 - \frac{n_1}{N_1}\right) = \frac{n_2}{N_2}$$

or

$$\frac{n_1}{N_1} + \frac{n_2}{N_2} = 1$$

It follows that, if the specimen experiences n_i cycles of stress amplitude S_i, for $i = 1, 2, \ldots$, the total cumulative damage fraction is taken to be

$$D = \sum_{i=1,2,\ldots} \left(\frac{n}{N}\right)_i \tag{11.20}$$

According to the Palmgren-Miner hypothesis, the specimen will undergo fatigue failure when D = 1. The order of application of stress is not specified here. This hypothesis is now assumed to hold when stress of varying frequency has an amplitude which changes randomly from cycle to cycle.

A Statistical Model for Fatigue Damage

The next assumption is that the maximum normal stress S(t) at a point on the surface of a structural member is represented by a stationary, narrow-band random trace. Such a trace could be obtained, for instance, from dynamic measurements of the bending strain on a cantilever beam subjected to random transverse loads. Figure 9.4 shows a typical response where z(t) = S(t) and where the envelopes of the peak values are shown as dotted lines. As was shown in Chapter X, this is the response that one would expect from a linear one degree of freedom system with a natural frequency ω_n. The stress is assumed to be linearly proportional to displacement, and the frequency band of the applied force is assumed to include the frequency ω_n. Although a vibrating beam has infinite degrees of freedom, the elastic bending stress in the beam could still approximately follow a narrow band trace if the beam's lowest few frequencies were excited and the lowest mode shape was the dominant one. In this case, the system could be approximated by a one degree of freedom model which vibrates at its lowest frequency ω_n.

The next assumption is that S(t) is stationary and Gaussian with zero mean. Thus, a statistical average frequency ω_s for the cycling of stress can be calculated. It can be shown that ω_s^2 is given by

$$\omega_s^2 = \frac{\int_{-\infty}^{\infty} \omega^2 \, S_s(\omega)\, d\omega}{\int_{-\infty}^{\infty} S_s(\omega)\, d\omega} = \left. \frac{\dfrac{d^2 R_s(\tau)}{d\tau^2}}{R_s(\tau)} \right|_{\tau=0} \tag{11.21}$$

where $S_s(\omega)$ and $R_s(\tau)$ are the power spectral density and autocorrelation function, respectively, for S(t).

A One Degree of Freedom System

Equation (11.21) can be evaluated for the one degree of freedom model where the displacement x(t) is a linear function of stress S(t). This occurs only when the power spectral density of the excitation force is ideally white. With this assumption, $S_x(\omega) = (\text{const.})\cdot S_S(\omega)$, and (11.21) has the form

$$\omega_s^2 = \frac{\int_{-\infty}^{\infty} \omega^2 S_x(\omega)\, d\omega}{\int_{-\infty}^{\infty} S_x(\omega)\, d\omega} \tag{11.22}$$

With the value of $S_x(\omega)$ from (10.29), the result is simply

$$\omega_s = \omega_n \tag{11.23}$$

This very important result means that the average frequency of a narrow band, stationary, Gaussian response is identical to the natural frequency of the vibrating system. Thus, the mean frequency of stress "cycles", defined as ν_0^+, is given by

$$\nu_0^+ = \frac{\omega_n}{2\pi} \tag{11.24}$$

The value of ν_0^+ is actually the average number of zero crossings of the $S(t) = 0$ axis with positive slope per unit time. The nomenclature reminds us of this. In time T, for instance, the expected number of cycles is $\nu_0^+ T$. These ideas will be used presently.

The Rayleigh Probability Density Function

Consider now the probability distribution of the peak amplitudes of stress as traced by the dotted envelope of Fig. 9.4. It is reasonable to base fatigue damage on the amplitudes of the stress envelope shown in this figure. If the probability density function for stress is Gaussian, that is, if

$$p(s) = \frac{1}{\sqrt{2\pi R_s(0)}}\, e^{-\frac{S^2(t)}{2R_s(0)}} \tag{11.25}$$

then it can be shown that

$$p(a) = \frac{a}{R_s(0)} e^{-\frac{a^2}{2 R_s(0)}} \tag{11.26}$$

where p(a) is the Rayleigh probability density function for the stress peaks at level a. The Rayleigh distribution is shown in Fig. 9.7 where, in this case, the response is designated S(t) instead of z(t). This distribution will now be used to find the mean or average damage to be expected from an ensemble of narrow-band stress histories of duration T.

The Average Time to Failure

It was mentioned that in time T the expected number of stress "cycles" is $\nu_0^+ T$. The expected fraction of these cycles whose stress amplitude lie between a and a + da is p(a) da. The expected number n(a) of such peaks is

$$n(a) = \nu_0^+ T \; p(a) \; da \tag{11.27}$$

A single peak of amplitude a causes an incremental damage of 1/N(a), according to the Palmgren-Miner hypothesis. N(a) is the number of cycles to failure in a constant-amplitude fatigue test with stress amplitude a. The expected fractional damage due to all "cycles" having peaks between a and a + da is

$$\frac{n(a)}{N(a)} = \nu_0^+ T \frac{p(a)}{N(a)} da \tag{11.28}$$

The total expected damage E(D) is the sum of contributions like (11.28) for all stress amplitudes encountered. That is

$$E(D) = \nu_0^+ T \int_0^\infty \frac{p(a)}{N(a)} da \tag{11.29}$$

With the empirical result for N = N(a) given by (11.19), with p(a) given by (11.26), and with ν_0^+ given by (11.24), (11.29) becomes

$$E(D) = \frac{\nu_0^+ T}{c R_s(0)} \int_0^\infty a^{b+1} e^{\frac{-a^2}{2R_s(0)}} da$$

$$= \frac{\omega_n}{2\pi} \frac{T}{c} \left(2R_s(0)\right)^{\frac{b}{2}} \Gamma\left(1 + \frac{b}{2}\right) \tag{11.30}$$

Here, $\Gamma(1 + b/2)$ is the Gamma function and can be evaluated from tables. Recall that $R_S(0) = \overline{S^2(t)}$ is the mean square value of the stress as a function of time and is given by the right side of (10.27) for the one degree of freedom system. The stress is related linearly to displacement as long as the excitation is ideally white. Thus

$$S(t) = \alpha\, x(t) \tag{11.31}$$

where α is the constant of proportionality. If follows that

$$R_S(0) = \overline{S^2(t)} = \int_{-\infty}^{\infty} \frac{\alpha^2}{k^2} |H(\omega)|^2 \; S_F(\omega)\; d\omega \tag{11.32}$$

where $|H(\omega)|$ is given by (10.4), k is the spring constant, and $S_F(\omega) = S_0$, the power spectral density of the applied random force. The procedure is: (1) evaluate α of (11.31) from a simple strength of materials analysis of static stress-deflection; (2) determine $S_F(\omega) = S_0$ for the force and thus $R_S(0)$ from (11.32); (3) determine the fatigue constants b and c that fit (10.19) and the data, as plotted in Fig. 11.18; (4) for the known value of ω_n, set $E(D) = 1$ in (11.30) and calculate T_F, the expected time to failure. This expected time to failure is then given by

$$T_F\Big|_{\text{AVE.}} = \frac{2\pi c}{\omega_n \left(2R_S(0)\right)^{\frac{b}{2}}} \cdot \frac{1}{\Gamma(1 + b/2)} \tag{11.33}$$

It should be emphasized that (11.33) predicts only an <u>average</u> time to failure for this one degree of freedom system. For some samples, the damage may be greater than one, while for others it may be less than one when failure occurs.

Variance in the Average Time to Failure

The present statistical model is based on a stationary-ergodic, Gaussian stress trace; a one degree of freedom linear system subjected to ideal white noise excitation; and light damping or a narrow-band stress frequency range. With these assumptions, Mark (1961) has derived a formula for the variance σ_D^2 of the expected accumulated damage E(D). With the mean damage E(D) after a time T, calculated from (11.30), the standard deviation σ_D of E(D) is found from

$$\sigma_D = \frac{E(D)}{\sqrt{\frac{\omega_n T}{2\pi}}} \left\{ \frac{f_1(b)}{\xi} - \frac{f_2(b)}{\xi^2 \frac{\omega_n T}{2\pi}} + \frac{f_3(b)}{\frac{\omega_n T}{2\pi}} \right\}^{\frac{1}{2}} \tag{11.34}$$

where the damping constant $\xi \leq 0.05$. Typical values of the function $f_1(b)$, $f_2(b)$ and $f_3(b)$ are tabulated as a function of the fatigue constant b in Table 11.5. As a consequence of this equation, one can calculate the maximum and minimum failure times during which an arbitrarily prescribed number of samples would fail. This idea, which should be helpful to a design engineer in determining confidence limits for a structural component, will be illustrated in the following example.

b	$f_1(b)$	$f_2(b)$	$f_3(b)$
1	0.0414	0.00323	0.0796
3	0.369	0.0290	0.212
5	1.280	0.0904	0.679
7	3.72	0.223	2.33
9	10.7	0.518	8.28
11	31.5	1.230	30.0
13	96.7	3.06	111.2
15	308.	8.11	415.

TABLE 11.5 Functions Needed to Evaluate the Variance in the Average Time to Failure (Crandall and Mark, 1963).

11.5 NUMERICAL EXAMPLES

Consider the single degree of freedom model of the fixed offshore structure described in the first part of Chapter X. A free body diagram of one of the four legs of Fig. 10.1 is shown in Fig. 11.19. As discussed in Chapter X, an approximate shape for the static deflection curve for elastic deformations is given by

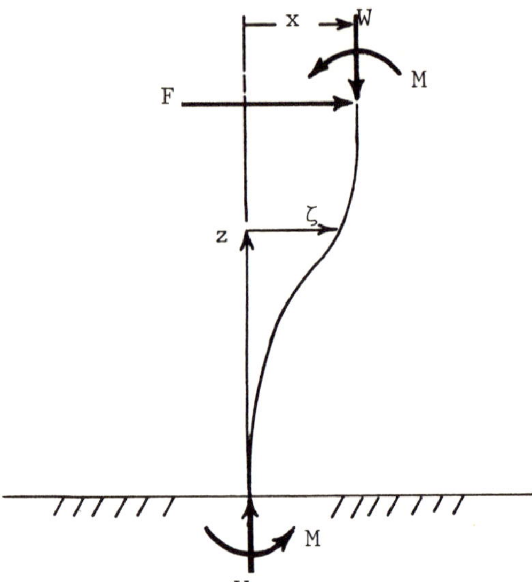

Figure 11.19. Forces Acting on One of the Four Legs of the Structure Shown in Fig. 10.1.

$$\zeta = \frac{F\ell^3}{24\ EI} \frac{1}{1 - \beta} \left(1 - \cos \frac{\pi z}{\ell}\right) \tag{11.35}$$

where

$$\beta = \frac{W\ell^2}{\pi^2\ EI}$$

When use is made of the symmetry of the deflection curve (11.35) it is deduced that the bending moments are equal at $z = 0$, the ocean floor, and at $z = \ell$, the platform-leg junction. This bending moment in each leg is also a maximum at these points. For static equilibrium, this moment is given by

$$M = \frac{1}{2} \ (Wx + F\ell) \tag{11.36}$$

which is found by equating the sum of the moments to zero about the point $z = 0$ in Fig. 11.19. With the value of x evaluated from (11.34) at $z = \ell$, (11.36) becomes

$$M = x \left[\frac{6\ EI}{\ell^2} - \left(\frac{12 - \pi^2}{2\pi^2}\right) W\right] \tag{11.37}$$

From elementary beam theory, the maximum bending stress occurs at the outer diameter of each leg at the locations of the maximum values of M. This stress is given by

$$S = \frac{Md}{2I} \tag{11.38}$$

where d is the leg diameter. With (11.37) this stress becomes

$$S = \frac{xd}{2I}\left[\frac{6\ EI}{\ell^2} - \left(\frac{12 - \pi^2}{2\pi^2}\right) W\right] \tag{11.39}$$

In this model, S is assumed to be the maximum fluctuating stress in the legs. Stress concentrations at the junctions $z = o, \ell$ and structural twisting caused by wave forces in planes other than the one shown in Fig. 11.19, are neglected. Assume also that the direct compressive stress caused by the platform weight has negligible effect on the fatigue damage accumulation in the legs as x fluctuates under forces of the waves. It was previously assumed that the dynamic deflection shape is similar to the static deflection shape given by (11.35). The constant α which relates the maximum stress response to the displacement response, or relation (11.31), can thus be deduced from (11.39), or

$$\alpha = \frac{d}{2I}\left[\frac{6\ EI}{\ell^2} - \left(\frac{12 - \pi^2}{2\pi^2}\right) W\right] \tag{11.40}$$

Assume further that the effective applied force $F = F(t)$ is stationary, ergodic and Gaussian with an ideally white power spectral density $S_F(\omega) = S_o$. The mean square value of the stress in terms of the mean square platform displacement, found by squaring and averaging (11.31) is given by

$$\overline{S^2(t)} = \alpha^2\ \overline{x^2(t)} \tag{11.41}$$

From (10.28) the result is

$$\overline{S^2(t)} = R_s(0) = \frac{\alpha^2\ \pi\ \omega_n\ S_o}{2\ \xi k^2} \tag{11.42}$$

As shown in Chapter X, (11.42) is also approximately valid if $S_F(\omega)$ is an ideal narrow band spectrum, provided that the restrictions on frequency and damping given by (10.33) are fulfilled.

Suppose that structural failure occurs when the expected accumulated damage in the section of a leg at the ocean floor becomes a maximum, or when E(D) = 1. The following procedure illustrates a way to calculate the average time to failure, T_F. For a structure of specified weight, geometry and material, W, $m_1 g$, ℓ, d, and EI are known. The value of α is given by (11.40). Experimental values are needed for the force spectrum S_o, for the damping factors ξ and C, and for the fatigue constants b and c. With these values, k, m and ω_n are calculated from (10.1), $R_S(0)$ is found from (11.42) and finally T_F is calculated from (11.33).

Consider the following specific case:

<u>Material:</u> mild steel, $E = 30 \times 10^6$ psi

<u>Fatigue Data:</u> (Fig. 11.18)	$b = 6$	
	$c = 10^{32}$ cycles $-\text{lb}^6/\text{in}^{12}$	
<u>The Structure:</u> (Figs. 10.1 and 11.19)	$\ell = 1800$ in.	leg length
	$d = 96$ in.	leg diameter
	$t = 0.5$ in.	leg wall thickness
	$W = 2.5 \times 10^6$ lb	¼ of platform weight
	$m_1 g = 78{,}600$ lb	weight of one leg
<u>Force Spectrum:</u>	$S_F(\omega) = S_o\ \text{lb}^2$ -sec/radian	
<u>Damping Factors:</u>	$\xi = 0.05$	Nath and Harleman (1967)
	$C = 0.70$	McLean, et al (1964), and Caldwell (1969)
<u>Calculations:</u>	$I \simeq \pi\left(\frac{d}{2}\right)^3 t = 1.74 \times 10^5\ \text{in}^4$	
	$\omega_n = 0.828$ radians/sec, (0.131 cps), Eq. (10.1)	
	$k = 9.03 \times 10^3$ lb/in	Eq. (10.1)
	$m = 1.32 \times 10^4\ \text{lb-sec}^2/\text{in}$	Eq. (10.1)
	$\alpha = 2.59 \times 10^3\ \text{lb/in}^3$	Eq. (11.40)

From (11.42) and (11.30), the results in terms of S_o are

$$\overline{S^2(t)} = R_S(0) = 2.14\, S_o \quad \text{lb}^2/\text{in}^4 \tag{11.43}$$

$$T = 1.60 \times 10^{30}\ E(D)/S_o^3 \quad \text{seconds}$$

$$= 5.07 \times 10^{22}\ E(D)/S_o^3 \quad \text{years} \tag{11.44}$$

The expected damage accumulation varies linearly with the average time required for that damage to occur, as shown by (11.44). Results for this numerical example are shown in Fig. 11.20 where S_o is assumed to be constant over the entire time until structural failure. The average times to failure are shown as circled points along the line E(D) = 1. For $S_o = 1.37 \times 10^7$ lb^2-sec/radian, for instance, T_F = 20 years.

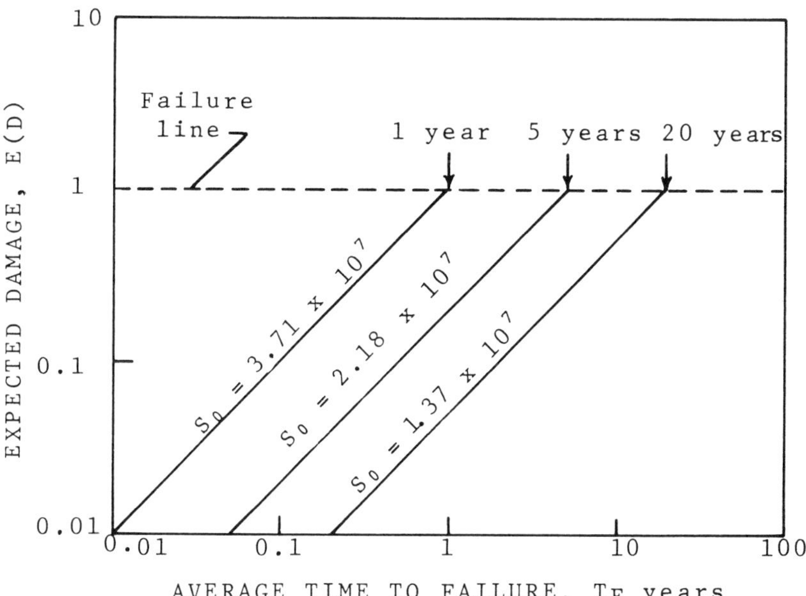

Figure 11.20 Damage Accumulation at the Ocean Floor in a Leg of a Fixed Offshore Structure.

It is important to observe that in experimental work a different unit of spectral density is often used. That is, cycles per unit time are used instead of radians per unit time, and positive frequencies only are considered in analyzing most experimental spectra. If $W_F(\omega)$ is the experimental power spectral density in cycles per unit time, then the values of $S_F(\omega) = S_o$, used to obtain plots such as Fig. 11.20 are obtained from the relationship

$$S_F(\omega) = \frac{W_F(\omega)}{4\pi} \tag{11.45}$$

where the factor of 2π converts the frequency from cycles per unit time to radians per unit time. The additional factor of two accounts for the conversion of positive frequencies only to positive and negative frequencies consistent with the theoretical development in which $S_F(\omega)$ is an even function of ω.

It is quite unrealistic to assume that S_o will remain relatively constant for more than a few days in an ocean environment, even over a relatively narrow frequency range which includes ω_n. Even so, after performing one of these calculations for which the value of S_o is based on the worst expected storm conditions, if T_F is found to be of the order of twenty years, the design would be conservative from the point of view of fatigue failure. This assumes, of course, that twenty years is all that is required of the structure. Suppose that $S_o = 1.37 \times 10^7$ lb^2 -sec/radian did represent the highest expected storm spectrum for the structure just described. The standard deviation in the accumulated damage after twenty years is given by (11.34), where $E(D) = 1$ and $T = T_F = 20$ years $= 6.32 \times 10^8$ sec, or

$$\sigma_D \simeq \left(\frac{2\pi \; f_1(b)}{\omega_n \; T_F \; \xi} \right)^{\frac{1}{2}} = 7.74 \times 10^{-4} \tag{11.46}$$

In this calculation, $f_1(b) = 2.5$ was found by a linear interpolation of the values in Table 11.5 for $b = 6$, and the terms involving $f_2(b)$ and $f_3(b)$ were found to be negligible in comparison to the $f_1(b)$ term. The standard deviation in the time to failure, σ_T, can be calculated from (11.44), where $T = T_F \pm \sigma_T$ and $E(D) = 1 \pm \sigma_D$. Thus

$$T_F \pm \sigma_T = 1.60 \times 10^{30} \quad (1 \pm \sigma_D)/S_o^3 \tag{11.47}$$

from which $\sigma_T = 5.64$ days when (11.46) is used. Since T_F will approach a Gaussian distribution as $T \to \infty$, (20 years is long enough here) there is a 68% chance that the structure will fail in 20 years plus or minus 5.64 days. This is because, for a Gaussian distribution, 68% of samples lie within the plus one and minus one sigma limits. From the practical viewpoint, the values of σ_D and σ_T are insignificant for the problem just stated, where $S_o = 1.37 \times 10^7$ lb^2-sec/radian.

Suppose that, after performing one of these calculations for which the value of S_o is based on the worst expected storm conditions, T_F is found to be of the order of one year. Suppose the structure is meant to last twenty years, over which time the worst storm conditions are estimated to occur only for a total time $T^{(1)}$ which is much less than the calculated one year period. If the storm spectrum, $S_F(\omega) = S_o^{(1)}$ and its time duration $T^{(1)}$ are given, $R_s(0)$

$= R_s^{(1)}(0)$ can be calculated as discussed above, and the accumulated damage can be calculated from (11.30), or

$$E_1(D) = \frac{\omega_n}{2\pi}\frac{T^{(1)}}{c}\left(2R_s^{(1)}(0)\right)^{\frac{b}{2}} \Gamma\left(1 + \frac{b}{2}\right) \tag{11.48}$$

where $E_1(D) < 1$. Now, if $S_F(\omega) = S_o^{(2)}$ is also given for the remainder of the lifetime of the structure, the corresponding value of $R_s(0) = R_s^{(2)}(0)$ can be calculated for this spectrum. Thus, the remaining lifetime $T^{(2)}$, for which the damage is $E\ (D) = 1 - E_2(D)$, can be found from (11.30) and (11.48), or

$$E_2(D) = \frac{\omega_n}{2\pi}\frac{T^{(2)}}{c}\left(2\ R_s^{(2)}(0)\right)^{\frac{b}{2}} \Gamma\left(1 + \frac{b}{2}\right) = 1 - E_1(D) \tag{11.49}$$

The average lifetime is thus given approximately by

$$T_F \simeq T^{(1)} + T^{(2)} \tag{11.50}$$

where $T^{(1)}$ is given and $T^{(2)}$ is calculated from (11.48) and (11.49)

One might be tempted to generalize this procedure to include more than two force spectrum values. Such calculations, however, would yield only crude approximations to T_F, as is also the case for the two component examples just considered. This is because the stationary hypothesis which was imposed on the spectrum in order to develop the above theory is violated if $S_F(\omega)$ is allowed to be a function of time, even a simple step function, from calm to storm conditions. If the transient effects of a time-varying force spectrum can be neglected, then one might have some confidence in this last type of calculation. This would be the case, for instance, if $S_F(\omega)$ were stationary for most of the structural lifetime, where the time of transition from storm to relative calm conditions was quite negligible. Finally, it is recalled that the Palmgren-Miner damage theory has not yet been proved experimentally. Until such proof is obtained, one should always view the results obtained from even the simplest of these calculations with some reservations.

11.6 REFERENCES

1. Caldwell, H. M. (1969), "Fluid Effects on Offshore Pipelines: A Static and Dynamic Model Study", M. S. Thesis, Department of Civil Engineering, Duke University.

2. Crandall, S. H. and Mark, W. D. (1963), Random Vibrations in Mechanical Systems, Academic Press, New York.

3. Eyring, H. and Ree, T. (1955), "A Generalized Theory of Plasticity Involving the Virial Theorem", Proceedings of the National Academy of Sciences, Vol. 41, p. 118.

4. Fontana, M. G. and Greene, N. D. (1967), Corrosion Engineering, McGraw-Hill Book Co., New York, p. 155.

5. Grover, H. J., Gordon, S. A., and Jackson, R. L. (1960), Fatigue of Metals and Structures, Battelle Memorial Institute, Bureau of Naval Weapons, U. S. Navy.

6. Lederman, P. B., and Kallas, D. H. (1968), "Materials: Key to Exploiting the Oceans", Chemical Engineering, June, p. 105.

7. Mark, W. D. (1961), "The Inherent Variation in Fatigue Damage Resulting from Random Vibration", Ph.D. Thesis, Department of Mechanical Engineering, M.I.T.

8. Maxwell, C. (1868), Philosophical Magazine, Vol. 35, p. 134.

9. McAdam, D. (1927), Proceedings of the American Society of Testing and Materials, Vol. 27, p. 102.

10. McLean, W. J., Laird, A. D., and Brewer, J. W. (1964), "Behavior of a Flexibly Supported Cylinder in a Fluid Stream", Report No. HPS-64-3, Institute of Engineering Research, University of California at Berkeley.

11. Miner, M. A. (1945), "Cumulative Damage in Fatigue", Journal of Applied Mechanics, ASME, Vol. 12, p. A159.

12. Nath, J. H., and Harleman, D. R. F. (1967), The Dynamic Response of Fixed Offshore Structures to Periodic and Random Waves, TR-102, Hydrodynamics Laboratory, M.I.T.

13. Palmgren, A. (1924), "Die Lebensdauer von Kugellagern", Feitschrift Verein Deutscher Ingenieure, Vol. 68, p. 339.

14. Pellini, W. S. (1954), "Symposium on Effect of Temperature on Brittle Behavior of Metals", ASTM Special Technical Publication No. 158.

15. Polakowski, N. H., and Ripling, E. J. (1966), Strength and Structure of Engineering Materials, Prentice-Hall, Inc., Englewood, New Jersey.

16. Porter, L. F. et al (1966), "Ultraservice Steels with Yield Strengths of 130 to 200 ksi", Metals Engineering Quarterly, August, p. 17.

17. Reinhart, F. M. (1965), "First Results - Deep Ocean Corrosion", Geo-Marine Technology, September, p. 15.

18. Russell, R., Chappell, and White (1927), Industrial Engineering Chemistry, Vol. 19, p. 65.

19. Shankman, A. D. (1968), "Materials for Pressure Hulls - Present and Future", Ocean Industry, Vol. 3, p. 29.

20. Tait, G. F. (1964), "Underwater Construction in Oceans Using Legged Platforms", Report of a one-day lecture, September, 1964, at the U. S. Naval Civil Engineering Laboratory, Port Hueneme, California.

21. Timoshenko, S. and MacCullough (1954), Elements of Strength of Materials, D. Van Nostrand Co., New York, p. 366.

22. Trouton, F. T. and Rankine, A. O. (1904), "On the Stretching and Torsion of Lead Wire beyond the Elastic Limit", Philosophical Magazine, Vol. 8, p. 538.

23. Uhlig, H. H. (1963), Corrosion and Corrosion Control, John Wiley and Sons, Inc., New York.

24. Uhlig, H. H. (1948), The Corrosion Handbook, Chapman and Hall, Ltd., London.

25. Uhlig, H. H., Triadis, D. and Stern, M. (1955), Journal of the Electrochemical Society, Vol. 102, p. 59.

26. Wilson, J. F. (1964), "Instrumentation to Control Strain Rate for Materials Undergoing Plastic Flow", Experimental Mechanics, SESA, Vol. 4, p. 11.

27. Wilson, J. F. and Garofalo, F. (1966), "Stress Relaxation in Metals Plastically Prestrained in Compression", Materials Research and Standards, ASTM, Vol. 6, p. 85.

28. Wilson, J. F. and Wilson, N. K. (1966), "Kinetics of Stress Relaxation in Metals", Transactions of the Society of Rheology, Vol. 10, p. 399.

29. Wray, P. J. and Richmond, O. (1968), "Experimental Approach to a Theory of Plasticity at Elevated Temperatures", Journal of Applied Physics, Vol. 39, p. 5754.

PART IV
APPLICATION TO OCEAN SYSTEMS

Perhaps the most important problem facing the design engineer is the determination of the reaction forces and/or motions of a proposed structure as excited by ocean waves, winds and currents. Again, as throughout these notes, the term "structure" is used in a very broad sense meaning all kinds of systems and devices ranging in size from small buoys to fixed platforms and on up to the supertankers and including the mooring (cable assemblage) and/or attachment (to a fixed point) arrangement, as well. The only restriction (or limitation) is that rigid bodies only are considered. The mooring or attachment arrangement is not limited thusly, however.

It turns out to be convenient (for reasons which will become self-illuminating) to classify these structures and corresponding attachment assemblies according to (i) relative size and (ii) a time parameter characterizing the method of attachment. By relative size, the structure can be considered to be less than equal to or greater than the dominant length (crest-to-crest, for example) of the exciting waves. In a similar manner, the time parameter characterizing the method of attachment can be considered as being less than, equal to, or greater than the period of the exciting waves. To illustrate, the importance of the latter classification category, it is instructive to review the characteristics and properties of various mechanical systems.

Following Stoker (1950), and excluding special cases, the response motions for linear and nonlinear systems may be compared as follows:

I. Linear Systems

A. Without damping

1. Free oscillation: Motion is simple harmonic.
2. Forced oscillation: Motion is a superposition of simple harmonic (free) and a forced oscillation having the same period as forcing function.

If free and forced periods are equal, resonance results.

B. With damping

1. Free oscillation: Motion is damped out exponentially.
2. Forced oscillation: Motion is a superposition of a free oscillation which is damped out exponentially and a forced oscillation with a frequency equal to that of the forcing function.

II. Nonlinear Systems

A. Without damping

1. Free oscillation: Motion is periodic but not simple harmonic.
2. Forced oscillation: General motion is unknown but many types of periodic motion (subharmonics, etc.) exist.

B. With damping

1. Free oscillation: Motion is damped out.
2. Forced oscillation: General motion is unknown.

Thus, in the case of linear systems without damping, the structure tends to: (1) oscillate about its own natural period when freely excited, or (2) oscillate about a superposition of its natural period and the forcing period under forced excitation. In the latter case, when damping is present, the craft tends to oscillate about a superposition of the free oscillation, which is damped out exponentially, and of the forced oscillation. As the time, t, increases, the response period approaches and eventually is equal to the period of the forcing function.

In the case of nonlinear forced systems with and without damping, the general motion is unknown, but many different types of periodic and non-periodic motions have been observed. Two particular motions of interest are those of subharmonic and superharmonic response, as described, for example, by von Kármán (1940), Den Hartog (1947) and Stoker (1950).

Fundamental concepts of the dynamics of linear spring-mass systems have been presented in Chapter X. Chapter XII presents a specific application of a moored-ship system as analyzed linearly. Chapter XIII discusses application of the nonlinear analysis to moored-ship systems, while Chapter XIV discusses specific nonlinear applications.

Chapter XII

LINEAR MOORED-SHIP SYSTEMS

Any floating body placed in a wave environment experiences a time-varying loading pattern which has two components. There is a linear part of the time-varying loads which causes the body to oscillate at frequencies in the region in which most of the surface gravity waves appear. In general, these loads have the most influence on roll, pitch and heave motions or alternatively those which have hydrostatic restoring forces (or moments). There is also a nonlinear part of the time-varying loads which may be very significant. For moored ships, this nonlinear part has an important effect on sway, surge and yaw motions particularly if it (the nonlinear part of the time-varying loads) coincides with one of the ships natural periods in these modes. For free-ships, this nonlinear part is responsible for the well-known drifting force. For a free ship, one could say that the sway, surge and yaw have zero frequency or infinite period.

This nonlinear part arises from two contributions. One contribution is that due to reflection and transmission of waves from the vessel, or simply the modification of the incident wave pattern due to the presence of the obstruction. The other part is due to the generation of waves by the vessel motions as the motions interact with the oncoming waves.

In the case of a multipoint mooring, the vessel's attitude remains relatively fixed with respect to the oncoming waves; thus, the generation of waves by the vessel's motions probably do not represent a significant contribution to the nonlinear loading history. On the other hand, for a single point mooring, it is likely that the reflection of waves is less important and that wave generation by the vessel becomes the major contribution to nonlinear effects.

In this chapter, we will consider only linear contributions to the loading pattern and we will develop in detail the appropriate equations for a multipoint mooring. This can be justified via the Froude-Kriloff assumption in that, for vessels which are small relative to the dominant wave length, the presence of the vessel does not materially alter the wave pattern. In principle, the same technique can be applied to a single point mooring except that (1) the mooring restoring forces have to be represented differently, and (2) the instantaneous attitude of the vessel relative to the oncoming wave pattern must be accounted for in evaluating the remaining coefficients of the equations of motion. This places some rather severe restrictions on use of the method for single point moorings, since the coupling terms are difficult to evaluate in any event and assume major importance for single point moorings.

As clearly indicated, the motion of a floating platform in a realistic seaway, either free or restrained, is a formidable problom for which a complete solution has not yet been achieved. Nevertheless, a considerable body of knowledge is available which may be used in conjunction with approximate techniques to yield estimates (of motions, forces and bending moments) which are sufficiently accurate for engineering analysis and design purposes. To summarize, our objective in this chapter is to review in detail the problem of estimating motions (and thereby forces, where appropriate) of floating platforms as induced by waves under the linearity restriction.

In contrast to the simple single degree of freedom system just introduced, a floating platform has the usual six (6) degrees of freedom (i.e., 3 translational (surge, sway and heave) and 3 rotational (roll, pitch and yaw); thus, it is necessary to formulate an appropriate set of differential equations analagous to that given in Chapter X. In the course of the analysis, the additional complexities appearing in the individual terms (inertia, damping, restoration, excitation) as well as the change in the basic relationships — which are required to be considered in the case of floating platforms — will receive treatment. It is to be emphasized that the basic assumption in the analysis to follow is linearity. Specifically, it is assumed that in the absence of the "right-hand side" excitation term, the platform motion can be described in terms of homogeneous, second-order, linear, differential equations with time as the independent variable. An excitation term is added to the homogeneous equations as an r.h.s. term which, in most instances, will be sinusoidal. The importance of the linearity assumption will emerge later. We begin by introducing the coordinate system.

12.1 COORDINATE SYSTEM

Following Korvin-Kroukovsky (1961) and others, it has been found convenient to stipulate two systems of axes in order to analyze ship motions in waves. One system of coordinate axis is fixed in the body and the other is fixed in space, as shown in the accompanying sketch.

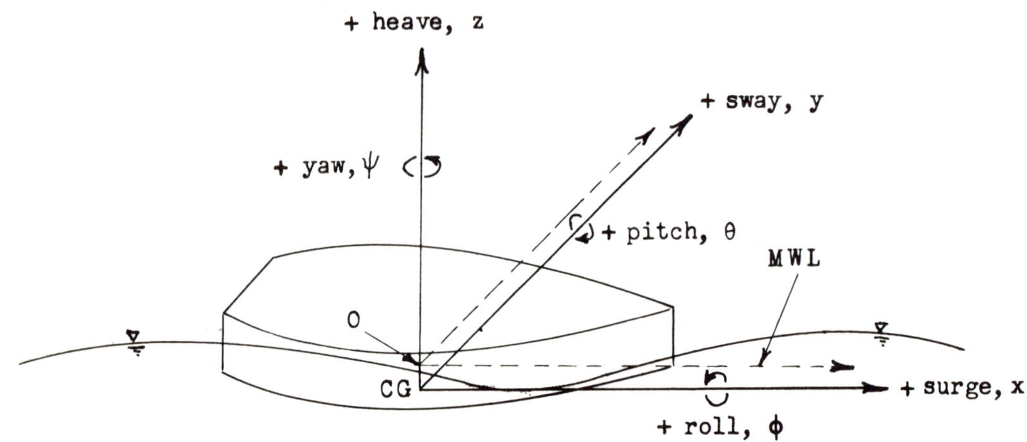

Definition sketch for co-ordinate system

CG = origin of "mean body" axes (i.e., that system fixed in body— to describe body motions)

O = origin of axes fixed in space (i.e., mean water level— to describe wave effects)

OG = distance between CG and mean water level

The system fixed in the body is a translating right-hand Cartesian system with the origin at the center of gravity of the body. The axes have a fixed orientation in that the origin translates with the center of gravity of the body but the axes cannot rotate. Thus, the axes may be considered to occupy the mean position of small angular oscillations. For this reason, some authors refer to them as "mean body" axes. With respect to the orientation of the barge (or ship), the x-axis is chosen positive toward the bow, the y-axis is positive to port, and the z-axis is positive upward. The dynamic variables are the linear displacements, surge (x), sway (y), and heave (z), along the respective axes, and the angular displacements, roll (ϕ), pitch (θ), and yaw (ψ), which are defined as positive in a direction of positive rotation about the x-, y-, and z-axes, respectively (i.e., port-upward, bow-downward, bow-portward). The positive direction of the forces and moments is similarly defined. The foregoing system of axes is sufficient to describe most of the motions of a platform in still water.

In order to evaluate the excitation effects, that is, the motion of a platform induced by the wave system, it has been found convenient to introduce another system of coordinate axes which is fixed in space. This is a nontranslating and nonrotating system and has its origin at the mean position of the free surface directly above mean position of the center of gravity of the platform. It is also a right-hand coordinate system.

Formulation of Equations of Motion: The equations of motion are formulated by equating to the motions the sum of the forces (and moments) acting on the body. The total force (or moment) acting on the body is the sum of the forces arising from dynamic body motions, F_{bm}, damping, F_d, hydrostatic restoring, F_h, mooring restoring, F_m, and excitation, F_w.

The force exerted by the body in accelerating the surrounding water gives rise to equal and opposite forces by the water on the ship. These forces (and moments) arise from the dynamic body motions; they are of an inertial nature, and being proportional to acceleration are usually expressed in terms of a fictitious added mass, which, in combination with the body's intrinsic mass, yields the commonly known "virtual mass". The total inertial force (moment) has components in all the translation (rotation) directions.

The damping forces involve the dissipation of energy and result from wave generation, viscosity, and eddy-making. Except for the roll motion, only damping due to wave generation has been evaluated up to the present. The total damping force (moment) has components in all the translation (rotation) directions.

The hydrostatic restoring forces (moments) arise from buoyancy effects induced by static displacement. The total hydrostatic restoring force has a component only in the vertical (z) direction, while the hydrostatic restoring moment has components in the roll (ϕ) and pitch (θ) directions.

The mooring restoring forces result from displacement of the body from the equilibrium mooring position. These forces will be assumed to be linear functions of displacement, an assumption which is usually justified for small displacements. Generally speaking, however, in most mooring situations for large displacements, the forces are usually nonlinear. The total mooring force is assumed to have components only in the surge (x) and sway (y) directions, while the total mooring moment is assumed to have a component only in the yaw (ψ) direction. A free or unmoored vessel, of course, has no mooring restoring forces.

Finally, the excitation forces (moments) are assumed to be due only to the waves. The effects of winds and currents are not considered. Later we will indicate how winds and currents may be treated. (One may consider the effect of wind and current is simply to displace the platform from its mean or equilibrium position. However, in so doing, the coupling or cross-multiplication terms arising when the orbital fluid particle motion of waves are parallel with those fluid particle motions from currents must be considered.) The wave forces (moments) have components in all the translation (rotation) directions. In view of the foregoing, the six equations of motion corresponding to the 6 degrees of physical freedom in which the body is permitted to move, may be written as follows:

Surge (x)

$$m\ddot{x} = X_i + X_d + X_m + X_w \tag{12.1.1}$$

Sway (y)

$$m\ddot{y} = Y_i + Y_d + Y_m + Y_w \tag{12.1.2}$$

Heave (z)

$$m\ddot{z} = Z_i + Z_d + Z_h + Z_w \tag{12.1.3}$$

Roll (ϕ)

$$I_x\ddot{\phi} = K_i + K_d + K_h + K_w \tag{12.1.4}$$

Pitch (θ)

$$I_y\ddot{\theta} = N_i + N_d + N_h + N_w \tag{12.1.5}$$

Yaw (ψ)

$$I_z\ddot{\psi} = M_i + M_d + M_m + M_w \tag{12.1.6}$$

where

m = mass of body

I_r = moment of inertia of ship about the r-axis

X, Y, Z = force in surge, sway and heave, respectively

K, N, M = moment in roll, pitch and yaw, respectively

with subscripts on the forces and moments indicating the type of force or moment, according to the following notation:

i = inertial

d = damping

h = hydrostatic restoring

m = mooring restoring

w = wave excitation

To sum up, the fundamental assumption of the present approach is that of linearity. Specifically, it is assumed that if the excitation terms in Equations (12.1.1) through (12.1.6) are neglected, the resulting equations are second-order linear differential equations with time t as the independent variable. The excitation terms merely add to the "right-hand side" a term of sinusoidal nature.

At this point, the suggested procedure is to evaluate the terms appearing in Equations (12.1.1) through (12.1.6), and then to solve the equations simultaneously. There are several possible approaches to carry out these evaluations. One which has been widely employed and applied successfully under varying conditions is the slender body theory. This is the approach which we will present in detail here and in the presentation, the meaning and nature of the terms will be clarified and evaluation by other methods will be suggested. We make one other simplification before proceeding to the evaluations. We assume (and this has been verified experimentally for ship-shape bodies) that there is little or no coupling between the variables in the two planes of motion, the longitudinal-vertical (surge, heave and pitch), and the horizontal-vertical (sway, roll and yaw). However, the longitudinal motions in the vertical plane are coupled to each other, and similarly, the lateral motions in the horizontal and transverse vertical planes are coupled to each other.

12.2 EVALUATION OF TERMS APPEARING IN EQUATIONS OF MOTION

As stated above, to find analytical expressions for force and moment terms appearing in Equations (12.1.1) through (12.1.6) an approximate method is employed which has been successfully applied to many studies of ship motion in waves. This method is the well-known slender body theory. It has been applied to submarines, destroyers, shallow-draft elongated barges and recently to supertankers. Essentially, this theory makes the assumption that, for an elongated body where a transverse dimension is small compared to its length, the flow at any cross-section is independent of the flow at any other section and, hence, the flow problem is reduced to a two-dimensional problem in the transverse plane. The forces at each section are found by this method, and the total force is found by carrying out an integration over the body length. The method was originally developed in the 1920's for the analysis of flow problems involving fuselage and airfoils. The technique has also been applied to wing sections with low ($\approx$1.0) aspect ratios, so that with proper modifications, it is possible that the technique has general application to a spectrum of hull forms.

Elaboration on Slender Body Theory: The following five sections are devoted to a detailed description of the force and moment components appearing in Equations (12.1.1) through (12.1.6). Particular attention is placed on the evaluation of these components for application to a broad range of hull shapes.

12.3 DYNAMIC FORCES AND MOMENTS DUE TO BODY MOTIONS

The hydrodynamic forces and moments of an inertial nature arise from the action of the forces and moments of the surrounding water on a platform, and are caused by the temporal displacement of the platform. The simplified result of slender body theory calculation is that the local force at any section is equal to the negative time rate of change of fluid momentum. The equations for the forces at each section then have the general form

$$\frac{dF_{bm}}{d\xi} = - \frac{D}{Dt}\left(\sum A'_{ij}\ \bar{q}_j\right) \qquad (12.3.1)$$

where the A'_{ij} denotes the sectional added mass along the ith direction due to motion in the jth direction; $\bar{q}_j$ is the effective average velocity at the platform in the jth direction, and D/Dt denotes the substansive or material derivative. Since the analysis is for zero speed, then D/Dt becomes simply $\partial/\partial t$.

The added mass terms, A'_{ij}, are frequency-dependent and thus take into account the effects of the free surface. The sum of the products $A'_{ij}\ \bar{q}_j$ insures that all couplings are included in the analysis subject to the limitation previously indicated (i.e., there are no couplings of any of the longitudinal motions with any of the lateral motions). To clarify the nature of the added mass tensor, the subscripts have the following meanings:

1 - Surge	4 - Roll
2 - Sway	5 - Pitch
3 - Heave	6 - Yaw

Additionally, it has been shown to be a symmetric tensor.

<u>Surge X_i</u>: Since pitch is the only longitudinal motion other than surge having a component in the surge direction, the summation indicated by Equation (12.3.1) is

$$A'_{11}\ \ddot{x} + A'_{15}\ \ddot{\theta}$$

Thus, the sectional surging force is

$$\frac{dX_i}{d\xi} = - (A'_{11}\ \ddot{x} + A'_{15}\ \ddot{\theta}) \qquad (12.3.2)$$

It will be noticed that yaw also has a component in the surge direction ($A'_{16}\Psi$) but we have already assumed that the longitudinal motions are uncoupled with the lateral motions. Thus, we neglect this component but it must be included if one is interested in coupling all six of the motion equations together. The added mass component A_{11} is known to be quite small for slender bodies and will be set equal to zero in our formulation herein. However, Sommet (date unknown) has investigated the problem theoretically and he presents the results in the form of integral equations which are applicable to platforms of any shape but which are too lengthy to reproduce herein. The added mass is shown to depend on the (i) hull outlines, (ii) depth of water, and (iii) characteristics of motion (frequency). In the case of a vertical sided prismatic hull whose waterplane area is in the form of a more or less elongated polygon, he gives the following expression for the added mass coefficient

$$A'_{11} = \frac{2}{\pi}\gamma\left[\frac{2}{\pi^2}\cdot\frac{h}{L-1}\left(1 - \frac{2h}{\pi^2 1}\right)\frac{\sin^2\pi\theta}{\theta} + \theta G(\gamma)\right]$$

where

$$\theta = \frac{\text{draft}}{\text{water depth}}$$

$$\delta = \frac{1}{L}$$

h = depth of water

L = length of vessel

γ = bow and stern angle with longitudinal vertical plane

$G(\gamma)$ = a function of the inverse length as shown in sketch on the following page

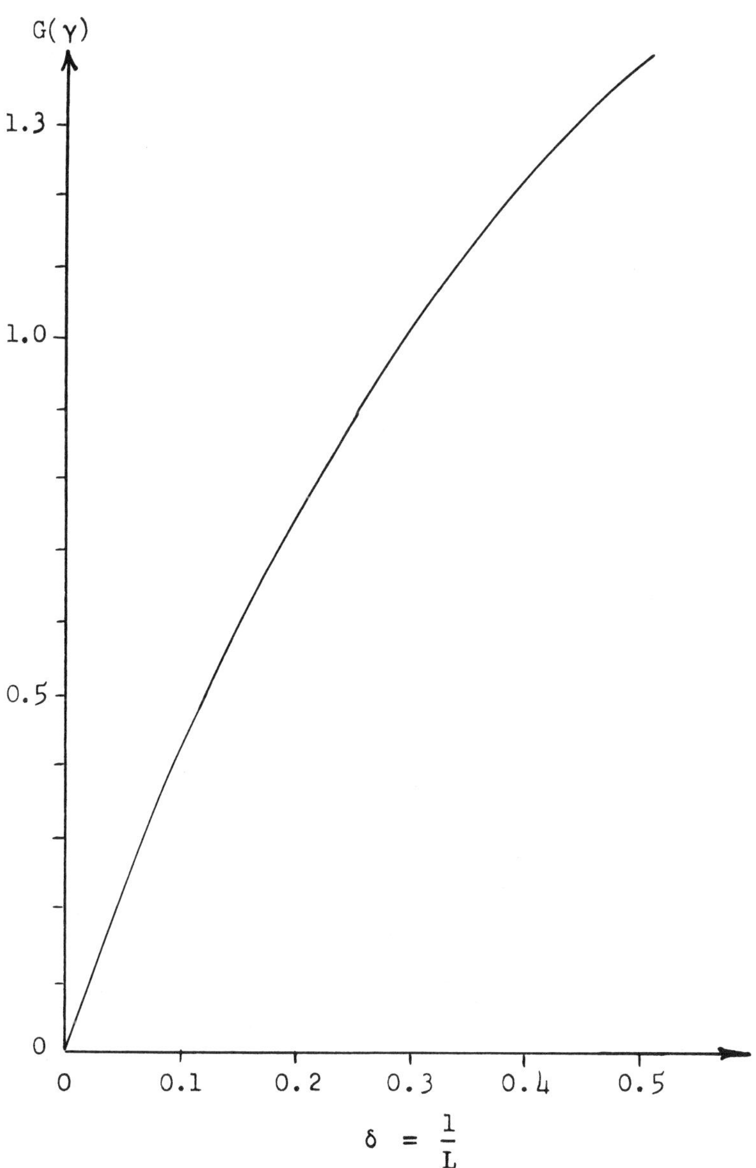

Relationship of $G(\gamma)$ to δ for determination of added mass in surge. (After Sommet, (date unknown)).

The important conclusions may be summarized as follows:

(1) Within given limits, the added mass coefficient for a given hull outline is inversely proportional to the water depth.

(2) The added mass coefficient is inversely proportional to the fineness ratio, being as much as 30% or 40% of the displacement mass for the blunter hull forms.

(3) The added mass increases with increasing frequency of oscillation.

Finally, where a hull is such that the basic assumptions given by Sommet are not confirmed for the calculations, the required information may be obtained by scale model tests. A small model is generally quite adequate for this type of test.

Returning to Equation (12.3.2), if we set the first term equal to zero, we can rewrite the equation as:

$$X_i = -\ddot{\theta}\int A'_{15}\, d\xi \qquad (12.3.3)$$

The symmetrical nature of the added mass tensor, A_{ij}, may be utilized to evaluate the integral appearing in Equation (12.3.3). The surging force is applied at the center of bouyancy of each elemental section or the total force may be considered to act at the center of bouyancy of the entire vessel. Thus, surging induces a pitching moment of magnitude $m\,|\overline{BG}|\,\ddot{x}$, where m is the platform mass, $\overline{BG}$ the vertical distance from the center of buoyancy to the center of gravity, and $\ddot{x}$ is the acceleration in the longitudinal direction (surge). Thus, the added mass in pitch due to motion in surge, A_{51}, is equal to the coefficient of the pitching moment, or

$$A_{51} = m\,\left|\overline{BG}\right| .$$

But from symmetry,

$$A_{15} = - A_{51} \, .$$

Thus, Equation (12.3.3) becomes

$$X_i = m \left| \overline{BG} \right| \ddot{\theta} \tag{12.3.4}$$

It is clear that the integral of the sectional added mass coefficients, A'_{15}, may be replaced by its composite A_{15}. The unprimed notation is employed to indicate that a summation has been carried out over the body length.

Sway, Y_i: The sectional dynamic force for lateral motion is formulated by reference to the coordinate axis fixed in space (i.e., at the mean free-water surface level). This axis is chosen because the roll axis coincides with the longitudinal axis passing through the CG of the platform and the necessity for determining individual CG's of each sectional element may be eliminated. Thus,

$$\frac{dY_{i_o}}{d\xi} = -\left(A'_{22}\, \ddot{y} + A'_{26}\, \ddot{\Psi} - A'_{24}\, \ddot{\phi} \right) \tag{12.3.5}$$

where the subscript "o" refers to the mean water surface level. The terms represent the added mass in sway resulting from motion in sway, yaw and roll, respectively. The difficulty of evaluating A'_{26} either theoretically or from known experiments requires that an appropriate substitution be made. It has been noted that since the quantity $\xi \ddot{\Psi}$ represents the linear acceleration of the CG of a sectional element in the sway direction due to motion in yaw, the use of A'_{22} in place of A'_{26} is justified. Here ξ is a variable distance along the x-axis. Further, since

$$A'_{24} = - A'_{42}$$

Equation (12.3.5) becomes, upon substitution,

$$\frac{dY_{i_o}}{d\xi} = -A'_{22}\ddot{y} - A'_{22}\xi\ddot{\psi} - A'_{42}\ddot{\phi} \qquad (12.3.6)$$

The last term in Equation (12.3.6) represents rolling motion about the free surface level. Rolling motion about the composite center of gravity may be geometrically decomposed into an equivalent linear velocity and a roll motion about the free surface level. Thus, in referring the sway force to the origin at the center of gravity of the body, it is necessary to add an equivalent linear velocity term. This equivalent term is $\overline{OG}\,A'_{22}$ o where $\overline{OG}$ is the distance from the center of gravity to the mean free surface level. Finally, Equation (12.3.6) becomes:

$$Y_i = -\ddot{y}\int_{\xi_s}^{\xi_b} A'_{22}\,d\xi - \ddot{\psi}\int_{\xi_s}^{\xi_b} A'_{22}\,\xi\,d\xi - \ddot{\phi}\int_{\xi_s}^{\xi_b}\left(A'_{42} + \overline{OG}\,A'_{22}\right)d\xi$$

(12.3.7)

The sectional added mass coefficient for lateral motions may be determined by a composite method which consists of using values appropriate for zero frequency and then modifying these values by a frequency-dependent function. The value of A'_{22} may be represented by

$$A'_{22} = \rho S k_y k'_4$$

where $\rho S k_y$ is the added mass for zero frequency and k'_4 is a frequency correction factor. The value of $S k_y$ is defined by

$$S\,k_y = C_z' \frac{\pi H^2}{2}$$

where H is the local section draft and C_z' is a factor taken from Prohaska's charts as shown on the following page. These charts were originally derived for vertical motions at infinite frequency. Thus, in using these charts to select a value of C_z' for an effective beam-draft ratio appropriate for lateral motions, the "picture" of the ship section must be turned by 90° and the beam-draft ratio interpreted as 4H/B. (See sketch.)

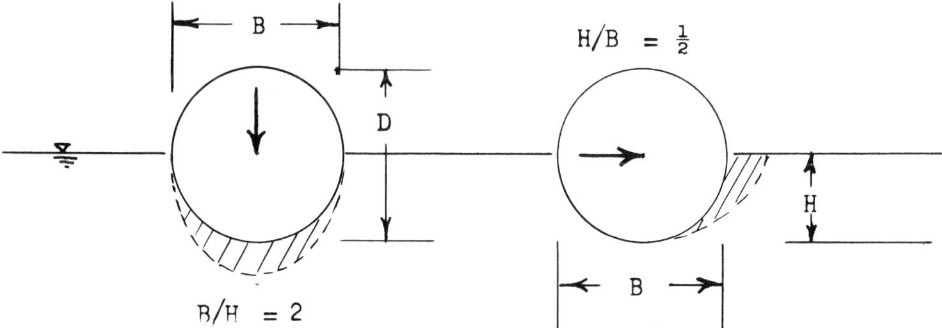

This takes into account the effect of the presence of the surface (i.e., the added mass for lateral motion is equal to one-half of the total added mass of the reflected body, about the free surface, for zero frequency). Prohaska's work included the testing of 10 representative (Lewis-form ship sections) model sections. The test measurements supported the conclusions obtained from an analytic investigation.

The frequency correction factor k_4' was derived (Kaplan and Ulc (1961)) as an approximation based on energy considerations and has been applied successfully to the analysis of bending moments of ships in waves. The form of the correction may not be precisely applicable to many different ship sections and its frequency variation may not be correct over the entire frequency range. However, in the absence of more precise information its use seems justified. The correction is given by

$$k_4' = 1 - \frac{2}{\pi}\,\frac{\omega^2 H}{g} = 1 - \frac{4H}{\lambda}$$

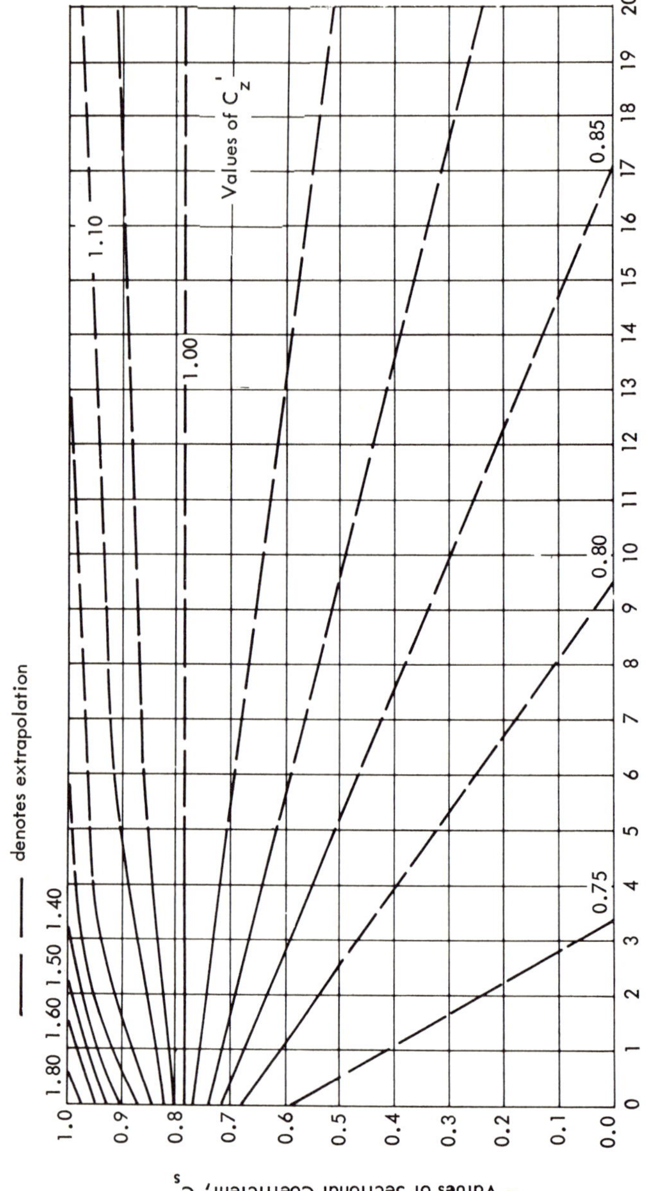

Extrapolation of coefficient C_z' from data by Prohaska (1947).

The added mass tensor component in sway A'_{42} due to motion in roll is given by Hu (1961) as

$$A'_{42} = C'_{42} \frac{\rho \pi H^3}{2}$$

Hu has computed values of C'_{42} for Lewis-form sections as functions of the beam-draft ratios, and section coefficients. Additional computations have been carried out by Kaplan and Putz (1962) to enlarge the range of parameters given by Hu. The values given by Hu were evaluated relative to the free surface level and are appropriate to the case of zero frequency.

Heave, Z_i: In the case of the vertical force, the displacement of the center of gravity of a sectional element is the resultant of pure heave and pitch motions. The sectional vertical force is

$$\frac{dZ_i}{d\xi} = - \frac{D}{Dt}\left(A'_{33}\, \dot{z} - A'_{35}\, \dot{\theta} \right) \qquad (12.3.8)$$

or

$$Z_i = - \ddot{z} \int_{\xi_s}^{\xi_b} A'_{33}\, d\xi + \ddot{\theta} \int_{\xi_s}^{\xi_b} A'_{35}\, d\xi \qquad (12.3.9)$$

The last term represents the heave contribution from pitching about the body center of gravity. The quantity $\xi \ddot{\theta}$, however, represents the linear acceleration of the CG of a sectional element in the heave direction due to motion in pitch. Thus, $A'_{33}\xi$ may be employed in place of A'_{35}, and Equation (12.3.9) becomes

$$Z_i = -\ddot{z}\int_{\xi_s}^{\xi_b} A'_{33}\, d\xi + \ddot{\theta}\int_{\xi_s}^{\xi_b} A'_{33}\,\xi\, d\xi \qquad (12.3.10)$$

The added mass coefficient A'_{33} is defined by

$$A'_{33} = C\rho\frac{\pi}{8}(B^*)^2$$

where values of C are given by Grim (1953) in the form of charts for Lewis-form sections as a function of the dimensionless frequency parameter

$$\frac{\omega^2 B^*}{2g} = \frac{\pi B^*}{\lambda}$$

for different values of beam-draft ratios and section coefficients. The method and resulting calculations include the effects of the free surface. An example of these charts is given in the following figure. To show that the format of the equation is correct, the hydrodynamic mass of a plate of width B^* and a normal dimension $\Delta\ell$, in an otherwise unlimited fluid is given by

$$A'_{33} = \frac{\rho\pi(B^*)^2}{4}\Delta\ell$$

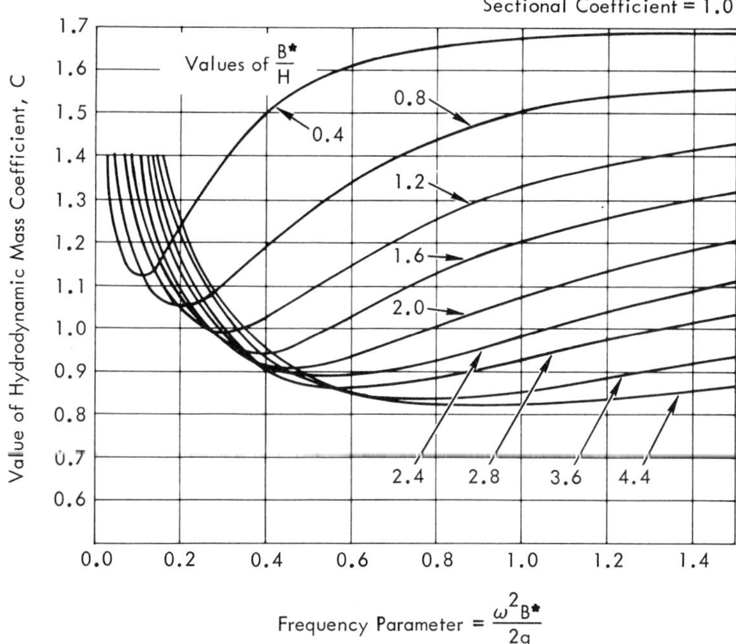

Hydrodynamic mass coefficient, C, for heaving motion. (From Grim, 1959a.)

For a unit normal dimension, $\Delta \ell = 1$, and we see then that the preceeding equation may appear as

$$A'_{33} = \frac{C}{2} \frac{\rho \pi (B^*)^2}{4}$$

where the factor $C/2$ takes into account the effects of the free surface, frequency of oscillation, hull shape and geometric dimensions normal to the direction of motion.

Pitch Moment, M_i: The values of the moments are easily obtained from the forces. Thus the sectional pitch moment is

$$\frac{dM_i}{d\xi} = -\xi \frac{dZ_i}{d\xi} + \zeta \frac{dX_i}{d\xi} \qquad (12.3.11)$$

where ζ is a dummy variable along the z-axis.

$$M_i = \int_{\xi_s}^{\xi_b} \xi \frac{dZ_i}{d\xi} d\xi + \int_{\xi_s}^{\xi_b} \zeta \frac{dX_i}{d\xi} d\xi \qquad (12.3.12)$$

The last expression in Equation (12.3.12) represents the contribution to pitch moment resulting from surging motion. Since the surging force acts through the center of buoyancy (distance BG below the center of gravity), this term can be evaluated separately and the need for determining an average ζ can be eliminated. Thus,

$$\int_{\xi_s}^{\xi_b} \zeta \frac{dX_i}{d\xi} d\xi = m \left| BG \right| \ddot{x} \qquad (12.3.13)$$

The pitch moment thus becomes

$$M_i = \ddot{z}\int_{\xi_s}^{\xi_b} A'_{33}\,\xi\, d\xi \; - \; \ddot{\theta}\int_{\xi_s}^{\xi_b} A'_{33}\,\xi^2\, d\xi \; + \; m\,|BG|\,\ddot{x} \qquad (12.3.14)$$

The evaluation of A'_{33} has been treated previously.

Yaw, N_i: The yaw moment is likewise obtained from the forces. The sectional yaw moment is

$$\frac{dN_i}{d\xi} = \xi\,\frac{dY_i}{d\xi} \qquad (12.3.15)$$

Substitution of Equation (12.3.7) into Equation (12.3.15), and integrating, yields the total yaw moment

$$N_i = -\ddot{y}\int_{\xi_s}^{\xi_b} A'_{22}\,\xi\, d\xi \; - \ddot{\psi}\int_{\xi_s}^{\xi_b} A'_{22}\,\xi^2\, d\xi$$

$$-\ddot{\phi}\int_{\xi_s}^{\xi_b}\left(A'_{42} + |OG|\; A'_{22}\right)\xi\; d\xi \qquad (12.3.16)$$

The evaluation of A'_{42} and A'_{22} has been treated previously.

Roll, K_i: The roll moment relative to the mean free surface level is formulated according to Equation (12.3.1) as

$$\frac{dK_{i_o}}{d\xi} = -\frac{D}{Dt}\left(A'_{42}\,\dot{y} + A'_{44}\,\dot{\phi} + A'_{46}\,\dot{\psi}\right) \qquad (12.3.17)$$

Reference to the coordinate axis whose origin is fixed in space eliminates the need for determining location of the sectional CG's. As in the case of sway, since the quantity $\xi\ddot{\psi}$ represents the linear acceleration in sway due to motion in yaw, Kaplan and Putz (1962) suggest substituting $A'_{42}\xi$ for A'_{46}. Equation (12.3.17) then becomes

$$K_{i_o} = \ddot{y}\int_{\xi_s}^{\xi_b} A'_{42}\,d\xi - \ddot{\psi}\int_{\xi_s}^{\xi_b} A'_{42}\,\xi\,d\xi - \ddot{\phi}\int_{\xi_s}^{\xi_b} A'_{44}\,d\xi \qquad (12.3.18)$$

The last integral in Equation (12.3.18) when combined with the body roll moment of inertia yields the total effective roll moment of inertia. The value of the latter quantity is obtained from prototype measurements of the natural roll period and metacentric heights. The roll moment may now be referred to the center of gravity of the platform by including the product of the lateral forces referred to the free surface level and an appropriate moment arm. Thus

$$K_i = K_{i_o} + \left|\overline{OG}\right| Y_{i_o} \qquad (12.3.19)$$

Substituting Equations (12.3.18) and (12.3.6) into Equation (12.3.19), the roll moment representation becomes

$$K_i = - \ddot{y} \int_{\xi_s}^{\xi_b} \left(A'_{42} + \left| \overline{OG} \right| A'_{22} \right) d\xi \; - \ddot{\psi} \int_s^b \left(A'_{42} + \left| \overline{OG} \right| A'_{22} \right) \xi \, d\xi$$

$$- \ddot{\phi} \int_{\xi_s}^{\xi_b} \left(A'_{44} - \left| \overline{OG} \right| A'_{42} \right) d\xi \qquad (12.3.20)$$

where the integral appearing in the final term represents the total added moment of inertia in roll. It may be computed from the well-known relationship $I_{xT} = mk_{xT}^2$ where k_{xT} is obtained from the relation

$$T = \frac{2\pi k_{xT}}{(g\,\overline{GM})^{\frac{1}{2}}}$$

Generally, $\overline{GM}$, the transverse metacentric height. is determined from static inclining experiments and the period T is the measured period of roll which includes the added mass effects. Thus, k_{xT} is seen to be the effective radius of gyration.

The representations for the forces and moments of an inertial nature presented in this section are seen to consist of linear combinations of terms proportional to acceleration.

12.4 DAMPING FORCES AND MOMENTS

The damping forces and moments arise from wave generation and viscous (drag) effects. For a first order approximation the damping forces and moments are assumed to be proportional to the body velocity and hence the problem reduces to a determination of the coefficients of velocity in the equations of motion. In accordance with the generalized formulations presented herein, the sectional damping terms appear as

$$\frac{dF_d}{d\xi} = \sum (CN')_{ij} \bar{q}_j \tag{12.4.1}$$

where $\bar{q}_j$ is the velocity component in the jth direction; $(CN')_{ij}$ is the sectional force per unit velocity along the jth direction due to motion in the ith direction and takes into account the three-dimensional effect on the damping force calculated from two-dimensional sections. Simply stated, the correction factor C, is the ratio of three-dimensional to two-dimensional damping in the jth direction when the motion is in the ith direction. They are termed "pure" damping factors and, with the exception of roll motion, are applied only when $j = i$. Otherwise, $C = 1$.

In the literature on ship motion studies, damping due to wave generation has been extensively treated and that due to viscous effects only slightly less. However, the voluminous background of information is mainly of an empirical nature and largely unusable for zero-speed studies. Thus, only a few selected references are cited herein.

Surge, X_d: The surge damping force has the form

$$\frac{dX_d}{d\xi} = C_{xx} N'_{xx} \dot{x} \tag{12.4.2}$$

$$X_d = C_{xx} \dot{x} \int_{\xi_s}^{\xi_b} N'_{xx} \, d\xi \tag{12.4.3}$$

For normal ship forms, the surge damping is known to be quite small since the surge motions do not produce appreciable waves. Viscous effects are also small. Newman (1961) points out that for three-dimensional damping of submerged ellipsoids at zero forward speed, the general behavior of surge and heave damping as functions of frequency is similar. Substituting the unit total damping force (unprimed) for the summation of the sectional damping forces, Equation (12.4.2) becomes

$$X_d = C_{xx} N_{xx} \dot{x}$$

Equating the surge damping coefficient to the heave damping coefficient and introducing the fractional product factors, $f(\omega)$, obtained from the results of Newman, the surge damping force becomes

$$X_d = f\, C_z\, N_z\, \dot{x} \qquad (12.4.4)$$

where C_z is the three-dimensional damping factor for pure heave and N_x is the total unit damping force in heave. The double-index subscript has been dropped in this notation. N_z is given by the relationship

$$N_z = \frac{\rho g^2}{\omega^3} \int_{\xi_s}^{\xi_b} (\bar{A}_z)^2 \, d\xi$$

where $\bar{A}_z$ is the ratio of amplitude of the heave-generated wave to the amplitude of heaving motion of the elementary section.

The fractional product factors $f(\omega)$ are small, being less than 10%. That is the surge damping due to wave generation for most ship forms is only about 10% of the corresponding heave damping. Values of $\bar{A}_z$ have been calculated by Grim (1959) for Lewis-form sections and are available as a function of

$$\frac{\omega^2 B^*}{2g} = \frac{\pi^2 B^*}{\lambda}$$

for different beam-draft ratios and section coefficients.

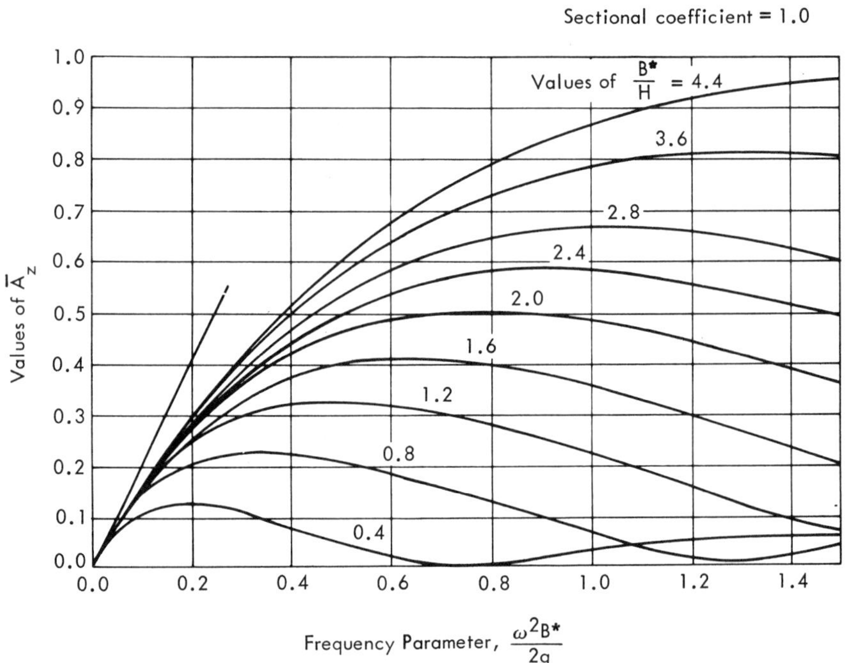

Ratio of amplitude of two-dimensional heave-generated wave to amplitude of heaving motion. (From Grim, (1959a.)

The three-dimensional damping factor in heave, C_z, along with that for pitch, C_θ, has been studied theoretically by Havelock and is available as a function of the frequency parameter

$$\frac{\omega^2 L}{g}$$

where L is the length of the vessel. The results were determined for a submerged spheroid at zero forward speed and are thus generally applicable only to representative surface-ship forms. The damping factors for heave and pitch are presented below.

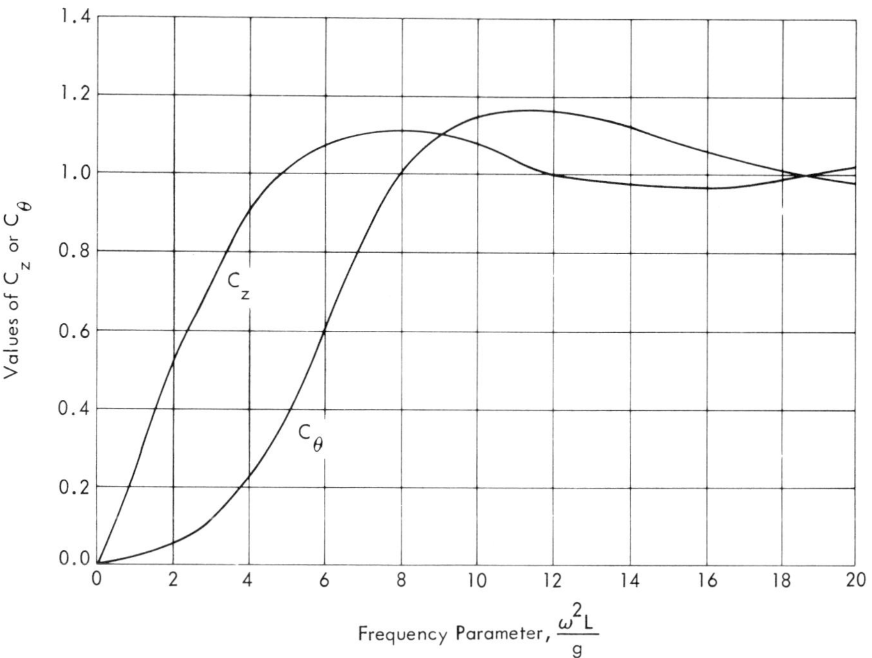

Three-dimensional damping factors in heave and pitch. (From Havelock, 1956.)

Sway, Y_d: For the case of lateral motion, the sway damping force at each section has the form

$$\frac{dY_d}{d\xi} = -C_{yy} N'_{yy} \dot{y} + C_{y\phi} N'_{y\phi} \dot{\phi} - C_{y\psi} N'_{y\psi} \dot{\psi} \qquad (12.4.5)$$

where N'_{yy} is the unit swaying force per unit lateral velocity. Equation (12.4.5) can be written

$$Y_d = -C_{yy} \dot{y} \int_{\xi_s}^{\xi_b} N'_{yy} \, d\xi + C_{y\phi} \dot{\phi} \int_{\xi_s}^{\xi_b} N'_{yo} \, d\xi - C_{y\psi} \dot{\psi} \int_{\xi_s}^{\xi_b} N'_{y\psi} \, d\xi \qquad (12.4.6)$$

Noting that $N'_{y\phi}$ is the unit swaying force per unit angular velocity due to roll and that rolling about the longitudinal axis induces a swaying force acting at the center of pressure (i.e., at a distance $|BG|$ from the CG) the quantity "$- |BG| N'_{yy}$" may be substituted for the term $N'_{y\phi}$. Kaplan and Putz (1962) state that the sway damping factor C_{yy} can also be applied to the lateral damping contribution from rolling motion. On that basis, and with $C_{y\psi} = 1$, the final form of the total sway damping force becomes

$$Y_d = -C_{yy} N_y \dot{y} - N_{y\psi} \dot{\psi} - C_{yy} |BG| N_y \dot{\phi} \qquad (12.4.7)$$

where N'_{yy} is given by the expression

$$N'_{yy} = \frac{\rho g^2}{\omega^3} \bar{A}_y^2 \tag{12.4.8}$$

The quantity $\bar{A}_y$ is the ratio of the amplitude of the sway-generated wave to the amplitude of swaying motion of the elementary section. Vossers (1960) provides a means of evaluating $\bar{A}_y$ from the relationship

$$\bar{A}_y = d_y \left(\frac{\omega^2 B^*}{2g} \right)^2 \tag{12.4.9}$$

This relation is due to Grim (1961) and is based on an asymptotic analysis of d_y. Substituting $\bar{A}_y$ into Equation (12.4.8) the lateral damping force per unit vertical velocity of the section is

$$N'_{yy} = \frac{\rho g^2}{\omega^3} \left(d_y \right)^2 \left(\frac{\omega^2 B^*}{2g} \right)^4$$

or

$$N'_{yy} = \frac{\rho g^2}{\omega^3} \left(d_y \right)^2 \left[\frac{\omega^8 (B^*)^4}{16 g^4} \right] = \frac{\rho \omega^5}{16 g^2} (B^*)^4 \left(d_y \right)^2 \tag{12.4.10}$$

Thus,

$$N_y = \int_{\xi_s}^{\xi_b} N'_{yy}\, d\xi = \frac{\rho\omega^2}{16\, g^2}\int_{\xi_s}^{\xi_b} (B^*)^4 \left(d_y\right)^2 d\xi \qquad (12.4.11)$$

and

$$N_{y\psi} = \int_{\xi_s}^{\xi_b} \xi N'_{yy}\, d\xi = \frac{\rho\omega^5}{16\, g^2}\int_{\xi_s}^{\xi_b} (B^*) \left(d_y\right)^2 \xi\, d\xi \qquad (12.4.12)$$

Vossers (1960) has derived an expression for d_y for Lewis-form sections and has presented the results in terms of the section coefficient, C_s, and values of the beam-draft ratio $B^*/2H$. The results are valid only for low frequencies where the parameter is $\omega^2 B^*/2g < 0.3$. At higher frequencies, it gives an over-estimate of the damping; however, for practical purposes this is not important since lateral motions are small for high length/beam ratio platforms. A sample of the coefficient d_y is given on the following page. (See Figure 12.4.1)

The three dimensional sway damping factor C_{yy} has been examined by Hu and Kaplan (1962) for the case of a submerged spheroid at zero forward speed. As with the case for the heave and pitch three-dimensional damping factors, their use is generally applicable only to representative surface-ship forms. The sway damping coefficient for a spheroid with a fineness ratio of $L/D=8$ is given in Figure 12.4.2.

Heave, Z_d: The vertical damping force at each section has the form

$$\frac{dZ_d}{d\xi} = - C_{zz} N'_{zz}\, \dot{z} + C_{z\theta}\, N'_{z\theta}\, \dot{\theta} \qquad (12.4.13)$$

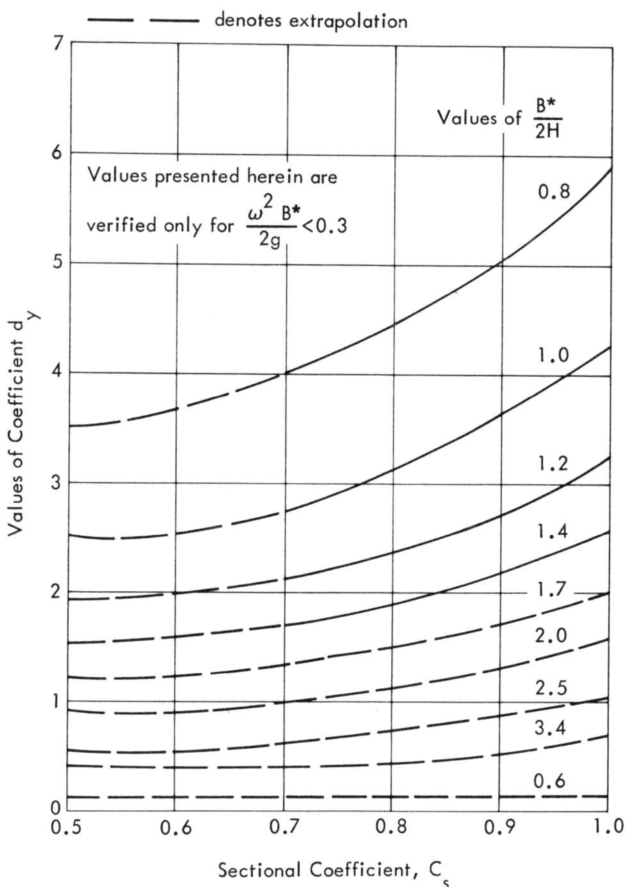

Figure 12.4.1 Extrapolation of coefficient d_y from data by Vossers (1960).

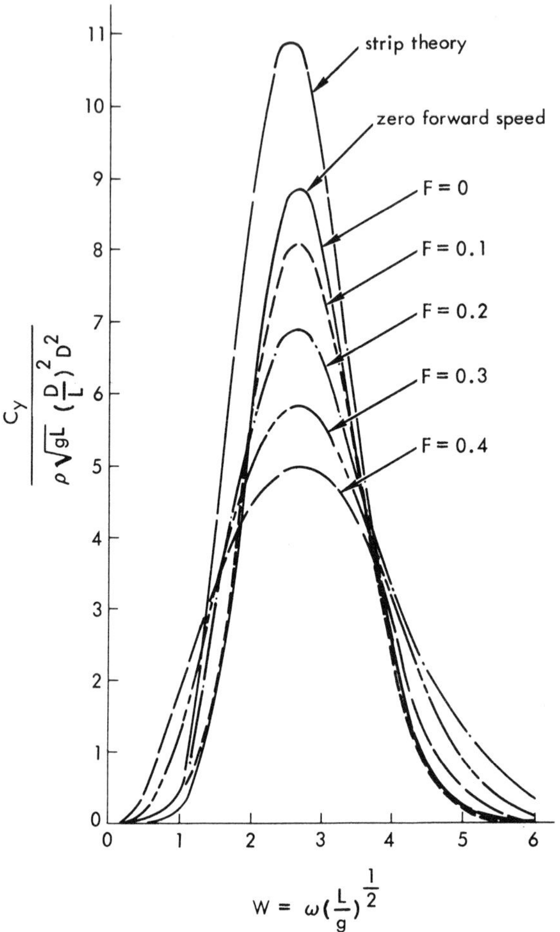

Figure 12.4.2 Sway damping coefficient for the spheroid $L/D = 8$, $L/f = 5$. (From Hu and Kaplan, 1962)

or

$$Z_d = -C_{zz}\,\dot{z}\int_{\xi_s}^{\xi_b} N'_{zz}\,d\xi + C_{z\theta}\,\dot{\theta}\int_{\xi_s}^{\xi_b} N'_{z\theta}\,d\xi \qquad (12.4.14)$$

where

$$N'_{zz} = \frac{\rho g^2 \left(\bar{A}_z\right)^2}{\omega^3} \qquad (12.4.15)$$

and

$$N_{z\theta} = \xi N'_{zz} \qquad (12.4.16)$$

Thus, since $C_{z\theta} = 1$, Equation (12.3.14) becomes

$$Z_d - C_{zz} N_z \dot{z} + N_{z\theta}\dot{\theta} \qquad (12.4.17)$$

where

$$N_z = \int_{\xi_s}^{\xi_b} N'_{zz}\,d\xi \quad \text{and} \quad N_{z\theta} = \int_{\xi_s}^{\xi_b} N'_{zz}\,\xi\,d\xi$$

Evaluation of $\bar{A}_z$ and C_{zz} have been discussed previously.

Pitch, M_d: The moments pitch, yaw, and roll are easily derived from the forces. The form of the sectional pitch damping moment is

$$\frac{dM_d}{d\xi} = -\xi\frac{dZ_d}{d\xi} = C_{\theta z}\, N'_{zz}\, \xi \dot{z} - C_{\theta\theta}\, N'_{zz}\, \xi^2 \dot{\theta} \quad (12.4.18)$$

or

$$M_d = C_{\theta z}\, \dot{z} \int_{\xi_s}^{\xi_b} N'_{zz}\, \xi\, d\xi - C_{\theta\theta}\, \dot{\theta} \int_{\xi_s}^{\xi_b} N'_{zz}\, \xi^2\, d\xi \quad (12.4.19)$$

Making the substitution that

$$N_\theta = \int_{\xi_s}^{\xi_b} N'_{zz}\, \xi^2\, d\xi\,,$$

and setting $C_{\theta z} = 1$, and $C_{\theta\theta} = C_{zz}$, the final form of the pitch moment becomes

$$M_d = N_{z\theta}\, \dot{z} - C_{zz}\, N_\theta\, \dot{\theta} \quad (12.4.20)$$

Again, as with the case for heave, the three-dimensional damping factor, C_{zz} and $\bar{A}_z$ have been discussed previously.

Yaw, N_d: The sectional yaw damping moment is

$$\frac{dN_d}{d\xi} = \xi \frac{dY_d}{d\xi} - C_{\psi y} N'_{yy} \xi \dot{y} - C_{\psi\psi} N'_{yy} \xi^2 \dot{\psi} - C_{\psi\dot{\phi}} |BG| N'_{yy} \xi \dot{\phi} \tag{12.4.21}$$

but, $C_{\psi y} = C_{\psi\phi} = 1$, and Equation (12.4.21) becomes

$$N_d = - \dot{y} \int_{\xi_s}^{\xi_b} N'_{yy} \xi d\xi - C_{\psi\psi} \dot{\psi} \int_{\xi_s}^{\xi_b} N'_{yy} \xi^2 d\xi - |BG| \dot{\phi} \int_{\xi_s}^{\xi_b} N'_{yy} \xi d\xi \tag{12.4.22}$$

Setting

$$N_\psi = \int_{\xi_s}^{\xi_b} N'_{yy} \xi^2 d\xi = \frac{\rho \omega^5}{16 g^2} \int_{\xi_s}^{\xi_b} (B^*)^4 \left| d_y \right|^2 \xi^2 \ d\xi$$

The final representation of the yaw moment is

$$N_d = - N_{y\psi} \dot{y} - C_{\psi\psi} N_\psi \dot{\psi} - |BG| N_{y\psi} \dot{\phi} \tag{12.4.23}$$

The parameter d_y has been discussed previously. The three-dimensional damping factor, C_ψ, has been studied by Hu and Kaplan (1962) and the results have been presented as a function of the parameter $\omega\sqrt{L/g}$ as shown in Figure 12.4.3 for the case of a submerged spheroid of fineness ratio $L/D = 8$. Thus, they are generally applicable only to representative surface ship forms.

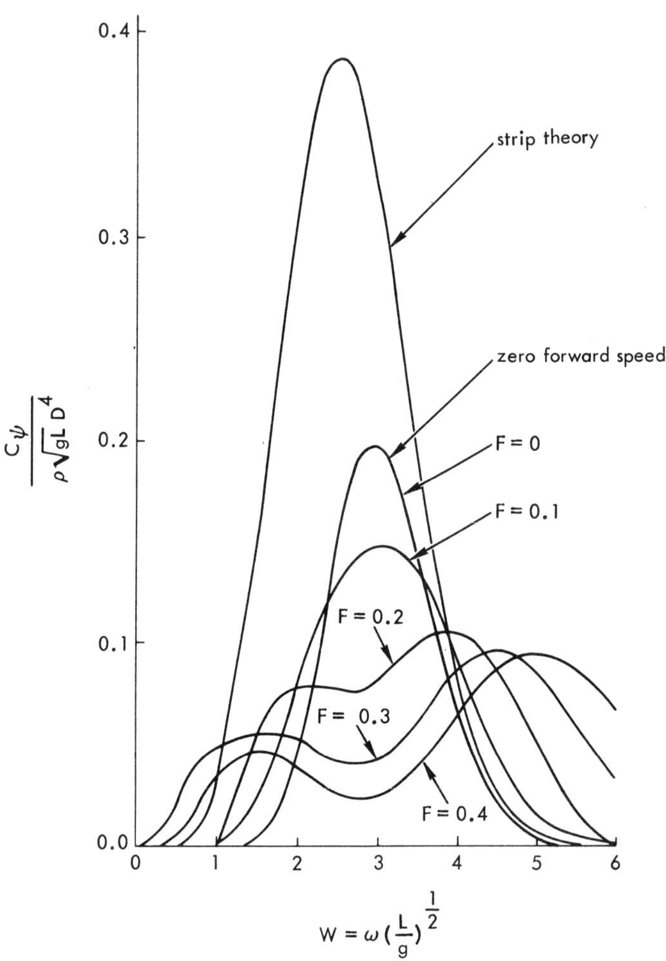

Figure 12.4.3 Yaw damping coefficient for the spheroid $L/D = 8$, $L/f = 5$. (From Hu and Kaplan, 1962)

Roll, K_d: The roll damping term due to lateral motions is approximated by the relation

$$K_d = |BG| \, Y_d \qquad (12.4.24)$$

where it is assumed that the center of action of the lateral force is at the center of buoyancy; thus, the total force due to roll damping is

$$K_d = - C_{yy} \, |BG| \, N_y \, \dot{y} - |BG| \, N_{y\psi} \, \dot{\psi} - C_{yy} \left(|BG| \right)^2 N_y \, \dot{\phi} \qquad (12.4.25)$$

Kaplan and Putz (1962) state that the quantity

$$C_{yy} \left(|BG| \right)^2 N_y$$

should be replaced by a pure roll damping term N_ϕ which may be obtained from experimental data. Specifically, N_ϕ is obtained from an examination of the roll decay curve. However, roll is a highly tuned motion and the damping value is only important near the natural roll period. Thus, it is suggested that the damping coefficient N_ϕ be calculated from the total roll moment of inertia, natural roll frequency, and the extinction characteristics of the roll decay curve. Specifically, N_ϕ may be calculated from the following relationship:

$$N_\phi = 2\mu \, I_{xT} \, \frac{2\pi}{T_\phi} \qquad (12.4.26)$$

where I_{xT} is the total roll moment of inertia, T_ϕ is the natural roll period, and μ is the dimensionless coefficient of decay, which is usually obtained from experimental data.

To elaborate, the roll damping coefficient N_ϕ appears as part of one of the coefficients in the uncoupled equation for damped harmonic oscillation, viz.

$$\ddot{\phi} + 2\mu\omega_\phi \dot{\phi} + \omega_\phi^2 \phi = 0$$

where ϕ is the roll amplitude, ω_ϕ is the circular frequency of the free undamped oscillations, and μ is termed the dimensionless coefficient of decay. The coefficient of decay μ, is the quantity which appears most frequently in the literature on roll extinction properties and thus it is a convenient basis for comparison. Korvin-Kroukovsky (1961) gives a comprehensive discussion on methods of obtaining values of μ and solutions to the above equation for normal ship forms. Empirical data on damping in roll is given by both Korvin-Kroukovsky and by Blagoveshchensky (1962). One simplified method of obtaining values of μ from model and/or prototype tests is the "logarithmic decrement" method widely employed in general vibratory problems. The logarithmic decrement is measured by the slope of log ϕ versus the number of cycles of oscillation. Thus,

$$2\pi\mu = \delta = \log\left(\frac{\phi_1}{\phi_2}\right)$$

where δ is the logarithmic decrement and ϕ_1 and ϕ_2 are the amplitudes of any two succeeding oscillations. Korvin-Kroukovsky reports that μ has a value of 0.08 for ships with bilge keels with a range of fluctuation of up to 40% from the mean value. Blagoveshchensky states that μ is highly dependent on the maximum amplitude of oscillation.

12.5 HYDROSTATIC RESTORING FORCES AND MOMENTS

The hydrostatic reactions are those caused by displacements of the platform from the equilibrium position in still water. These restoring effects, which exist only for the heave, pitch and roll motions, are well known, being derived in detail in a number of important references (e.g., St. Denis and Pierson (1953)) and therefore only the simplified results are presented herein.

Assuming that the angular displacements in roll are small and that the effective buoyancy change is due to total immersion of the local cross-sectional element (whose sides are assumed to be almost parallel near the intersection with the free surface), the sectional vertical force due to net vertical displacements is

$$\frac{dZ_h}{d\xi} = -\rho g B^* (z - \xi\theta) \tag{12.5.1}$$

Similarly, the hydrostatic restoring pitch moment is

$$\frac{dM_h}{d\xi} = -\xi \frac{dZ_h}{d\theta} \tag{12.5.2}$$

From Equations (12.5.1) and (12.5.2) the final forms of the hydrostatic restoring heave force and pitch moment are, respectively,

$$Z_h = -\rho g z \int_{\xi_s}^{\xi_b} B^* \, d\xi + \rho g \theta \int_{\xi_s}^{\xi_b} B^* \xi \, d\xi \tag{12.5.3}$$

and

$$M_h = \rho g z \int_{\xi_s}^{\xi_b} B^* \xi \, d\xi - \rho g \theta \int_{\xi_s}^{\xi_b} B^* \xi^2 \, d\xi \tag{12.5.4}$$

The righting moment for roll motion is the well-known expression

$$K_h = -\rho g \nabla \left|GM\right| \phi = - W \left|GM\right| \phi \qquad (12.5.5)$$

where ∇ is the displaced volume; $|GM|$ is the transverse metacentric height, which is the distance between the center of gravity and the metacenter (a point on the vertical axis of symmetry intersected by the resultant of the displaced center of buoyancy caused by small roll motions); and W is the total platform displacement.

12.6 MOORING FORCES AND MOMENTS

Restoring forces for moored vessels have greatest influence on the displacememts in the horizontal plane, (i.e., surge, sway and yaw). The mooring restoring forces are not effective for the heave, pitch and roll motions, since these motions have very large hydrostatic restoring forces. Any changes in the mooring forces occur as a result of changes in the energy state of the cable system resulting from platform displacements. Any mooring configuration can be replaced by an equivalent mooring system which coincides with the ship coordinate axis system as shown below.

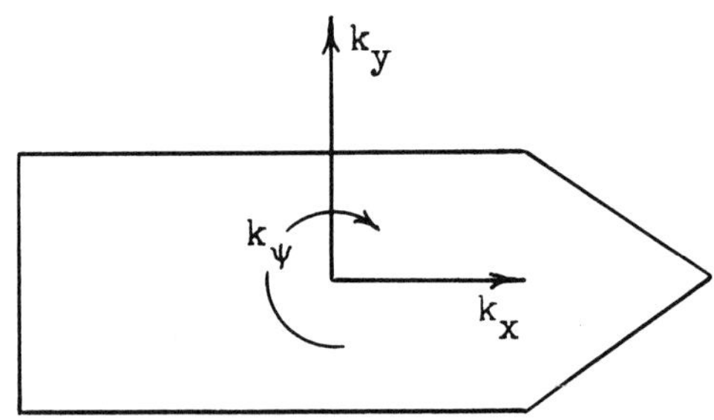

This includes even a single point mooring system. (This will be discussed in detail later.) The ship (or platform) can be displaced analytically in the surge and sway directions and can be rotated in yaw mode and from these calculations, the load-displacement and/or moment-rotational diagrams can be constructed. If the restoring force is proportional to the displacement (i.e., linear), then the 'spring constants', for surge (k_x), sway (k_y) and yaw (k_ψ) can be determined. Even though the load-displacement diagrams are nonlinear, often the displacements induced by waves are small enough so that they can be considered linear. The effect of current and wind, generally, is to displace the craft to a new equilibrium position. The introduction of the existence of natural periods in surge, sway and yaw (with possible large motions associated with resonances in these degrees of freedom) is the main characteristic which distinguishes the motions of free ships from those of moored ships in waves.

In summary, the variations in the forces in the moorings are due to motions of the platform and they can be easily found since they are related kinematically to the motions. Even if the moorings are nonlinear, the mooring restoring forces can still be determined as a function of the platform displacements but a solution other than the one which will be illustrated must be considered.

12.7 EVALUATION OF WAVE-EXCITING FORCES AND MOMENTS

The hydrodynamic forces and moments induced by waves are obtained by application of slender body theory. Kaplan (1957) states that:

> "A consequence of the use of slender body theory is that the crossflow in each cross-section is independent of the flow at other sections, and the connection of this two-dimensional flow with the actual three-dimensional problem is through the boundary condition, which contains the longitudinal coordinate ξ as a parameter. At each section the boundary condition is a function of x, since it is dependent on the body geometry. The total force acting on the body is found by integrating the elemental forces at each section over the length of the body."

In a general survey of the application of strip theory to wave-induced effects for slender bodies, Korvin-Kroukovsky (1961) states that the wave-induced force acting on an individual ship section of the length $d\xi$ is composed of inertial, damping and displacement components. The inertial component is proportional to acceleration; the damping component is proportional to velocity, and the displacement component is, of course, proportional to a displacement term.

Thus, the total wave-induced force acting on an elemental section is

$$\frac{dF_w}{d\xi} = \frac{dF_w^{(1)}}{d\xi} + \frac{dF_w^{(2)}}{d\xi} + \frac{dF_w^{(3)}}{d\xi} \qquad (12.7.1)$$

where 1 indicates an inertial contribution, 2 indicates a damping contribution, and 3 indicates a displacement contribution.

Inertial Terms: Kaplan and Putz (1962) state that the wave forces on a surface ship can be represented in the same manner as for a submerged body, but with the frequency dependence of the added mass terms included in the final representation. On this basis, the inertial contribution to the wave-exciting forces for the present case of zero speed is represented by

$$\frac{dF_w^{(1)}}{d\xi} = \rho S + \sum A_{ij} \frac{D\bar{q}_{oi}}{Dt} \qquad (12.7.2)$$

where $dF_W^{(1)}$ is the inertial wave force acting on the elemental section, $d\xi$; ρ is the mass density of sea water; S is the cross-sectional area; A_{ij} is the sectional added mass coefficient along the i direction due to its motion in the j direction; D/Dt is an operator denoting the substansive or material derivative; and $\bar{q}_{oi}$ is a component of the orbital velocity evaluated at some reference point on the platform. The reference point chosen for this analysis is the intersection of the vertical plane containing the longitudinal centerline (due to symmetry) and the horizontal plane representing the mean half draft of the platform. The latter is an approximation to allow for the influence of the exponential decay of wave particle velocity. The wave particle velocity, $\bar{q}_{oi}$, is obtained from the velocity potential given by Lamb (1945) and compatible with the adopted coordinate system. This wave potential is

$$\Phi_w = a\,c \exp\left[\frac{2\pi z}{\lambda}\right] \cos \frac{2\pi}{\lambda} (x \cos\beta + y \sin\beta - ct) \quad (12.7.3)$$

where a is the wave amplitude; c is the propagation velocity, $C^2 = (g\lambda/2\pi) \tanh (2\pi d/\lambda)$; λ is the wave length; z is the vertical coordinate denoting the point at which the velocity will be evaluated and will be set equal to the mean half draft of the platform, $-\bar{h}$; β is the angle lying between the x-axis and the normal to the wave crests and lying in the range $-\pi \leq \beta \leq \pi$; and d is the mean water depth. Note that the terms $x \cos\beta$ and $y \sin\beta$ are merely the projections of $\sqrt{x^2 + y^2}$ on the lines $\cos\beta$ and $\sin\beta$, respectively. The surface wave elevation, η, is

$$\eta = \frac{1}{g} \left.\frac{\partial \phi_w}{\partial t}\right|_{z=0}$$

$$\eta = \frac{1}{g}\, a\, c^2 \frac{2\pi}{\lambda} \sin \frac{2\pi}{\lambda} (x \cos\beta + y \sin\beta - ct) \quad (12.7.4)$$

or, substituting,

$$\eta = a \tanh \frac{2\pi d}{\lambda} \sin \frac{2\pi}{\lambda} (x \cos\beta + y \sin\beta - ct) \quad (12.7.5)$$

For deep water, $\tanh 2\pi d/\lambda \approx 1$, and Equation (12.7.5) becomes

$$\eta = a \sin \frac{2\pi}{\lambda} (x \cos\beta + y \sin\beta - ct) \tag{12.7.6}$$

Damping Terms: Damping forces arising from wave action are due to the difference between the body velocities and the wave orbital velocities. These forces should not be confused with damping due to motion of the body in otherwise still water, discussed elsewhere. Kaplan and Putz (1962) state that the "basis for including such a representation in the wave-induced effects for slender bodies as in the recent thesis by Vossers(1962) while it has been used in other studies purely on the basis of empirical reasoning." Thus, formulating an expression for the damping forces, we have

$$\frac{dF_w^{(2)}}{d\xi} = C_i N_{ij} \bar{q}_{oi} \tag{12.7.7}$$

where $\bar{q}_{oi}$ is as defined previously, C_i is a three-dimensional damping factor, and N_{ij} is the sectional (elemental) damping coefficient.

Displacement Terms: The displacement terms are due to the buoyant effect of the waves as they pass the platform. Thus, the displacement terms are included only in the vertical force and may be represented by

$$\frac{dF_w^{(3)}}{d\xi} = \rho g \beta^* \eta(\xi, t) \tag{12.7.8}$$

where β^* is the local section beam. The wave force expressions for surge, X_W; sway, Y_W; heave, Z_W; pitch, M_W; yaw, N_W; and roll, K_W, may now be derived in accordance with Equations (12.7.1), (12.7.2), (12.7.7) and (12.7.8).

Surge X_w: Evaluating the wave orbital velocity, q_{ox}, from the wave potential, Equation (12.7.3) gives

$$q_{ox} = u = -\frac{\partial \phi_w}{\partial x} = -ac\frac{2\pi}{\lambda}\exp\left[\frac{2\pi z}{\lambda}\right]\cos\beta\left[\sin\frac{2\pi}{\lambda}(x\cos\beta + \sin\beta - ct)\right] \tag{12.7.9}$$

The substansive derivative,

$$\frac{D(u)}{Dt} = \frac{\partial(u)}{\partial t} + (u)\frac{\partial(u)}{\partial x} + (v)\frac{\partial(u)}{\partial y} + (w)\frac{\partial(u)}{\partial z} \tag{12.7.10}$$

where the first term on the right-hand side represents the temporal acceleration and the last three terms represent the convective acceleration due to convergence of the flow. Since one of the basic assumptions of slender body theory as used herein is that the convergence of the wave motion flow pattern resulting from presence of the body is negligible, the convective terms may be set equal to zero. Thus,

$$u\frac{\partial u}{\partial x} + v\frac{\partial u}{\partial y} + w\frac{\partial u}{\partial z} = 0 \tag{12.7.11}$$

and

$$\frac{Du}{Dt} = \frac{\partial u}{\partial t}$$

So,

$$\frac{\partial u}{\partial t} = -\, a\, c^2 \left(\frac{2\pi}{\lambda}\right)^2 \exp\left[\frac{2\pi z}{\lambda}\right] \cos\beta \left[\cos \frac{2\pi}{\lambda}\, (x \cos\beta + y \sin\beta - ct)\right] \tag{12.7.12}$$

Recall that

$$\cos (A - B) = \cos A \cos B + \sin A \sin B \tag{12.7.13}$$

and

$$\frac{Du}{Dt} = \frac{\partial u}{\partial t} = -a\, c^2 \left(\frac{2\pi}{\lambda}\right)^2 \exp\left[\frac{2\pi z}{\lambda}\right] \cos\beta \left\{\left[\cos \frac{2\pi}{\lambda}\, (x \cos\beta + y \sin\beta)\right] \frac{2\pi ct}{\lambda}\right.$$

$$\left. + \left[\sin \frac{2\pi}{\lambda}\, (x \cos\beta + y \sin\beta)\right] \sin \frac{2\pi ct}{\lambda}\right\} \tag{12.7.14}$$

From Equation (12.7.2),

$$X_w^{(1)} = \int_{\xi_s}^{\xi_b} \left(\rho S + \sum A'_{ij}\right) \frac{Du}{Dt}\, d\xi$$

Since $\sum A_{ij}$ is the sum of the added masses in surge and is approximately equal to zero for a slender body, then

$$X_w^{(1)} = - a c^2 \left(\frac{2\pi}{\lambda}\right)^2 \exp\left[\frac{2\pi z}{\lambda}\right] \cos\beta \left[\cos\omega t \int_{\xi_s}^{\xi_b} S \cos\frac{2\pi}{\lambda}(x\cos\beta + y\sin\beta)\, d\xi + \sin\omega t \int_{\xi_s}^{\xi_b} S \sin\frac{2\pi}{\lambda}(x\cos\beta + y\sin\beta)\, d\xi\right] \tag{12.7.15}$$

where $S = f(\xi)$, and $c = \lambda/T = \lambda\omega/2\pi$ by definition.

In order to compute $X_w^{(1)}$, the right-hand side must be evaluated at some reference point $\bar{z}$, y. The reference point chosen in this analysis is $z = -\bar{h}$, $y = 0$. The mean half draft $\bar{h}$, is chosen to approximate the influence of the exponential decay of wave particle velocity. The vertical centerline is chosen because of symmetry. However, since most platforms are not really slender, and since the beam of most present day platforms of interest are not small relative to some of the wave lengths, an additional correction must be introduced. Kaplan and Putz (1962) state that:

"Since the various integrals are really of the form

$$\int_{\xi_s}^{\xi_b} f(\xi) \begin{matrix}\cos\\ \sin\end{matrix}\left(\frac{2\pi x}{\lambda}\cos\beta + \frac{2\pi y}{\lambda}\sin\beta\right) d\xi$$

a proper move would be to represent the influence of this additional term by finding the mean value of the beam. Using the maximum beam, B, as a base, since the platform is close to a full planform shape, the quantity of interest is

$$\frac{1}{B}\int_{-\frac{B}{2}}^{\frac{B}{2}} dy \int_{\xi_s}^{\xi_b} f(\xi) \begin{matrix}\cos\\ \sin\end{matrix}\left(\frac{2\pi x}{\lambda}\cos\beta + \frac{2\pi y}{\lambda}\sin\beta\right) d\xi$$

which is equal to

$$\frac{\sin\left(\frac{\pi B}{\lambda}\sin\beta\right)}{\frac{\pi B}{\lambda}} \int_{\xi_s}^{\xi_b} f(\xi) \begin{matrix}\cos\\ \sin\end{matrix}\left(\frac{2\pi x}{\lambda}\cos\beta\right) d\xi$$

which accounts for the influence of the beam-wave length ratio in a more rational manner (although not the most precise method) than simply setting $y = 0$."

Therefore, setting $z = -\bar{h}$, $y = 0$ and introducing the product factor

$$\frac{\sin\left(\frac{\pi B}{\lambda}\sin\beta\right)}{\frac{\pi B}{\lambda}\sin\beta} = \Lambda$$

the expression for the surging force $X_w^{(1)}$ becomes

$$X_w^{(1)} = -\rho a\, c^2\left(\frac{2\pi}{\lambda}\right)^2 \exp\left[-\frac{2\pi\bar{h}}{\lambda}\right] \Lambda \cos\beta\left[\cos\omega t\int_{\xi_s}^{\xi_b} S\cos\frac{2\pi}{\lambda}(x\cos\beta)\,d\xi\right.$$

$$\left. + \sin\omega t\int_{\xi_s}^{\xi_b} S\sin\frac{2\pi}{\lambda}(x\cos\beta)\,d\xi\right] \qquad (12.7.16)$$

where $c^2 = (g\lambda/2\pi)\tanh(2\pi d/\lambda)$.

Surge damping due to wave generation is generally quite small and in the absence of more detailed information, is neglected. Thus,

$$X_w^{(2)} = 0 \qquad (12.7.17)$$

Adding Equations (12.7.16) and (12.7.17), gives the general expression for evaluating the surging force in deep or shallow water. In terms of ω and λ, it may be written as

$$X_w = -\rho a\,\omega^2 \exp\left[-\frac{2\pi\bar{h}}{\lambda}\right] \Lambda \cos\beta\left[\cos\omega t\int_{\xi_s}^{\xi_b} S\cos\frac{2\pi}{\lambda}(x\cos\beta)\,d\xi\right.$$

$$\left. + \sin\omega t\int_{\xi_s}^{\xi_b} S\sin\frac{2\pi}{\lambda}(x\cos\beta)\,d\xi\right] \qquad (12.7.18)$$

where the only variable is ω, since λ may be found from the relationship

$$\omega^2 = \left(\frac{2\pi}{\lambda}\right) g \tanh \frac{2\pi d}{\lambda} .$$

For head-on waves, $\beta = +\pi$, $\cos\beta = -1$, $\sin\beta = 0$, and Equation (12.7.18) becomes

$$X_w = +\rho a \omega^2 \exp\left[-\frac{2\pi \bar{h}}{\lambda}\right] \left[\cos \omega t \int_{\xi_s}^{\xi_b} S \cos\left(-\frac{2\pi x}{\lambda}\right) d\xi \right.$$

$$\left. + \sin \omega t \int_{\xi_s}^{\xi_b} S \sin\left(-\frac{2\pi x}{\lambda}\right) d\xi \right] \qquad (12.7.19)$$

Making use of the trigonometric properties,

$$X_w = \rho a \omega^2 \exp\left[-\frac{2\pi \bar{h}}{\lambda}\right] \left[\cos \omega t \int_{\xi_s}^{\xi_b} S \cos\left(\frac{2\pi x}{\lambda}\right) d\xi \right.$$

$$\left. - \sin \omega t \int_{\xi_s}^{\xi_b} S \sin\left(\frac{2\pi x}{\lambda}\right) d\xi \right] \qquad (12.7.20)$$

For deep water, $2\pi/\lambda = \omega^2/g = g/c^2$ and Equation (12.7.20) may be written

$$X_w = a\rho\omega^2 \exp\left[-\frac{\omega^2}{g}\bar{h}\right]\left[\cos\omega t \int_{\xi_s}^{\xi_b} S \cos\left(\frac{\omega^2}{g}x\right) d\xi\right.$$

$$\left. - \sin\omega t \int_{\xi_s}^{\xi_b} S \sin\frac{\omega^2}{g} x\, d\xi\right] \qquad (12.7.21)$$

For beam-on waves, $B = \pi/2$, $\cos\beta = 0$, and $\sin\beta = 1$, and Equation (12.7.18) becomes

$$X_w = 0 \qquad (12.7.22)$$

which is the result in all depths of water.

Sway, Y_w: Evaluating the wave orbital velocity, q_{oy}, from the wave potential, Equation (12.7.3) gives

$$q_{oy} = v = -\frac{\partial\phi_w}{\partial y} = a\, c^2\left(\frac{2\pi}{\lambda}\right)\exp\left[\frac{2\pi z}{\lambda}\right]\sin\beta\left[\sin\frac{2\pi}{\lambda}(x\cos\beta + y\sin\beta - ct)\right]$$

$$(12.7.23)$$

Again, the convective acceleration terms are set equal to zero, and

$$\frac{Dv}{Dt} = \frac{\partial v}{\partial t} = -\, a\, c^2 \left(\frac{2\pi}{\lambda}\right)^2 \exp\left[\frac{2\pi z}{\lambda}\right] \sin\beta \left[\cos\frac{2\pi}{\lambda}(x\cos\beta + y\sin\beta - ct)\right] \tag{12.7.24}$$

or

$$\frac{\partial v}{\partial t} = -\, a\,\omega^2 \exp\left[\frac{2\pi z}{\lambda}\right] \sin\beta \left\{(\cos\omega t)\left[\cos\frac{2\pi}{\lambda}(x\cos\beta + y\sin\beta)\right] + (\sin\omega t)\left[\sin\frac{2\pi}{\lambda}(x\cos\beta + y\sin\beta)\right]\right\} \tag{12.7.25}$$

From Equation (12.7.2),

$$Y_w^{(1)} = \int_{\xi_s}^{\xi_b} \left(\rho S + \sum A'_{ij}\right) \frac{Dv}{Dt}\, d\xi$$

but, $\sum A_{ij} = A'_{22} - A'_{42}\,(2\pi/\lambda)$ for sway motions, where A'_{22} is the sectional added mass in sway due to motion in sway and A'_{42} is the sectional added mass in roll due to motion in sway. Justification for including the latter term, although not germane to the immediate problem, is to take account of the possible effect of relative dimension of draft to wave length. Since the orbital velocities in deep water waves decay with depth exponentially, and since the velocities are evaluated at the mean half draft, it can be shown that there is in addition a "roll moment" which is substantially accounted for by including the next highest order inertial term in powers of wave number $(2\pi/\lambda)$.

The sway force, $Y_W^{(1)}$, then becomes

$$Y_W^{(1)} = -\rho\omega^2 a \exp\left[\frac{2\pi z}{\lambda}\right] \sin\beta \left\{\cos\omega t \int_{\xi_s}^{\xi_b} \left(S + \frac{A'_{22}}{\rho}\right.\right.$$

$$\left. + \frac{2\pi A'_{42}}{\lambda\rho}\right) \cos\left[\left(\frac{2\pi}{\lambda}\right)(x\cos\beta + y\sin\beta)\right] d\xi$$

$$+ \sin\omega t \int_{\xi_s}^{\xi_b} \left(S + \frac{A'_{22}}{\rho} + \frac{2\pi A'_{42}}{\lambda\rho}\right)$$

$$\left. \cdot \sin\left[\left(\frac{2\pi}{\lambda}\right)(x\cos\beta + y\sin\beta)\right] d\xi\right\} \qquad (12.7.26)$$

Evaluating the force at the mean half draft, -h, and at the centerline reference position, $y = 0$, and introducing the correction for the relative dimensions of the platform beam and wave lengths results in

$$Y_w^{(1)} = -\rho\omega^2 \text{ a } \exp\left[-\frac{2\pi h}{\lambda}\right]\Lambda \sin\beta\left\{\cos \omega t\int_{\xi_s}^{\xi_b}\left(S + \frac{A_{22}'}{\rho}\right.\right.$$

$$\left.+\frac{2\pi A_{42}'}{\lambda\rho}\right)\cos\left[\left(\frac{2\pi}{\lambda}\right)(x \cos\beta)\right]d\xi + \sin \omega t\int_{\xi_s}^{\xi_b}\left(S + \frac{A_{22}'}{\rho}\right.$$

$$\left.\left.+\frac{2\pi A_{42}'}{\lambda\rho}\right)\sin\left[\frac{2\pi}{\lambda}(x \cos\beta)\right]d\xi\right. \qquad (12.7.27)$$

The damping force, $Y_w^{(2)}$, from Equation (12.7.7) is

$$Y_w^{(2)} = \int_{\xi_s}^{\xi_b} C_y N_{iy}' q_{oy}\, d\xi = C_y\int_{\xi_s}^{\xi_b} N_{yy}' v\, d\xi \qquad (12.7.28)$$

or

$$Y_w^{(2)} = C_y \text{ a}\,\omega \exp\left[\frac{2\pi z}{\lambda}\right]\sin\beta\left[\cos \omega t\int_{\xi_s}^{\xi_b} N_{yy}' \sin\frac{2\pi}{\lambda}(x \cos\beta + y \sin\beta)d\xi\right.$$

$$\left.-\sin \omega t\int_{\xi_s}^{\xi_b} N_{yy}' \cos\frac{2\pi}{\lambda}(x \cos\beta + y \sin\beta)\, d\xi\right] \qquad (12.7.29)$$

Again evaluating at the mean half draft, $-\bar{h}$, and at the centerline reference, $y = 0$, yields

$$Y_w^{(2)} = C_y\ a\,\omega \exp\left[-\frac{2\pi\bar{h}}{\lambda}\right]\Lambda \sin\beta\left[\cos\omega t\int_{\xi_s}^{\xi_b} N'_{yy}\sin\left(\frac{2\pi}{\lambda}x\cos\beta\right)d\xi\right.$$

$$\left. - \sin\omega t\int_{\xi_s}^{\xi_b} N'_{yy}\cos\left(\frac{2\pi}{\lambda}x\cos\beta\right)d\xi\right] \qquad (12.7.30)$$

The swaying force, Y_w, is equal to the sum of $Y_w^{(1)}$ and $Y_w^{(2)}$.

$$Y_w = -\rho\omega^2\ a\exp\left[-\frac{2\pi\bar{h}}{\lambda}\right]\Lambda\ \sin\beta\left[\cos\omega t\int_{\xi_s}^{\xi_b}\left(S + \frac{A'_{22}}{\rho} + \frac{2\pi A'_{42}}{\lambda\rho}\right)\right.$$

$$\cdot\cos\left(\frac{2\pi}{\lambda}x\cos\beta\right) + \sin\omega t\int_{\xi_s}^{\xi_b}\left(S + \frac{A'_{22}}{\rho} + \frac{2\pi A'_{42}}{\lambda\rho}\right)$$

$$\left.\cdot\sin\left(\frac{2\pi}{\lambda}x\cos\beta\right)d\xi\right] + C_y\ a\,\omega\ \exp\left[-\frac{2\pi\bar{h}}{\lambda}\right]\Lambda$$

$$\cdot\sin\beta\left[\cos\omega t\int_{\xi_s}^{\xi_b} N'_{yy}\sin\left(\frac{2\pi}{\lambda}x\cos\beta\right)d\xi\right.$$

$$\left. - \sin\omega t\int_{\xi_s}^{\xi_b} N'_{yy}\cos\left(\frac{2\pi}{\lambda}x\cos\beta\right)d\xi\right] \qquad (12.7.31)$$

which is the general expression for evaluating the swaying force.

For head-on waves, $\beta = \pi$, $\cos\beta = -1$, $\sin\beta = 0$, and Equation (12.7.31) reduces to

$$Y_w = 0 \qquad (12.7.32)$$

which is the result for both deep and shallow water waves.

For beam-on waves, $\beta = \pi/2$, $\cos\beta = 0$, $\sin\beta = 1$, and Equation (12.7.31) reduces to

$$Y_w = -\rho\omega^2 \, a \exp\left[-\frac{2\pi\bar{h}}{\lambda}\right] \frac{\sin\frac{\pi B}{\lambda}}{\frac{\pi B}{\lambda}} \left[\cos\omega t \int_{\xi_s}^{\xi_b} \left(S + \frac{A'_{22}}{\rho} + \frac{2\,A'_{42}}{\lambda\rho}\right) d\xi\right]$$

$$- C_y \, a\omega \exp\left[-\frac{2\pi\bar{h}}{\lambda}\right] \frac{\sin\frac{\pi B}{\lambda}}{\frac{\pi B}{\lambda}} \left(\sin\omega t \int_{\xi_s}^{\xi_b} N'_{yy} \, d\xi\right) \qquad (12.7.33)$$

which is valid for deep water waves and approximately so for shallow water waves.

Heave Z_w: Evaluating the wave orbital velocity, q_{oz}, from the wave potential, Equation (12.7.3) gives

$$q_{oz} = w = -\frac{\partial\phi_w}{\partial z} = -a\,c\,\frac{2\pi}{\lambda}\exp\left[\frac{2\pi z}{\lambda}\right]\cos\frac{2\pi}{\lambda}(x\cos\beta + y\sin\beta - ct) \qquad (12.7.34)$$

Again, the convective acceleration terms are set equal to zero, and

$$\frac{Dw}{Dt} = \frac{\partial w}{\partial t} = - a\, c^2 \left(\frac{2\pi}{\lambda}\right)^2 \exp\left[\frac{2\pi z}{\lambda}\right] \sin \frac{2\pi}{\lambda} (x \cos\beta + y \sin\beta - ct) \tag{12.7.35}$$

Making use of the trigonometric identities

$$\frac{\partial w}{\partial t} = - a\, c^2 \left(\frac{2\pi}{\lambda}\right)^2 \exp\left[\frac{2\pi z}{\lambda}\right] \left[\cos \omega t \sin \frac{2\pi}{\lambda} (x \cos\beta + y \sin\beta) - \sin \omega t \cos \frac{2\pi}{\lambda} (x \cos\beta + y \sin\beta)\right] \tag{12.7.36}$$

From Equation (12.7.2),

$$Z_w^{(1)} = \int_{\xi_s}^{\xi_b} \left(\rho S + \sum A_{ij}\right) \frac{Dw}{Dt}\, d\xi \tag{12.7.37}$$

and the only added mass tensor of interest is A'_{33}, which is the sectional added mass coefficient in heave due to motion in heave.

Therefore,

$$\sum A_{ij} = A'_{33}$$

and

$$Z_w^{(1)} = \int_{\xi_s}^{\xi_b} \left(\rho S + A'_{33}\right) \frac{Dw}{Dt}\, d\xi \tag{12.7.38}$$

so that

$$Z_W^{(1)} = -\rho a \omega^2 \exp\left[\frac{2\pi z}{\lambda}\right]\left[\cos \omega t \int_{\xi_s}^{\xi_b} \left(S + \frac{A'_{33}}{\rho}\right) \sin \frac{2\pi}{\lambda} \quad \cos\beta + y \sin\beta) d\xi\right.$$

$$\left. - \sin \omega t \int_{\xi_s}^{\xi_b} \left(S + \frac{A'_{33}}{\rho}\right) \cos \frac{2\pi}{\lambda} (x \cos\beta + y \sin\beta)\, d\xi\right] \tag{12.7.39}$$

Evaluating the force at the mean half draft, $-\bar{h}$, and at the center-line, $y = 0$, after introducing the proper corrective term, results in

$$Z_W^{(1)} = -\rho a \omega^2 \exp\left[-\frac{2\pi\bar{h}}{\lambda}\right] \Lambda \left[\cos \omega t \int_{\xi_s}^{\xi_b} \left(S + \frac{A'_{33}}{\rho}\right) \sin\left(\frac{2\pi}{\lambda} x \cos\beta\right) d\xi\right.$$

$$\left. - \sin \omega t \int_{\xi_s}^{\xi_b} \left(S + \frac{A'_{33}}{\rho}\right) \cos\left(\frac{2\pi}{\lambda} x \cos\beta\right) d\xi\right] \tag{12.7.40}$$

The damping force, $A_W^{(2)}$, from Equation (12.7.7) is

$$Z_W^{(2)} = \int_{\xi_s}^{b} C_z N'_{iz} q_{oz}\, d\xi = C_z \int_{\xi_s}^{\xi_b} N'_{zz} w\, d\xi \tag{12.7.41}$$

$$Z_w^{(2)} = - C_z \, a\omega \exp\left[\frac{2\pi z}{\lambda}\right]\left[\cos \omega t \int_{\xi_s}^{\xi_b} N_{zz}' \cos \frac{2\pi}{\lambda}(x \cos\beta + y \sin\beta) d\xi\right.$$

$$\left. + \sin \omega t \int_{\xi_s}^{\xi_b} N_{zz}' \sin \frac{2\pi}{\lambda}(x \cos\beta + y \sin\beta)\, d\xi\right] \qquad (12.7.42)$$

which becomes upon evaluation

$$Z_w^{(2)} = - C_z \, a\omega \exp\left[-\frac{2\pi\bar{h}}{\lambda}\right] \Lambda \left[\cos \omega t \int_{\xi_s}^{\xi_b} N_{zz}' \cos\left(\frac{2\pi}{\lambda} x \cos\beta\right) d\xi\right.$$

$$\left. + \sin \omega t \int_{\xi_s}^{\xi_b} N_{zz}' \sin\left(\frac{2\pi}{\lambda} x \cos\beta\right) d\xi\right] \qquad (12.7.43)$$

The displacement force, $Z_w^{(3)}$, is due to the periodic buoyancy alterations of the platform as the wave passes along the hull, and is therefore proportional to the displacement of the water surface. Thus,

$$Z_w^{(3)} = \int_{\xi_s}^{\xi_b} \rho \, g \, B^* \, \eta(\xi, t) \, d\xi \qquad (12.7.44)$$

where B^* is the local section beam and $\eta(\xi,t)$ is the water surface elevation.

Since

$$\eta = a \left[\tanh \frac{2\pi d}{\lambda}\right] \sin \frac{2\pi}{\lambda} (x \cos\beta + y \sin\beta - ct)$$

then

$$Z_w^{(3)} = \rho g a \left(\tanh \frac{2\pi d}{\lambda}\right) \left[\cos \omega t \int_{\xi}^{\xi_b} B^* \sin \frac{2\pi}{\lambda} (x \cos\beta + y \sin\beta) d\xi\right.$$

$$\left. - \sin \omega t \int_{\xi_s}^{\xi_b} B^* \cos \frac{2\pi}{\lambda} (x \cos\beta + y \sin\beta)\, d\xi \right] \qquad (12.7.45)$$

or after evaluation

$$Z_w^{(3)} = \rho g a \left(\tanh \frac{2\pi d}{\lambda}\right) \Lambda \left[\cos \omega t \int_{\xi_s}^{\xi_b} B^* \sin\left(\frac{2\pi}{\lambda} x \cos\beta\right) d\xi\right.$$

$$\left. - \sin \omega t \int_{\xi_s}^{\xi_b} B^* \cos\left(\frac{2\pi}{\lambda} x \cos\beta\right) d\xi \right] \qquad (12.7.46)$$

The total heaving force, Z_w, is equal to the sum of $Z_w^{(1)}$, $Z_w^{(2)}$, and $Z_w^{(3)}$, or

$$Z_w = -\rho a \omega^2 \exp\left[-\frac{2\pi \bar{h}}{\lambda}\right] \Lambda \left[\cos \omega t \int_{\xi_s}^{\xi_b} \left(S + \frac{A'_{33}}{\rho}\right) \sin\left(\frac{2\pi}{\lambda} x \cos \beta\right) d\xi\right.$$

$$\left. - \sin \omega t \int_{\xi_s}^{\xi_b} \left(S + \frac{A'_{33}}{\rho}\right) \cos\left(\frac{2\pi}{\lambda} x \cos \beta\right) d\xi\right]$$

$$- C_z a\omega \exp\left[-\frac{2\pi \bar{h}}{\lambda}\right] \Lambda \left[\cos \omega t \int_{\xi_s}^{\xi_b} N'_{zz} \cos\left(\frac{2\pi}{\lambda} x \cos \beta\right) d\xi\right.$$

$$\left. + \sin \omega t \int_{\xi_s}^{\xi_b} N'_{zz} \sin\left(\frac{2\pi}{\lambda} x \cos \beta\right) d\xi\right]$$

$$+ \rho g a \tanh \frac{2\pi d}{\lambda} \Lambda \left[\cos \omega t \int_{\xi_s}^{\xi_b} B^* \sin\left(\frac{2\pi}{\lambda} x \cos \beta\right) d\xi\right.$$

$$\left. - \sin \omega t \int_{\xi_s}^{\xi_b} B^* \cos\left(\frac{2\pi}{\lambda} x \cos \beta\right) d\xi\right] \qquad (12.7.47)$$

For head-on waves, $\beta = \pi$, $\cos\beta = -1$, $\sin\beta = 0$, and Equation (12.7.47) becomes

$$Z_w = +\rho a \omega^2 \exp\left[-\frac{2\pi\bar{h}}{\lambda}\right]\left[\cos\ \ t\int_{\xi_s}^{\xi_b}\left(S + \frac{A'_{33}}{\rho}\right)\sin\left(\frac{2\pi}{\lambda}x\right)\ d\xi\right.$$

$$+ \sin\omega t\int_{\xi_s}^{\xi_b}\left(S + \frac{A'_{33}}{\rho}\right)\cos\left(\frac{2\pi}{\lambda}x\right)\ d\xi\Bigg]$$

$$- C_z a \omega \exp\left[-\frac{2\pi\bar{h}}{\lambda}\right]\left[\cos\omega t\int_{\xi_s}^{\xi_b} N'_{zz}\cos\left(\frac{2\pi}{\lambda}x\right)\ d\xi\right.$$

$$- \sin\omega t\int_{\xi_s}^{\xi_b} N'_{zz}\sin\left(\frac{2\pi}{\lambda}x\right)\ d\xi\Bigg]$$

$$- \rho g\ a\ \tanh\left(\frac{2\pi d}{\lambda}\right)\left[\cos\omega t\int_{\xi_s}^{\xi_b} B^*\sin\left(\frac{2\pi}{\lambda}x\right)\ d\xi\right.$$

$$+ \sin\omega t\int_{\xi_s}^{\xi_b} B^*\cos\left(\frac{2\pi}{\lambda}x\right)\ d\xi\Bigg] \qquad (12.7.48)$$

For beam-on waves, $\beta = \pi/2$, $\cos\beta = 0$, $\sin\beta = 1$, and

$$Z_w = \rho a \omega^2 \exp\left[-\frac{2\pi\bar{h}}{\lambda}\right] \frac{\sin\frac{\pi B}{\lambda}}{\frac{\pi B}{\lambda}} \sin\omega t \int_{\xi_s}^{\xi_b} \left(S + \frac{A'_{33}}{\rho}\right) d\xi$$

$$- C_z a \omega \exp\left[-\frac{2\pi\bar{h}}{\lambda}\right] \frac{\sin\frac{\pi B}{\lambda}}{\frac{\pi B}{\lambda}} \cos\omega t \int_{\xi_s}^{\xi_b} N'_{zz}\, d\xi$$

$$- g a \tanh\left(\frac{2\pi d}{\lambda}\right) \frac{\sin\frac{\pi B}{\lambda}}{\frac{\pi B}{\lambda}} \sin\omega t \int_{\xi_s}^{\xi_b} B^{*}\, d\xi \qquad (12.7.49)$$

Equations (12.7.48) and (12.7.49) are appropriate expressions for deep water waves but only approximate expressions for shallow water waves.

Yaw, N_w: The wave-induced yaw moment, N_w, is readily obtained from the swaying force, Y_w, since

$$N_w = \int_{\xi_s}^{\xi_b} \xi \, dY_w \, d\xi \qquad (12.7.50)$$

where ξ is the moment arm measured from the center of gravity. Thus, the general expression for the yaw moment, N_w, is

$$
\begin{aligned}
N_w = & -\rho\omega^2 \, a \, \exp\left[-\frac{2\pi\bar{h}}{\lambda}\right] \Lambda \, \sin\beta \left[\cos\omega t \int_{\xi_s}^{\xi_b} \left(S + \frac{A'_{22}}{\rho}\right.\right. \\
& \left. + \frac{2\pi A'_{42}}{\lambda\rho}\right)(\xi) \cos\left(\frac{2\pi}{\lambda} x \cos\beta\right) d\xi + \sin \; t \int_{\xi_s}^{\xi_b} \left(S + \frac{A'_{22}}{\rho}\right. \\
& \left.\left. + \frac{2\pi A'_{42}}{\lambda\rho}\right)(\xi) \sin\left(\frac{2\pi}{\lambda} x \cos\beta\right) d\xi\right] + C_y \, a \, \exp\left[-\frac{2\pi\bar{h}}{\lambda}\right] \\
& \cdot \Lambda \, \sin\beta \left[\cos\omega t \int_{\xi_s}^{\xi_b} N'_{yy}(\xi) \cos\left(\frac{2\pi}{\lambda} x \cos\beta\right) d\xi\right. \\
& \left. - \sin\omega t \int_{\xi_s}^{\xi_b} N'_{yy}(\xi) \sin\left(\frac{2\pi}{\lambda} x \cos\beta\right) d\xi\right]
\end{aligned}
\qquad (12.7.51)
$$

In the case of head-on waves, $\beta = \pi$, $\sin\beta = 0$, and Equation (12.7.51) becomes

$$
N_w = 0 . \qquad (12.7.52)
$$

In the case of beam-on waves, $\beta = \pi/2$, $\sin\beta = 1$, $\cos\beta = 0$, and Equation (12.7.51) becomes

$$N_w = -\rho\omega^2 a \exp\left[-\frac{2\pi\bar{h}}{\lambda}\right]\frac{\sin\frac{\pi B}{\lambda}}{\frac{\pi B}{\lambda}}\left[\cos\omega t\int_{\xi_s}^{\xi_b}\left(S + \frac{A'_{22}}{\rho}\right.\right.$$

$$\left.\left.+ \frac{2\pi A'_{42}}{\lambda\rho}\right)(\xi)\cos\left(\frac{2\pi}{\lambda}x\right)d\xi\right]$$

$$- C_y a\omega \exp\left[-\frac{2\pi\bar{h}}{\lambda}\right]\frac{\sin\frac{\pi B}{\lambda}}{\frac{\pi B}{\lambda}}\left[\cos\omega t\int_{\xi_s}^{\xi_b} N'_{yy}(\xi)\,d\xi\right]$$

(12.7.53)

Equations (12.7.52) and (12.7.53) are appropriate expressions for deep water waves but only approximate expressions for shallow water waves.

<u>Pitch, M_w</u>: The wave-induced pitch moment, M_w, is readily obtained from the heaving force, Z_w, since

$$M_w = -\int_{\xi_s}^{\xi_b} \xi \, dZ_w \, d\xi \qquad (12.7.54)$$

where ξ is the moment arm measured from the center of gravity and the minus sign is necessary to preserve the right-hand notation. Thus, the general expression for the pitch moment is

$$M_W = +\rho a \omega^2 \exp\left[-\frac{2\pi\bar{h}}{\lambda}\right] \Lambda \left[\cos \omega t \int_{\xi_s}^{\xi_b} \left(S + \frac{A'_{33}}{\rho}\right)(\xi) \left(\frac{2\pi}{\lambda} x \cos \beta\right) d\xi\right.$$

$$\left. - \sin \omega t \int_{\xi_s}^{\xi_b} \left(S + \frac{A'_{33}}{\rho}\right)(\xi) \cos\left(\frac{2\pi}{\lambda} x \cos \beta\right) d\xi\right]$$

$$+ C_z a \exp\left[-\frac{2\pi\bar{h}}{\lambda}\right] \Lambda \left[\cos \omega t \int_{\xi_s}^{\xi_b} N'_{zz}(\xi) \cos\left(\frac{2\pi}{\lambda} x \cos \beta\right) d\xi\right.$$

$$\left. + \sin \omega t \int_{\xi_s}^{\xi_b} N'_{zz}(\xi) \sin\left(\frac{2\pi}{\lambda} x \cos \beta\right) d\xi\right]$$

$$- \rho g a \tanh \frac{2\pi d}{\lambda} \Lambda \left[\cos \omega t \int_{\xi_s}^{\xi_b} B^*(\xi) \sin\left(\frac{2\pi}{\lambda} x \cos \beta\right) d\xi\right.$$

$$\left. - \sin \omega t \int_{\xi_s}^{\xi_b} B^*(\xi) \cos\left(\frac{2\pi}{\lambda} x \cos \beta\right) d\xi\right] \qquad (12.7.55)$$

For head-on waves, $\beta = \pi$, $\cos\beta = 1$, $\sin\beta = 0$, and the appropriate expression for the wave-induced pitch moment becomes

$$M_W = -\rho a \omega^2 \exp\left[-\frac{2\pi\bar{h}}{\lambda}\right]\left[\cos\omega t\int_{\xi_s}^{\xi_b}\left(S + \frac{A'_{33}}{\rho}\right)(\xi)\sin\left(\frac{2\pi}{\lambda}x\right)d\xi\right.$$

$$+ \left.\sin\omega t\int_{\xi_s}^{\xi_b}\left(S + \frac{A'_{33}}{\rho}\right)(\xi)\cos\left(\frac{2\pi}{\lambda}x\right)d\xi\right]$$

$$+ C_z a \omega \exp\left[-\frac{2\pi\bar{h}}{\lambda}\right]\left[\cos\omega t\int_{\xi_s}^{\xi_b} N'_{zz}(\xi)\cos\left(\frac{2\pi x}{\lambda}\right)d\xi\right.$$

$$- \left.\sin\omega t\int_{\xi_s}^{\xi_b} N'_{zz}(\xi)\sin\left(\frac{2\pi x}{\lambda}\right)d\xi\right]$$

$$+ \rho g a \tanh\frac{2\pi d}{\lambda}\left[\cos\omega t\int_{\xi_s}^{\xi_b} B^*(\xi)\sin\left(\frac{2\pi}{\lambda}x\right)d\xi\right.$$

$$+ \left.\sin\omega t\int_{\xi_s}^{\xi_b} B^*(\xi)\cos\left(\frac{2\pi}{\lambda}x\right)d\xi\right] \qquad (12.7.56)$$

For beam-on waves, $\beta = \pi/2$, $\cos\beta = 0$, and $\sin\beta = 1$, and Equation (12.7.56) becomes

$$M_w = -\rho g a \omega^2 \exp\left[-\frac{2\pi\bar{h}}{\lambda}\right] \frac{\sin\frac{\pi B}{\lambda}}{\frac{\pi B}{\lambda}} \left[\sin\omega t \int_{\xi_s}^{\xi_b} \left(S + \frac{A_{33}'}{\rho}\right)(\xi)\, d\xi\right]$$

$$+ C_z a \omega \exp\left[-\frac{2\pi\bar{h}}{\lambda}\right] \frac{\sin\frac{\pi B}{\lambda}}{\frac{\pi B}{\lambda}} \left[\cos\omega t \int_{\xi_s}^{\xi_b} N_{zz}'(\xi)\, d\xi\right]$$

$$+ \rho g a \tanh\frac{2\pi d}{\lambda} \frac{\sin\frac{\pi B}{\lambda}}{\frac{\pi B}{\lambda}} \left[\sin\omega t \int_{\xi_s}^{\xi_b} B^*(\xi)\, d\xi\right]$$

(12.7.57)

Equations (12.7.55), (12.7.56), and (12.7.57) are appropriate expressions for deep water waves but only approximate expressions for shallow water waves.

Roll, K_w: The roll moment, $K_o^{(1)}$, referring to the free surface lever, is given by Hu (1961) and Kaplan and Putz (1962) as

$$K_{ow}^{(1)} = \int_{\xi_s}^{\xi_b} \left(-\rho S \bar{z}_{cb} + A_{42}' + A_{44}' \frac{2\pi}{\lambda}\right) \frac{D\bar{q}_{oy}}{Dt}\, d\xi \qquad (12.7.58)$$

where $\bar{z}_{cb}$ is the location of the center of buoyance of each section $(\bar{z}_{cb} < 0)$, and is measured from the free surface level; A_{42}' is the sectional added moment of inertia in roll due to motion in sway; A_{44}' is the sectional added moment of inertia in roll due to motion in roll.

Since damping in roll is known to be quite small, then $K_o^{(2)} = 0$. Hu also gives the roll moment, $K_o^{(3)}$, as

$$K_{ow}^{(3)} = \int_{\xi_s}^{\xi_b} \left[-\frac{\rho}{12} (B^*)^3 \frac{D\bar{q}_{oy}}{Dt} \right] d\xi \qquad (12.7.59)$$

Adding Equations (12.7.58) and (12.7.59) gives

$$K_{ow} = \int_{\xi_s}^{\xi_b} \left[-\rho S \bar{z}_{cb} - \frac{\rho}{12} (B^*)^3 + A_{42}' + \frac{2\pi}{\lambda} A_{44}' \right] \frac{D\bar{q}_{oy}}{Dt} d\xi$$

(12.7.60)

Since Kaplan and Putz state that "the added moment of inertia in roll, A_{44}', is quite difficult to compute, and with little faith in the validity of such a result for motions of this type," an approximation is necessary to assist in the evaluation of the additional roll moment. Therefore, disregarding the term $(2\pi/\lambda)A_{44}'$ for the moment, the remaining terms are evaluated as

$$K_{ow}' = \int_{\xi_s}^{\xi_b} \left[A_{42}' - \rho S \bar{z}_{cb} - \frac{\rho}{12} (B^*)^3 \right] \frac{Dv}{Dt} d\xi$$

$$K'_{ow} = -a\omega^2 \exp\left[-\frac{2\pi\bar{h}}{\lambda}\right] \Lambda \sin\beta$$

$$\cdot\left\{\cos\omega t\int_{\xi_s}^{\xi_b}\left[A'_{42} - \rho S\bar{z}_{cb} - \frac{\rho}{12}(B^*)^3\right]\cos\left(\frac{2\pi}{\lambda}x\cos\beta\right)d\xi\right.$$

$$\left.+\sin\omega t\int_{\xi_s}^{\xi_b}\left[A'_{42} - \rho S\bar{z}_{cb} - \frac{\rho}{12}(B^*)^3\right]\sin\left(\frac{2\pi}{\lambda}x\cos\beta\right)d\xi\right\}$$

(12.7.61)

Kaplan and Putz point out that the added moment of inertia, A'_{44}, is actually that of the submerged part of the platform rolling about a point either at the free surface or at the mean half depth. They give an effective total added roll inertia of the form

$$K''_{ow} = -\frac{A_{44}}{L}a\omega^2\frac{2\pi}{\lambda}\exp\left[-\frac{2\pi\bar{h}}{\lambda}\right]\Lambda\sin\beta\left[\cos\omega t\int_{\xi_s}^{\xi_b}\cos\left(\frac{2\pi}{\lambda}x\cos\beta\right)d\xi\right.$$

$$\left.+\sin\omega t\int_{\xi_s}^{\xi_b}\sin\left(\frac{2\pi}{\lambda}x\cos\beta\right)d\xi\right]$$

(12.7.62)

where L is the length of the platform and A_{44} is the added moment of inertia of the platform due to rotation about the center of gravity. If the waterline length

$$\left(|\xi_b| + |\xi_s| \right)$$

and platform length, L, are approximately equal, then Equation (12.7.62) can be simplified to

$$K_{ow}'' = -A_{44}\, a\, \omega^2 \left(\frac{2\pi}{\lambda}\right) \exp\left[-\frac{2\pi \bar{h}}{\lambda}\right] \Lambda \quad \sin\beta \cos\omega t \; \frac{\sin\left(\frac{\pi L}{\lambda}\cos\beta\right)}{\frac{\pi L}{\lambda}\cos\beta} \tag{12.7.63}$$

The total roll moment, K_{ow}, about the free surface is, therefore,

$$K_{ow} = K_{ow}' + K_{ow}'' \tag{12.7.64}$$

Referring the roll moment to the center of gravity results in

$$K_w = K_{ow} + \overline{OG}\, Y_w = K_{ow}' + K_{ow}'' + \overline{OG}\, Y_w \tag{12.7.65}$$

where $\overline{OG}$ is the moment arm from the center of gravity to the free surface. The general expression for the roll moment thus becomes

$$K_W = - \; a\,\omega^2 \exp\left[-\frac{2\pi\bar{h}}{\lambda}\right] \Lambda \;\sin\beta$$

$$\cdot\left\{\cos\omega t \int_s^b \left[\frac{A'_{42}}{\rho} - S\,\bar{z}_{cb} - \frac{(B^*)^3}{12}\right] \cos\left(\frac{2\pi}{\lambda}\,x\cos\beta\right) d\xi\right.$$

$$\left. + \;\sin\omega t \int_s^b \left[\frac{A'_{42}}{\rho} - S\,\bar{z}_{cb} - \frac{(B^*)^3}{12}\right] \sin\left(\frac{2\pi}{\lambda}\,x\cos\beta\right) d\xi\right\}$$

$$- \;\frac{A_{44}}{L}\, a\,\omega^2 \;\frac{2\pi}{\lambda} \;\exp\left[-\frac{2\pi\bar{h}}{\lambda}\right] \Lambda \;\sin\beta$$

$$\cdot\left[\cos\omega t \int_{\xi_s}^{\xi_b} \cos\left(\frac{2\pi}{\lambda}\,x\cos\beta\right) d\xi\right.$$

$$\left. + \;\sin\omega t \int_{\xi_s}^{\xi_b} \sin\left(\frac{2\pi}{\lambda}\,x\cos\beta\right) d\xi\right] + \overline{OG}\,Y_W \qquad (12.7.66)$$

For head-on waves, $\beta = \pi$, $\sin\beta = 0$, and $K_W = 0$.

For beam-on waves, $\beta = \pi/2$, $\sin\beta = 1$, and $\cos\beta = 0$, and K_w becomes

$$K_w = -\rho a \omega^2 \exp\left[-\frac{2\pi\bar{h}}{\lambda}\right]\left(\frac{\sin\frac{\pi B}{\lambda}}{\frac{\pi B}{\lambda}}\right)\left\{\cos\omega t\int_{\xi_s}^{\xi_b}\left[\frac{A'_{42}}{\rho} - S\,\bar{z}_{cb} - \frac{(B^*)^3}{12}\right]d\xi\right\}$$

$$-\frac{A'_{44}}{L}\, a\,\omega^2\left(\frac{2\pi}{\lambda}\right)\exp\left[-\frac{2\pi\bar{h}}{\lambda}\right]\left(\frac{\sin\frac{\pi B}{\lambda}}{\frac{\pi B}{\lambda}}\right)\left[\cos\omega t\int_{\xi_s}^{\xi_b}\cos\left(\frac{2\pi x}{\lambda}\right)d\xi\right]$$

$$-\overline{OG}\left\{\rho\omega^2 a\exp\left[-\frac{2\pi\bar{h}}{\lambda}\right]\left(\frac{\sin\frac{\pi B}{\lambda}}{\frac{\pi B}{\lambda}}\right)\left[\cos\omega t\int_{\xi_s}^{\xi_b}\left(S + \frac{A'_{22}}{\rho} + \frac{2\pi A'_{42}}{\lambda\rho}\right)d\xi\right]\right.$$

$$\left.-C_y\, a\,\omega\exp\left[-\frac{2\pi\bar{h}}{\lambda}\right]\left(\frac{\sin\frac{\pi B}{\lambda}}{\frac{\pi B}{\lambda}}\right)\left(\sin\omega t\int_{\xi_s}^{\xi_b}N'_{yy}\,d\xi\right)\right\} \qquad (12.7.67)$$

Summary: The wave exciting forces and moments are seen to consist of expressions of the form

$$F_r^w \quad \text{or} \quad M_r^w = A_r^w\cos\omega t + B_r^w\sin\omega t \qquad (12.7.68)$$

for the rth motion which are frequency-and heading-dependent and assume a surface wave elevation η of

$$\eta(x, y, t) = a\cos\left[\omega t + \frac{\pi}{2} - \left(\frac{2\pi}{\lambda}\right)(x\cos\beta + y\sin\beta)\right]$$

where a is the wave amplitude and β is the angle between the x-direction and the direction of propagation of the waves. Assuming a unit amplitude (a = 1), and coefficients A_r^w and B_r^w of the general expressions are as follows:

Surge:

$$A_x^w = -\rho\omega^2 \exp\left[-\frac{2\pi\bar{h}}{\lambda}\right] \Lambda \cos\beta \int_{\xi_s}^{\xi_b} S \cos\left(\frac{2\pi}{\lambda} x \cos\beta\right) d\xi \quad (12.7.69)$$

$$B_x^w = -\rho\omega^2 \exp\left[-\frac{2\pi\bar{h}}{\lambda}\right] \Lambda \cos\beta \int_{\xi_s}^{\xi_b} S \sin\left(\frac{2\pi}{\lambda} x \cos\beta\right) d\xi \quad (12.7.70)$$

Sway:

$$A_y^w = -\rho\omega^2 \exp\left[-\frac{2\pi\bar{h}}{\lambda}\right] \Lambda \sin\beta \int_{\xi_s}^{\xi_b} \left(S + \frac{A'_{22}}{} + \frac{2\pi A'_{42}}{\lambda}\right) \cos\left(\frac{2\pi}{\lambda} x \cos\beta\right) d\xi$$

$$+ C_y\omega \exp\left[-\frac{2\pi\bar{h}}{\lambda}\right] \Lambda \sin\beta \int_{\xi_s}^{\xi_b} N'_{yy} \sin\left(\frac{2\pi}{\lambda} x \cos\beta\right) d\xi$$

(12.7.71)

$$B_y^w = -\rho\omega^2 \exp\left[-\frac{2\pi\bar{h}}{\lambda}\right] \Lambda \sin\beta \int_{\xi_s}^{\xi_b} \left(S + \frac{A'_{22}}{} + \frac{2\pi A'_{42}}{\lambda}\right) \sin\left(\frac{2\pi}{\lambda} x \cos\beta\right) d\xi$$

$$- C_y\omega \exp\left[-\frac{2\pi\bar{h}}{\lambda}\right] \Lambda \sin\beta \int_{\xi_s}^{\xi_b} N'_{yy} \cos\left(\frac{2\pi}{\lambda} x \cos\beta\right) d\xi$$

(12.7.72)

Heave:

$$A_z{}^W = -\rho\omega^2 \exp\left[-\frac{2\pi\bar{h}}{\lambda}\right] \Lambda \int_{\xi_s}^{\xi_b} \left(S + \frac{A'_{33}}{\rho}\right) \sin\left(\frac{2\pi}{\lambda} x \cos\beta\right) d\xi$$

$$- C_z\omega \exp\left[-\frac{2\pi\bar{h}}{\lambda}\right] \Lambda \int_{\xi_s}^{\xi_b} N'_{zz} \cos\left(\frac{2\pi}{\lambda} x \cos\beta\right) d\xi$$

$$+ \rho g \tanh\frac{2\pi d}{\lambda} \Lambda \int_{\xi_s}^{\xi_b} B^* \sin\left(\frac{2\pi}{\lambda} x \cos\beta\right) d\xi \tag{12.7.73}$$

$$B_z{}^W = \rho\omega^2 \exp\left[-\frac{2\pi\bar{h}}{\lambda}\right] \Lambda \int_{\xi_s}^{\xi_b} \left(S + \frac{A'_{33}}{\rho}\right) \cos\left(\frac{2\pi}{\lambda} x \cos\beta\right) d\xi$$

$$- C_z\omega \exp\left[-\frac{2\pi\bar{h}}{\lambda}\right] \Lambda \int_{\xi_s}^{\xi_b} N'_{zz} \sin\left(\frac{2\pi}{\lambda} x \cos\beta\right) d\xi$$

$$+ \rho g \tanh\frac{2\pi d}{\lambda} \Lambda \int_{\xi_s}^{\xi_b} B^* \cos\left(\frac{2\pi}{\lambda} x \cos\beta\right) d\xi \tag{12.7.74}$$

Yaw:

$$A_{\psi}^{w} = -\rho\omega^2 \exp\left[-\frac{2\pi\bar{h}}{\lambda}\right]\Lambda \sin\beta\int_{\xi_s}^{\xi_b}\left(S + \frac{A'_{22}}{\rho} + \frac{2\pi A'_{42}}{\lambda\rho}\right)(\xi)\cos\left(\frac{2\pi}{\lambda} x \cos\beta\right) d\xi$$

$$+ C_y\omega \exp\left[-\frac{2\pi\bar{h}}{\lambda}\right] \Lambda \sin\beta\int_{\xi_s}^{\xi_b} N'_{yy}(\xi) \cos\left(\frac{2\pi}{\lambda} x \cos\beta\right) d\xi$$

(12.7.75)

$$B_{\psi}^{w} = -\rho\omega^2 \exp\left[-\frac{2\pi\bar{h}}{\lambda}\right]\Lambda \sin\beta\int_{\xi_s}^{\xi_b}\left(S + \frac{A'_{22}}{\rho} + \frac{2\pi A'_{42}}{\lambda\rho}\right)(\xi) \sin\left(\frac{2\pi}{\lambda} x \cos\beta\right) d\xi$$

$$- C_y\omega \exp\left[-\frac{2\pi\bar{h}}{\lambda}\right] \Lambda \sin\beta\int_{\xi_s}^{\xi_b} N'_{yy}(\xi) \sin\left(\frac{2\pi}{\lambda} x \cos\beta\right) d\xi$$

(12.7.76)

Pitch:

$$A_\theta^{\,w} = \rho\omega^2 \exp\left[-\frac{2\pi\bar{h}}{\lambda}\right] \Lambda \int_{\xi_s}^{\xi_b} \left(S + \frac{A_{33}'}{\rho}\right)(\xi) \sin\left(\frac{2\pi}{\lambda} x \cos\beta\right) d\xi$$

$$+ C_z\omega \exp\left[-\frac{2\pi\bar{h}}{\lambda}\right] \Lambda \int_{\xi_s}^{\xi_b} N_{zz}'(\xi) \cos\left(\frac{2\pi}{\lambda} x \cos\beta\right) d\xi$$

$$- \rho g \tanh\frac{2\pi d}{\lambda} \Lambda \int_{\xi_s}^{\xi_b} B^*(\xi) \sin\left(\frac{2\pi}{\lambda} x \cos\beta\right) d\xi$$

(12.7.77)

$$B_\theta^{\,w} = -\rho\omega^2 \exp\left[-\frac{2\pi\bar{h}}{\lambda}\right] \Lambda \int_{\xi_s}^{\xi_b} \left(S + \frac{A_{33}'}{\rho}\right)(\xi) \cos\left(\frac{2\pi}{\lambda} x \cos\beta\right) d\xi$$

$$+ C_z\omega \exp\left[-\frac{2\pi\bar{h}}{\lambda}\right] \Lambda \int_{\xi_s}^{\xi_b} N_{zz}'(\xi) \sin\left(\frac{2\pi}{\lambda} x \cos\beta\right) d\xi$$

$$+ \rho g \tanh\frac{2\pi d}{\lambda} \Lambda \int_{\xi_s}^{\xi_b} B^*(\xi) \cos\left(\frac{2\pi}{\lambda} x \cos\beta\right) d\xi$$

(12.7.78)

Roll:

$$A_\phi^W = -\rho\omega^2 \exp\left[-\frac{2\pi\bar{h}}{\lambda}\right] \Lambda \sin\beta \int_{\xi_s}^{\xi_b} \left[\frac{A'_{42}}{\rho} - S\,\bar{z}_{cb} - \frac{(B^*)^3}{12}\right] \cos\left(\frac{2\pi}{\lambda} x \cos\beta\right) d\xi$$

$$- \frac{A_{44}}{L}\,\omega^2 \left(\frac{2\pi}{\lambda}\right) \exp\left[-\frac{2\pi\bar{h}}{\lambda}\right] \Lambda \sin\beta \int_{\xi_s}^{\xi_b} \cos\left(\frac{2\pi}{\lambda} x \cos\beta\right) d\xi$$

$$- \overline{OG}\rho\omega^2 \exp\left[-\frac{2\pi\bar{h}}{\lambda}\right] \Lambda \sin\beta \int_{\xi_s}^{\xi_b} \left(S + \frac{A'_{22}}{\rho} + \frac{2\pi A'_{42}}{\lambda\rho}\right) \cos\left(\frac{2\pi}{\lambda} x \cos\beta\right) d\xi$$

$$+ \overline{OG}\,C_y\,\omega \exp\left[-\frac{2\pi\bar{h}}{\lambda}\right] \Lambda \sin\beta \int_{\xi_s}^{\xi_b} N'_{yy} \sin\left(\frac{2\pi}{\lambda} x \cos\beta\right) d\xi$$

(12.7.79)

$$B_\phi^{\,w} = -\rho\omega^2 \exp\left[-\frac{2\pi\bar{h}}{\lambda}\right]\Lambda \sin\beta \int_{\xi_s}^{\xi_b}\left[\frac{A_{42}'}{\rho} - S\bar{z}_{cb} - \frac{(B^*)^3}{12}\right]\sin\left(\frac{2\pi}{\lambda}\, x \cos\beta\right)d\xi$$

$$-\frac{A_{44}}{L}\,\omega^2\left(\frac{2\pi}{\lambda}\right)\exp\left[-\frac{2\pi\bar{h}}{\lambda}\right]\Lambda \sin\beta \int_{\xi_s}^{\xi_b}\sin\left(\frac{2\pi}{\lambda}\, x\cos\beta\right)d\xi$$

$$-\overline{OG}\,\rho\omega^2 \exp\left[-\frac{2\pi\bar{h}}{\lambda}\right]\Lambda \sin\beta \int_{\xi_s}^{\xi_b}\left(S + \frac{A_{22}'}{\rho} + \frac{2\pi A_{42}'}{\lambda\rho}\right)\sin\left(\frac{2\pi}{\lambda}\, x\cos\beta\right)d\xi$$

$$-\overline{OG}\, C_y\,\omega \exp\left[-\frac{2\pi\bar{h}}{\lambda}\right]\Lambda \sin\beta \int_{\xi_s}^{\xi_b} N_{yy}' \cos\left(\frac{2\pi}{\lambda}\, x\cos\beta\right)d\xi$$

(12.7.80)

12.8 SOLUTION OF THE EQUATIONS OF MOTION

The equations of motion are thus formulated as linear combinations of terms proportional to accelerations, terms proportional to velocities, terms proportional to displacements, and terms sinusoidal in time. The method of undetermined coefficients is employed to obtain solutions to the equations of motion. Specifically, it is assumed that since the exciting forces are sinusoidal the responses (i.e., the motions) will also consist of sinusoidal oscillations of the same frequency, but not necessarily in phase.

Recall that the time history of the surface wave elevation is

$$\eta(x, y, t) = a \cos\left[\omega t + \frac{\pi}{2} - \frac{2\pi}{\lambda}(x\cos\beta + y\sin\beta)\right]$$

which reduces to

$$\eta = (0,\ 0,\ t) \quad = \quad \cos\left(\omega t + \frac{\pi}{2}\right)$$

when evaluated at the center of gravity of the platform. Next, the rth component of the (known) excitation force or moment which is of the form

$$F_r^{w} \quad \text{or} \quad M_r^{w} = A_r^{w} \cos\ \omega t + B_r^{w} \sin\ \omega t$$

can also be written as

$$F_r^{w} \quad \text{or} \quad M_r^{w} = \sqrt{\left(A_r^{w}\right)^2 + \left(B_r^{w}\right)^2} \quad \cos\ (\ \omega t + \gamma_r)$$

where

$$\gamma_r \quad = \tan^{-1}\left(-\frac{B_r^{w}}{A_r^{w}}\right)$$

Since the motions are also sinusoidal, they are assumed to have the form

$$r(t) \quad = \quad a_r \cos\left(\omega t + \frac{\pi}{2} + \delta_r\right)$$

where $r(t)$ represents any of the six motions. Since the exciting waves are assumed to have unit amplitude and phase angle $\pi/2$, a_r and δ_r are by definition the amplitude-response operator and phase-response operator, respectively. Thus, the problem is reduced to finding a_r and δ_r for each of the motions.

The procedure is to substitute into the equations of motion the assumed solution and their derivatives, and then to solve for the amplitude-response and phase-response operators. Introducing the complex notation

$$\bar{X}_w = \sqrt{\left(A_x^w\right)^2 + \left(B_x^w\right)^2}\ \exp\left[i\gamma_x\right] \qquad \bar{Y}_w = \sqrt{\left(A_y^w\right)^2 + \left(B_y^w\right)^2}\ \exp\left[i\gamma_y\right]$$

$$\bar{Z}_w = \sqrt{\left(A_z^w\right)^2 + \left(B_z^w\right)^2}\ \exp\left[i\gamma_z\right] \qquad \bar{\Phi}_w = \sqrt{\left(A_\phi^w\right)^2 + \left(B_\phi^w\right)^2}\ \exp\left[i\gamma_\phi\right]$$

$$\bar{\Theta}_w = \sqrt{\left(A_\theta^w\right)^2 + \left(B_\theta^w\right)^2}\ \exp\left[i\gamma_\theta\right] \qquad \bar{\Psi}_w = \sqrt{\left(A_\psi^w\right)^2 + \left(B_\psi^w\right)^2}\ \exp\left[i\gamma_\psi\right]$$

$$\bar{x} = a_x \exp\left[i\left(\delta_x + \frac{\pi}{2}\right)\right] \qquad \bar{y} = a_y \exp\left[i\left(\delta_y + \frac{\pi}{2}\right)\right] \qquad \bar{z} = a_z \exp\left[i\left(\delta_z + \frac{\pi}{2}\right)\right]$$

$$\bar{\phi} = a_\phi \exp\left[i\left(\delta_o + \frac{\pi}{2}\right)\right] \qquad \bar{\theta} = a_\theta \exp\left[i\left(\delta_\theta + \frac{\pi}{2}\right)\right] \qquad \bar{\psi} = a_\psi \exp\left[i\left(\delta_\psi + \frac{\pi}{2}\right)\right]$$

so that

$$F_x^{W}(t) = \text{Real}\left(\bar{X}_W \exp\left[i\omega t\right]\right)$$

$$F_y^{W}(t) = \text{Real}\left(\bar{Y}_W \exp\left[i\omega t\right]\right)$$

$$F_z^{W}(t) = \text{Real}\left(\bar{Z}_W \exp\left[i\omega t\right]\right)$$

$$M_\phi^{W}(t) = \text{Real}\left(\bar{\Phi}_W \exp\left[i\omega t\right]\right)$$

$$M_\theta^{W}(t) = \text{Real}\left(\bar{\theta}_W \exp\left[i\omega t\right]\right)$$

$$M_\psi^{W}(t) = \text{Real}\left(\bar{\Psi}_W \exp\left[i\omega t\right]\right)$$

and similarly, the assumed solutions are

$$x(t) = \text{Real}\left(\bar{x} \exp\left[i\omega t\right]\right) \qquad \phi(t) = \text{Real}\left(\bar{\Phi} \exp\left[i\omega t\right]\right)$$

$$y(t) = \text{Real}\left(\bar{y} \exp\left[i\omega t\right]\right) \qquad \theta(t) = \text{Real}\left(\bar{\theta} \exp\left[i\omega t\right]\right)$$

$$z(t) = \text{Real}\left(\bar{z} \exp\left[i\omega t\right]\right) \qquad \psi(t) = \text{Real}\left(\bar{\Psi} \exp\left[i\omega t\right]\right)$$

Thus, the known real functions,

$$F_x^W, \quad F_y^W, \quad F_z^W, \quad M_o^W, \quad M_\theta^W, \quad M_\psi^W$$

and the unknown real functions, x, y, z, o, θ, and ψ, may be replaced by both real and imaginary parts of the complex functions,

$$\bar{X}_W \exp\left[i\omega t\right] \qquad \bar{Y}_W \exp\left[i\omega t\right] \qquad \bar{Z}_W \exp\left[i\omega t\right]$$

$$\bar{\Phi}_W \exp\left[i\omega t\right] \qquad \bar{\Theta}_W \exp\left[i\omega t\right] \qquad \bar{\Psi}_W \exp\left[i\omega t\right]$$

The reason for substituting the complex functions is that their first and second time derivatives are just equal to the functions themselves multiplied by $i\omega$ and $-\omega^2$, respectively. After making the substitutions and solving for the unknowns $\bar{x}$, $\bar{y}$, $\bar{z}$, $\bar{o}$, $\bar{\theta}$, and $\bar{\Psi}$ (and thus for a_x, δ_x, a_y, δ_y, a_z, δ_z, a_ϕ, δ_ϕ, a_θ, δ_θ, a_ψ, δ_ψ), the correct final solution can be found by taking the real parts of

$$\bar{x} \exp\left[i\omega t\right] \qquad \bar{y} \exp\left[i\omega t\right] \qquad \bar{z} \exp\left[i\omega t\right]$$

$$\bar{\phi} \exp\left[i\omega t\right] \qquad \bar{\theta} \exp\left[i\omega t\right] \qquad \bar{\psi} \exp\left[i\omega t\right]$$

By substituting the assumed complex function solutions and their appropriate time derivatives, canceling the factor $\exp[i\omega t]$, and transposing all reaction terms to the left-hand side, the equations of motion are reduced to

$$\overline{X}_w = - m\omega^2 \,\overline{x} + m\,|BG|\,\omega^2 \,\overline{\theta} + i\,\omega\, N_x \,\overline{x} + K_x \,\overline{x}$$

$$\overline{Y}_w = - m\omega^2 \,\overline{y} - \omega^2 \rho \int_{\xi_s}^{\xi_b} \frac{A'_{22}}{\rho}\, d\xi \,\overline{y} - \omega^2 \rho \int_{\xi_s}^{\xi_b} \frac{A'_{22}}{\rho}\, \xi \, d\xi \,\overline{\Psi}$$

$$- \omega^2 \rho \int_{\xi_s}^{\xi_b} \frac{A'_{42}}{\rho}\, d\xi \,\overline{\phi} - \omega^2 \,|\overline{OG}|\, \rho \int_{\xi_s}^{\xi_b} \frac{A'_{22}}{\rho}\, d\xi \,\overline{\phi}$$

$$+ i\,\omega\, C_y\, N_y \,\overline{y} + i\,\omega\, N_{y\psi} \,\overline{\Psi} + i\,\omega\, C_y\, N_y \,|BG|\, \overline{\phi} + k_y \,\overline{y}$$

$$\overline{Z}_w = - m\omega^2 \,\overline{z} - \omega^2 \rho \int_{\xi_s}^{\xi_b} \frac{A'_{33}}{\rho}\, d\xi \,\overline{z} + \omega^2 \rho \int_{\xi_s}^{\xi_b} \frac{A'_{33}}{\rho}\, \xi \, d\xi \,\overline{\theta}$$

$$+ i\,\omega\, C_z\, N_z \,\overline{z} - i\omega N_{z\theta} \,\overline{\theta} + \rho g \int_{\xi_s}^{\xi_b} B^* \, d\xi \,\overline{z} - \rho g \int_{\xi_s}^{\xi_b} B^* \xi \, d\xi \,\overline{\theta}$$

$$\bar{\Phi}_W = -\omega^2 I_{xT}\,\bar{\phi} - \omega^2\rho\int_{\xi_s}^{\xi_b} \frac{A'_{42}}{\rho}\,d\xi\,\bar{y} - \omega^2(\overline{OG})\,\rho\int_{\xi_s}^{\xi_b} \frac{A'_{22}}{\rho}\,d\xi\,\bar{y}$$

$$-\omega^2\rho\int_{\xi_s}^{\xi_b} \frac{A'_{42}}{\rho}\,\xi\,d\xi\,\bar{\psi} - \omega^2(\overline{OG})\,\rho\int_{\xi_s}^{\xi_b} \frac{A'_{22}}{\rho}\,\xi\,d\xi\,\bar{\Psi} + i\omega N_\phi\,\bar{\phi}$$

$$+ i\omega C_y N_y\,|BG|\,\bar{y} + i\omega N_{y\psi}\,|BG|\,\bar{\Psi} + W\,|GM|\,\bar{\phi}$$

$$\bar{\Theta}_W = -\omega^2 I_y\,\bar{\theta} + \omega^2\rho\int_{s}^{b} \frac{A'_{33}}{\rho}\,\xi\,d\xi\,\bar{z} - \omega^2\rho\int_{\xi_s}^{\xi_b} \frac{A'_{33}}{\rho}\,\xi^2\,d\xi\,\bar{\theta}$$

$$+ m\,|BG|\,\omega^2\,\bar{x} - i\omega N_{z\theta}\,\bar{z} + i\omega C_\theta N_\theta\,\bar{\theta}$$

$$-\rho g\int_{\xi_s}^{\xi_b} B^*\,\xi\,d\xi\,\bar{z} + \rho g\int_{\xi_s}^{\xi_b} B^*\,\xi^2\,d\xi\,\bar{\theta}$$

$$\bar{\Psi}_w = -\omega^2 I_z \bar{\Psi} - \omega^2 \rho \int_{\xi_s}^{\xi_b} \frac{A'_{22}}{\rho} \xi \, d\xi \; \bar{y} - \omega^2 \rho \int_{\xi_s}^{\xi_b} \frac{A'_{22}}{\rho} \xi \, d\xi \; \bar{\Psi}$$

$$- \omega^2 \rho \int_{\xi_s}^{\xi_b} \frac{A'_{42}}{\rho} \xi \, d\xi \; \bar{\phi} - \omega^2 (\overline{OG}) \rho \int_{\xi_s}^{\xi_b} \frac{A'_{22}}{\rho} \xi \, d\xi \; \bar{\phi}$$

$$+ \; i \omega N_{y\psi} \bar{y} + i \omega C_{\psi} N_{\psi} \bar{\Psi} + i \omega N_{y\psi} |BG| \bar{\phi} + k_{\psi} \bar{\Psi}$$

The complex equations are conveniently represented in matrix form as

$$\begin{vmatrix} a_{11} & a_{12} & a_{13} \\ a_{21} & a_{22} & a_{23} \\ a_{31} & a_{32} & a_{33} \end{vmatrix} \begin{vmatrix} \bar{x} \\ \bar{z} \\ \bar{\theta} \end{vmatrix} = \begin{vmatrix} \bar{X}_w \\ \bar{Z}_w \\ \bar{M}_w \end{vmatrix}$$

for longitudinal motions and

$$\begin{vmatrix} b_{11} & b_{12} & b_{13} \\ b_{21} & b_{22} & b_{23} \\ b_{31} & b_{32} & b_{33} \end{vmatrix} \begin{vmatrix} \bar{y} \\ \bar{\Psi} \\ \bar{\phi} \end{vmatrix} = \begin{vmatrix} \bar{Y}_w \\ \bar{\Psi}_w \\ \bar{\Phi}_w \end{vmatrix}$$

$$a_{11} = \left(-m\,\omega^2 + k_x\right) + i\left(\omega N_x\right)$$

$$a_{12} = a_{21} = 0$$

$$a_{13} = a_{31} = \left(m\,|BG|\,\omega^2\right)$$

$$a_{22} = \left(-m\,\omega^2 - \omega^2\rho\int_{\xi_s}^{\xi_b} \frac{A'_{33}}{\rho}\,d\xi + \rho g\int_{\xi_s}^{\xi_b} B^*\,d\xi\right) + i\left(\omega C_z\ N_z\right)$$

$$a_{23} = a_{32} = \left(\omega^2\rho\int_{\xi_s}^{\xi_b} \frac{A'_{33}}{\rho}\,\xi\,d\xi - \rho g\int_{\xi_s}^{\xi_b} B^*\,\xi\,d\xi\right) + i\left(-\omega N_{z\theta}\right)$$

$$a_{33} = \left(-\omega^2\,I_y - \omega^2\rho\int_{\xi_s}^{\xi_b} \frac{A'_{33}}{\rho}\,\xi^2\,d\xi + \rho g\int_{\xi_s}^{\xi_b} B^*\,\xi^2\,d\xi\right) + i\left(\omega C_\theta\ N_\theta\right)$$

$$b_{11} = \left(-m\,\omega^2 - \omega^2\rho\int_{\xi_s}^{\xi_b} \frac{A'_{22}}{\rho}\,d\xi + k_y\right) + i\left(\omega C_y\ N_y\right)$$

$$b_{12} = b_{21}\left(\omega^2\rho\int_{\xi_s}^{\xi_b} \frac{A'_{22}}{\rho}\,\xi\,d\xi\right) + i\left(\omega N_{y\psi}\right)$$

$$b_{13} = b_{31}\left[-\omega^2 \rho\int_{\xi_s}^{\xi_b} \frac{A'_{42}}{\rho}\, d\xi - \omega^2(\overline{OG})\, \rho\int_{\xi_s}^{\xi_b} \frac{A'_{22}}{\rho}\, d\xi\right] + i\left(\omega C_y N_y \left|BG\right|\right)$$

$$b_{22} = \left(-\omega^2 I_z - \omega^2 \rho\int_{\xi_s}^{\xi_b} \frac{A'_{22}}{\rho}\, \xi^2\, d\xi - k_\psi\right) + i\left(\omega C_\psi N_\psi\right)$$

$$b_{23} = b_{32} = \left[-\omega^2 \rho\int_{\xi_s}^{\xi_b} \frac{A'_{42}}{\rho}\, \xi\, d\xi - \omega^2 (\overline{OG})\, \rho\int_{\xi_s}^{\xi_b} \frac{A'_{22}}{\rho}\, \xi\, d\xi\right]$$

$$+ i\left(\omega N_{y\psi} \left|BG\right|\right)$$

$$b_{33} = \left(-\omega^2 I_{xT} + W \left|GM\right|\right) + i\left(\omega N_\phi\right)$$

In solving for x, y, z, ϕ, θ, and ψ, it is convenient to multiply both sides by the appropriate inverse matrix. Thus, if (A) and (B) represent the longitudinal and lateral component matrices, respectively, then $(A)^{-1}$ and $(B)^{-1}$ represent their corresponding inverse matrices. Finally, the equations may be represented in the form

$$\begin{pmatrix} \bar{x} \\ \bar{z} \\ \bar{\theta} \end{pmatrix} = (A)^{-1} \begin{pmatrix} \bar{X}_w \\ \bar{Z}_w \\ \bar{\Theta}_w \end{pmatrix}$$

$$\begin{vmatrix} \bar{y} \\ \bar{\Psi} \\ \bar{\phi} \end{vmatrix} = (B)^{-1} \begin{vmatrix} \bar{Y}_w \\ \bar{\Psi}_w \\ \bar{\Phi}_w \end{vmatrix}$$

The inverse matrices are obtained by application of the well-known formula from matrix theory,

$$(A)^{-1} = \frac{\text{Adj }(A)}{\text{Det }(A)}$$

where Det (A) is the determinant of the matrix (A), and Adj (A) ("adjunct of (A)") is the matrix whose elements are the cofactors (i.e., the signed minors) of the elements of (A), but placed in transposed position.

In order to carry out the matrix multiplication indicated, it is convenient to express the known (complex) quantities

$$X_w,\ Z_w,\ \Theta_w,\ Y_w,\ \Psi_w,\ \Phi_w$$

in the form (u - iv) instead of the form

$$\sqrt{\left(A_r^w\right)^2 + \left(B_r^w\right)^2}\ \exp\left[i\,\gamma_r\right]$$

in which they were defined. This is done by applying the identity

$$\sqrt{\left(A_r^W\right)^2 + \left(B_r^W\right)^2} \; \exp\left[i\gamma_r\right] \quad \left(A_r^W\right) + i\left(-B_r^W\right)$$

which is easily proved from trigonometric equivalent of $\exp\left[i\gamma_r\right]$ and the definition of δ_r.

The results are thus obtained in the form (u - iv). They are easily converted to imaginary exponential form, from which the desired amplitude-response operators a_r and phase-response operators δ_r are obtained. These quantities are functions of both frequency and wave propagation direction for the various motions. From the response operators, the various motions are given by the formula

$$r(t) = a_r \cos\left(\omega t + \frac{\pi}{2} + \delta_r\right)$$

for imposed exciting waves of unit amplitude and phase angle $\pi/2$. For waves of amplitude a_o and phase angle δ_n, the motions will be given by

$$r(t) = a_o \, a_r \cos\left(\omega t + \delta_n + \delta_r\right)$$

It may be seen that a positive δ_r means that the peak amplitude of the motion lags behind the peak amplitude of the surface water elevation, and conversely, a negative δ_r means that the peak amplitude of the motion leads the peak amplitude of the surface elevation.

12.9 REFERENCES

1. Blagoveshchensky, S. N. (1962). "Theory of Ship Motion," translated from the first Russian edition by Theodor and Leonilla Strelkoff under the editorship of Louis Landweber, Iowa Institute of Hydraulic Research, 2 vol., Dover Publications, Inc., New York.

2. Den Hartog, J. P. (1957). "Mechanical Vibrations," McGraw-Hill Book Company, Inc., New York.

3. Grim, O. (1953). "Bereshnung der durch schwingungen eines Schiffskorper erzeugten hydrodynamischen Krafte," Jahrbuch der Schiffbautechnischen Gesellschaft 47, Band 1953, pp. 277-299.

4. Grim, O. (1959a). "Die Schwingungen von schwimmenden, zweidimensionalen Korpern," Hamburgische Schiffbau-Versuchsanstalt Gesselschaft, Report No. 1171, September.

5. Grim, O. (1959b). "The Hydrodynamic Forces in Roll Research," translated from German by Stevens Institute of Technology, Davidson Laboratory, Note 533, May.

6. Havelock, T. H., Sir (1956). "The Damping of Heave and Pitch: A Comparison of Two-Dimensional and Three-Dimensional Calculations," Transactions, The Institution of Naval Architects, Vol. 98, No. 4, pp. 464-468, October.

7. Hu, Pung Nien (1961). "Lateral Forces and Moments on Ships in Oblique Waves," Stevens Institute of Technology, Davidson Laboratory Report 831, June.

8. Hu, Pung Nien, and Kaplan, P. (1962). "On the Lateral Damping Coefficients of Submerged Slender Bodies of Revolution," Stevens Institute of Technology, Davidson Laboratory Report No. 830, February.

9. Kaplan, Paul (1957). "Application of Slender-Body Theory to the Forces Acting on Submerged Bodies and Surface Ships in Regular Waves," Journal of Ship Research, Vol. 1, No. 3, pp. 40-49, November.

10. Kaplan, Paul, and Putz, R. R. (1962). "The Motions of a Moored Construction Type Barge in Irregular Waves and Their Influence on Construction Operation," Contract Nby-32206, an investigation conducted by Marine Advisers, Inc., La Jolla, California, for U. S. Naval Civil Engineering Laboratory, Port Hueneme, California, August. (Unpublished)

11. Kaplan, Paul, and Ulc, Stanley (1961). "A Dimensional Method for Calculating Lateral Bending Moments on Ships in Oblique Waves," Technical Research Group, Inc., Report TRG-147-SR-1, November. (Unpublished)

12. Korvin-Kroukovsky, B. V. (1961). Theory of Seakeeping, Society of Naval Architects and Marine Engineers, New York.

13. Lewis, F. M. (1929). "The Inertia of the Water Surrounding a Vibrating Ship," Transactions, The Society of Naval Architects and Marine Engineers, Vol. 37, pp. 1-20.

14. Newman, J. N. (1962). "The Damping of an Oscillating Ellipsoid Near a Free Surface," Journal of Ship Research, Vol. 5, No. 3, pp. 44-59, December.

15. Prohaska, C. W., (1947). "The Vertical Vibration of Ships," The Shipbuilder and Marine Engine—Builder, pp. 542-546, 593-599, October-November.

16. St. Denis, M., and Pierson, W. J. (1953). "On the Motions of Ships in Confused Seas," Transactions, Society of Naval Architects and Marine Engineers, Vol. 61, pp. 280-357.

17. Sommet, J. (Date Unknown). "The Added Mass of a Ship Oscillating with Longitudinal Motion, Sogreah, Grenoble, France.

18. Stoker, J. J. (1950). "Non-Linear Vibrations in Mechanical and Electrical Systems," Interscience Publishers, Inc., New York.

19. von Kármán, Th. (1940). "The Engineer Grapples with Non-Linear Problems," Bulletin, American Mathematical Society, Vol. 46, pp. 615-683.

20. Vossers, G. (1960). "Fundamentals of the Behavior of Ships in Waves," International Shipbuilding Progress, Vol. 7, No. 65 pp. 28-46.

21. Vossers, G. (1962). "Some Applications of the Slender-Body Theory in Ship Hydrodynamics," The Netherlands Ship Model Basin, Publication No. 214, H. Veenman and Zonen N. V. Wageningen, The Netherlands.

Chapter XIII

NONLINEAR RESPONSE OF A MOORED SHIP TO SEA OSCILLATIONS

13.1 THE MATHEMATICAL MODEL

A possible mooring line suspension system for a large vessel is shown in Figure 13.1. Consider the longitudinal motion of this

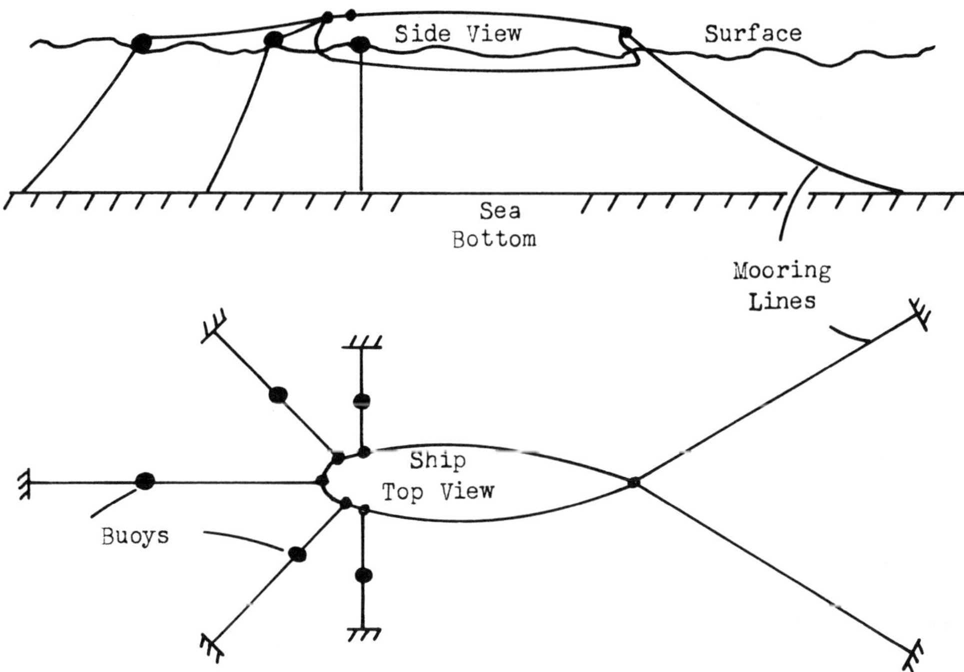

Figure 13.1 Schematic Views of a Moored Ship

ship in a harbor basin under a simple harmonic disturbing force $F(t) = F \cos \omega t$. This disturbing force, which is caused by the harbor waves impinging on the ship's hull, often has a frequency ω which is close to the natural frequency of oscillation of the water within the harbor basin. This problem is also discussed by Wilson (1951) and Abramson (1955). In reality, neither ω or the force amplitude F are constant; ω may cover a spectrum of frequencies; and both may vary with time. In the following analysis, however, the motion of the ship in the direction of $F(t)$ only is considered, where ω and F are constant. In addition, it is assumed that the composite effect of all seven mooring lines on ship's motion is essentially the same as that for just two equivalent horizontal lines, as shown in Figure 13.2. It will be

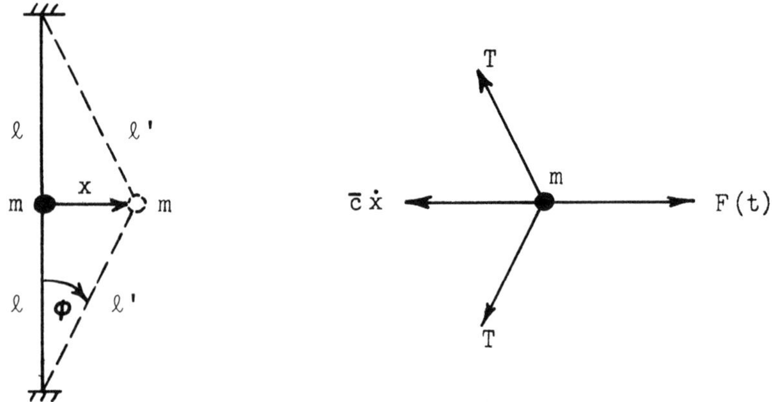

Figure 13.2 Simplified Model of a Moored Ship

shown that these assumptions lead to a nonlinear differential equation of motion which is a special form of the more general one derived by Wilson (1951). With approximate solutions to the special form, one can gain some physical insight into the actual behavior of a moored ship. Of ultimate concern, of course, are the tension forces in the mooring lines. With displacement solutions to the simplified nonlinear differential equation for the ship's motion, crude estimates of forces in the lines of the more complex case shown in Figure 13.1 can be made.

Consider the simplified model shown in Figure 13.2 where the two mooring lines are linear and elastic, to first approximation. That is, the stretch in each line for a longitudinal displacement x, is given by

$$(\ell' - \ell) = \frac{\Delta T \ \ell}{AE}$$

where ΔT is the change in tension force from the position $x = 0$; ℓ is the initial length; A is the cross-sectional area; and E is the apparent Young's modulus for each of the two equivalent mooring lines. Thus, if T_o is the initial line tension, the force at a displacement x is given by

$$T = T_o + \Delta T = T_o + AE\left(\frac{\ell' - \ell}{\ell}\right)$$

With the above two equations and the dimensions of Figure 13.2, the component of T in the plus x direction is expressed as

$$T_x = -T \sin\phi = -T\frac{x}{\ell'} = -\frac{x}{\ell'}\left[T_o + AE\left(\frac{\ell' - \ell}{\ell}\right)\right]$$

$$= \frac{-T_o x}{\ell} \frac{1}{\sqrt{1 + \left(\frac{x}{\ell}\right)^2}} - AE\left(\frac{x}{\ell}\right)\left[1 - \frac{1}{\sqrt{1 + \left(\frac{x}{\ell}\right)^2}}\right]$$

The expression for T_x is simplified by making use of the binomial expansion

$$\frac{1}{\sqrt{1 + \left(\frac{x}{\ell}\right)^2}} = 1 + \left(-\frac{1}{2}\right)\left(\frac{x}{\ell}\right)^2 + \frac{1}{2!}\left(-\frac{1}{2}\right)\left(-\frac{3}{2}\right)\left(\frac{x}{\ell}\right)^4 + \cdots$$

where terms up to $(x/\ell)^2$ only are retained. Thus, the expression for T_x for small values of longitudinal displacements x is given by

$$T_x = -T_o \frac{x}{\ell} - \frac{AE}{2\ell^3}\left[1 - \frac{T_o}{AE}\right] x^3$$

In most cases, the initial tension in mooring lines is small compared to the quantity AE, so that

$$\frac{T_o}{AE} << 1$$

With this assumption, the horizontal component of tension in one line is

$$T_x = -T_o\left(\frac{x}{\ell}\right) - \frac{AE}{2\ell^3} x^3$$

Newton's second law of motion can now be applied to the free body diagram of the ship of mass m shown in Figure 13.2, or

$$2T_x - \bar{c}\dot{x} + F(t) = m\ddot{x}$$

It is assumed that the damping force, $\bar{c}\dot{x}$, is linear to first approximation. Thus, the differential equation of motion can be written as

$$m\ddot{x} + \bar{c}\dot{x} + \left(\frac{2T_o}{\ell}\right) x + \left(\frac{AE}{\ell^3}\right) x^3 = F(t) \qquad (13.1)$$

where $\bar{c}$ is the damping coefficient. Closed form solutions to (13.1) have not yet been found, even for the cases when $F(t)$ is harmonic, a constant or zero. Approximate solutions are available however, some of which are worked in detail by Stoker (1950). Special solutions will now be discussed.

13.2 UNDAMPED, FREE VIBRATIONS

In the case where $\bar{c} = 0$ and $F(t) = 0$, (13.1) becomes

$$\ddot{x} + \alpha x + \beta x^3 = 0 \tag{13.2}$$

where

$$\alpha = \frac{2T_o}{m\ell} \; ; \quad \beta = \frac{AE}{m\ell^3}$$

The acceleration $\ddot{x}$ can be written in terms of velocity $\dot{x} = v$ or

$$\ddot{x} = \frac{d^2x}{dt^2} = \frac{dv}{dt} = \frac{dv}{dx}\frac{dx}{dt} = v\frac{dv}{dx}$$

Thus, (13.2) becomes

$$v\frac{dv}{dx} = -(\alpha x + \beta x^3) \tag{13.3}$$

in which the variables are separable. Integrating this equation between the limits of $x = 0$ at $v = v_{max} = v_o$ and $x = x_{max} = a$ at $v = 0$, the result is

$$\int_{v_o}^{o} v\, dv = - \int_{o}^{a} (\alpha x + \beta x^3)\, dx$$

or

$$\frac{v_o^2}{2} = \alpha \frac{a^2}{2} + \beta \frac{a^4}{4}$$

The amplitude a of oscillation is thus seen to satisfy the relationship

$$a^2 = \frac{1}{\beta}\left[- \alpha + (\alpha^2 + 2\beta v_o^2)^{\frac{1}{2}} \right] \qquad (13.4)$$

The time $\overline{T}$ to complete one cycle of oscillation (in which m follows the path $x = 0, + a, 0, -a, 0$), is found by integrating (13.3) twice, or

$$\int_{v_o}^{v} v\, dv = - \int_{o}^{x} (\alpha x + \beta x^3)\, dx$$

$$v^2 - v_o^2 = - \alpha x^2 - \beta \frac{x^4}{2}$$

After the variables are separated, where $v = dx/dt$, a second integration gives the result

$$\frac{\overline{T}}{4} = \int_{o}^{\overline{T}/4} dt = \int_{o}^{a} \left(v_o^2 - \alpha x^2 - \beta \frac{x^4}{2} \right)^{-\frac{1}{2}} dx \qquad (13.5)$$

The upper limits in the last equation correspond to the time required for 1/4 of the total oscillation. By making use of (13.4) and of the transformation

$$x = a \sin \theta$$

(13.5) can be written as

$$\bar{T} = 4\sqrt{2} \int_0^{\pi/2} \left(2\alpha + \beta a^2 + \beta a^2 \sin^2\theta \right)^{-\frac{1}{2}} d\theta \qquad (13.6)$$

Equation (13.6) shows that the period $\bar{T}$ depends on the amplitude a of oscillation, as long as $\beta \neq 0$. When $\beta = 0$, the period is independent of a, and (13.6) gives the familiar result for the period of a linear harmonic oscillator,

$$T = \frac{2\pi}{\sqrt{\alpha}}$$

The natural frequency of the nonlinear system (13.2) is simply

$$\omega_n = \frac{2\pi}{\bar{T}} \qquad (13.7)$$

where $\bar{T}$, given by (13.6), can be found by numerical integration. Equation (13.7) is sketched in Figure 13.3 for a fixed value of α, for $\beta = 0$ and for $\beta > 0$. We note that the frequency is independent of amplitude for $\beta = 0$, but that ω_n increases with amplitude when $\beta > 0$. The amplitude dependence of frequency is characteristic of a moored vessel's response to a sudden gust, whether or not damping is present.

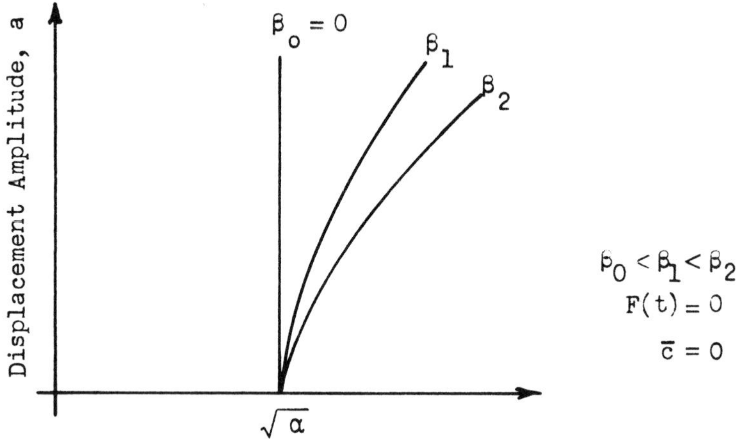

Figure 13.3 Dependency of the Vibration Amplitude of the Ship with Frequency

13.3 UNDAMPED, FORCED VIBRATIONS

When damping is zero and the impressed force is $F \cos \omega t$, (13.1) has the form

$$\ddot{x} = -\alpha x - \beta x^3 + \frac{F}{m} \cos \omega t \tag{13.8}$$

This differential equation can be solved approximately by an iteration procedure due to Duffing, as discussed by Stoker (1950). In this procedure, a solution is first assumed in the form

$$x = x_o = a \cos \omega t$$

for which the right-hand side of (13.8) becomes

$$\ddot{x} = \ddot{x}_1 = -\left(\alpha a + \frac{3}{4}\beta a^3 - \frac{F}{m}\right)\cos\omega t$$

$$-\frac{1}{4}\beta a^3 \cos 3\omega t \qquad (13.9)$$

where use was made of the identity

$$\cos^3\omega t = \frac{3}{4}\cos\omega t + \frac{1}{4}\cos 3\omega t$$

When (13.9) is integrated twice, the result for zero constants of integration is

$$x = x_1 = \frac{1}{\omega^2}\left(\alpha a + \frac{3}{4}\beta a^3 - \frac{F}{m}\right)\cos\omega t$$

$$+ \frac{1}{36}\frac{\beta A^3}{\omega^2}\cos 3\omega t$$

For small values of β and a small force F, Duffing argued that the amplitude a of oscillations could be approximated by the coefficient of $\cos\omega t$. Thus

$$a = \frac{1}{\omega^2}\left(\alpha a + \frac{3}{4}\beta a^3 - \frac{F}{m}\right)$$

or

$$\omega^2 = \alpha + \frac{3}{4} \beta a^2 - \frac{F}{ma} \qquad (13.10)$$

In this case, the amplitude of oscillation depends not only on α and β, but also upon the forcing frequency ω. The absolute value of the amplitude $|a|$ from (13.10) is shown in Figure (13.4) as a function of ω. Here, values of α and β are fixed for various values of the applied force amplitude F. When $|a|$ is compared with $|H(\omega)|$ of Figure 10.2, there is an important conclusion: the response curves in the nonlinear case where $\beta > 0$ can be thought of as arising from those for the linear case where $\beta = 0$ by bending the latter to the right.

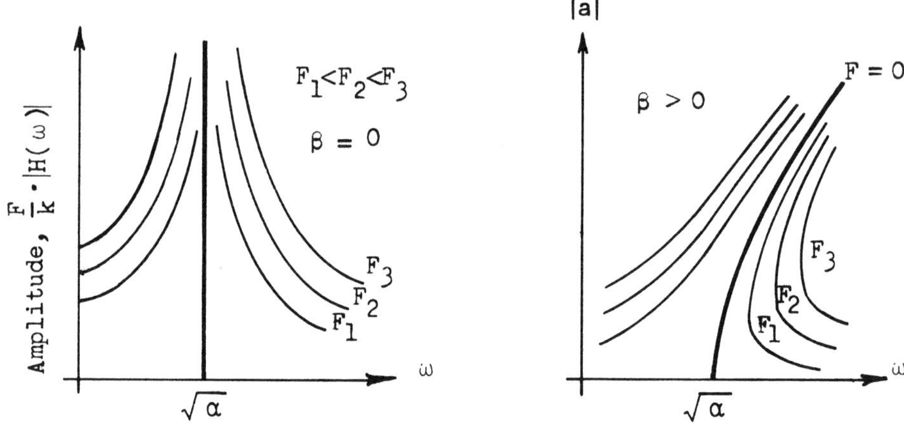

Figure 13.4 Comparison of the Response Curves for the Linear and Nonlinear Cases

13.4 DAMPED, FORCED VIBRATIONS

If damping is neglected, there is an oscillation of the form $x = a \cos \omega t$ either in phase with the impressed force $F \cos \omega t$ or 180° out of phase with it. If damping is present, however, the displacement and the impressed force can be expected to be out of phase, just as in the corresponding case when $\beta = 0$. In order to take into account this difference in phase, the impressed force could be prescribed and the phase of the solution left to be determined. In the present case, however, it is more convenient to fix the phase of the solution and leave the phase of the impressed force as a quantity to be determined. Thus, (13.1) is taken in the form

$$\ddot{x} + c\dot{x} + (\alpha x + \beta x^3) = H \cos \omega t - G \sin \omega t$$

where

$$c = \frac{\bar{c}}{m} \; ; \qquad \frac{F}{m} = \sqrt{H^2 + G^2} \tag{13.11}$$

Here, (F/m) is fixed but the phase angle is left to be determined. As in the previous iteration procedure, the quantities c, β, G and H are all assumed to be small, and a first approximate solution is chosen as $x_o = a \cos \omega t$. When this is inserted into (13.11), the coefficients of like trigonometric functions are equated, the results are

$$(\alpha - \omega^2) a + \frac{3}{2} \beta a^3 = H$$

$$a c \omega = G \tag{13.12}$$

In arriving at (13.12), the coefficient of $\cos 3\omega t$ was considered very small, and was thus neglected. This leads to an important result: if damping exists, there can be no periodic motion unless an impressed force is present; that is, $H = G = 0$ implies $a = 0$ if $c \neq 0$.

By squaring and adding (13.12), the amplitude of the response can be obtained in terms of the magnitude of the impressed force F and the forcing frequency ω. That is

$$\left(\frac{F}{m}\right)^2 = \left[(\alpha - \omega^2)\, a + \frac{3}{2}\beta a^3\right]^2 + c^2 a^2 \omega^2$$

$$= H^2 + G^2 \tag{13.13}$$

Figure (13.5) shows the response $|a|$ as a function of ω, for several values of (F/m), where α and β are constant. Again, it is observed that these curves could be considered as arising from the linear response curves ($\beta = 0$) by bending the latter to the right.

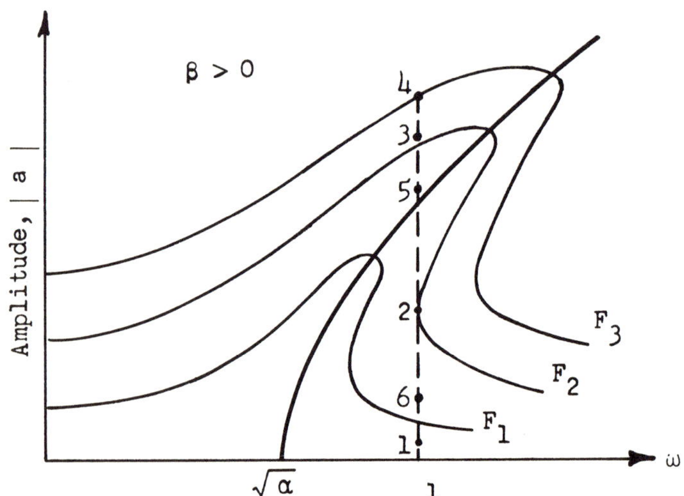

Figure 13.5 Displacement Response Curves for Damped, Forced Vibrations

13.5 THE JUMP PHENOMENA

The amplitude response curves of Figure 13.5 lead to some important conclusions about the possible behavior of a moored ship. Suppose that damping is constant; that $\omega = \omega_1$ = constant for a period of time; and that the magnitude of the harmonic force F increases slowly. At first, the vessel responds by oscillating out of phase with F (point 1. on Figure 13.5). As F increases, the amplitude of the ship response does also, until a critical value F_2 is reached (point 2.). There, the ship motion changes phase and jumps to a higher amplitude of motion, the value F_2 at point 3.. This happens unless F decreases, in which case the amplitude of ship's motion decreases also. A further increase in F brings the amplitude of the ship's response to point 4. on Figure 13.5.

With a decline in F, values of the ship's amplitude also decrease, with motions in phase with F. At point 5., where F is very small, the ship's amplitude again jumps to point 6., where F is also very small. From point 6., the oscillations die out due to damping and finally reach point 1. again. It is assumed that the growth and decay of the wave forces, at nearly constant frequency, are over a time period which is long compared to the frequency of the ship's oscillation.

Jumps in $|a|$ can also occur if the wave force amplitude is constant and the forcing frequency slowly changes. This path is traced out in Figure 13.6 as points 7.—11. for increasing ω and $F = F_2$. A jump in $|a|$ occurs between points 9. and 10. As ω decreases from point 11., however, the path is 11., 12., 13., 8. and 7., where the jump is from 13. to 8..

In reality, both the force amplitude and the forcing frequency may vary with time. One may also look upon the resultant force on the ship as composed of many separate harmonic forces, each with a particular frequency and each with a particular amplitude. Thus, one can visualize very complex motions and erratic jumps in the motions of a moored ship, especially when it has many mooring lines. It is hoped that this simplified analysis has given the reader some physical insight into the phenomena, as well as a view of the mathematical complexities. Analytical methods still need to be developed in order to optimize the design of safe mooring lines.

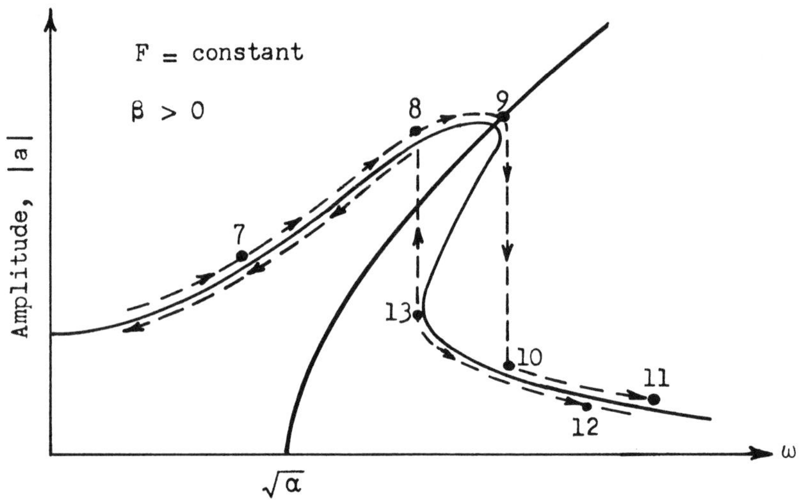

Figure 13.6 The Jump Phenomena at Constant Amplitude of Force

13.6 SUBHARMONICS

In the above discussion, the only solutions for the displacement response of the moored ship to external harmonic excitation were those with frequencies ω coinciding with the exciting force. It is recalled that a continuous, linear system such as a simply supported beam can undergo transverse vibrations at frequencies equal to integral multiples (2, 3, ...n) of the exciting frequency, but vibrations at fractional frequencies of ω do not exist. However, in single degree of freedom, nonlinear systems such as those described by (13.8) or (13.11), vibrations at frequencies equal to fractions (1/2, 1/3, ...1/n) of the exciting frequency can sometimes be sustained. Such vibrations which occur at smaller frequencies than the exciting force are called subharmonic vibrations. Stoker (1950) discusses subharmonic oscillations of frequency $\omega/3$ for the undamped case, (13.8), and the damped case, (13.11). The essential features of this solution to (13.8) will now be summarized.

To show that (13.8) can have at least one subharmonic, a solution of the following form is assumed.

$$x = A_{1/3} \cos\left(\frac{\omega t}{3}\right) + A_1 \cos \omega t \qquad (13.14)$$

When (13.14) is substituted into (13.8) and trigometric identities are used, two algebraic equations in α, β, A_1, A and F are found by equating coefficients of $\cos(\omega t/3)$ and $\cos \omega t$, respectively. From these results, it is found that for $\omega = 3\sqrt{\alpha}$, the amplitude $A_{1/3}$ is nonzero if β is vanishingly small. Also, $A_1 = -F/(8\alpha m)$ for this value of ω and for $\beta = 0$. An approximate solution of (13.8) is next found by an iteration procedure, in which the above values, $\omega = 3\sqrt{\alpha}$, $\beta = 0$, and $A_1 = A$ are prescribed initially. The results are

$$\omega^2 = 9\alpha + \frac{27}{4}\beta\left(A_{1/3}^2 + A_{1/3}A + 2A^2\right)$$

$$A_1 = A + \frac{1}{32}\frac{\beta}{\alpha}\left(A_{1/3}^3 - 21A_{1/3}^2 A - 27A_{1/3}A^2 - 51A^3\right)$$

where

$$A = -\frac{F}{8\alpha m} \qquad (13.15)$$

An inspection of (13.15) leads to several conclusions about the nature of the subharmonic response. Discussion is limited to the case for $\beta > 0$. The first of these equations represents a hyperbola in the ω, A plane. The forcing frequency has a minimum value ω_{min} when $A_{1/3} = -A/2$, which is given by

$$\omega_{min} = 3\sqrt{\alpha + \frac{21}{16}\beta A^2} \qquad (13.16)$$

Thus, only if $\omega > \omega_{min}$ can a subharmonic of frequency $\omega/3$ exist. Otherwise, $A_{1/3} = 0$ to satisfy (13.15) for real values of this amplitude. The response curve for $A_{1/3} = 0$ and $A_1 = a$ has already been presented in Figure 13.4, and is repeated in Figure 13.7 for one value of $F = F_1$. Also shown on Figure 13.7 is the amplitude

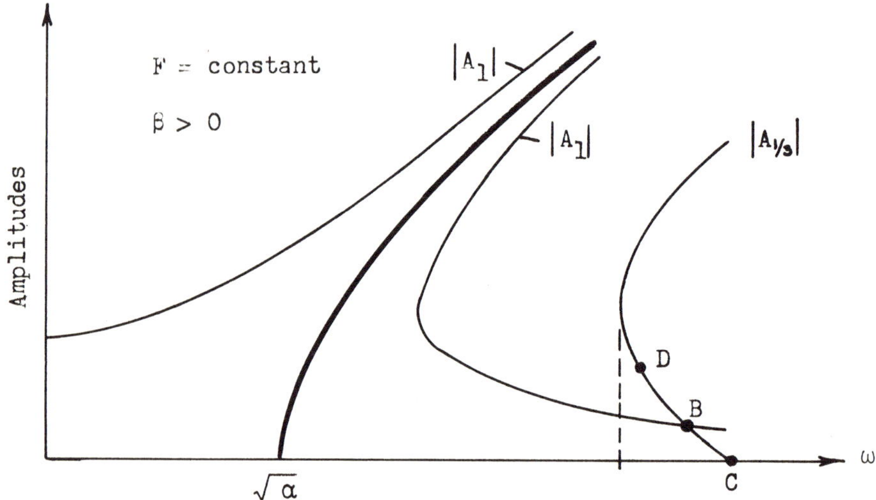

Figure 13.7 Response Amplitudes Including a Subharmonic Oscillation

of the subharmonic response, $A_{1/3}$. The intersection point B of the two amplitudes occurs at

$$A_1 = A - \frac{51}{32} \beta A^3 \qquad (13.17)$$

where $A = -F/(8\alpha m)$.

An inspection of Figure 13.7 for constant F will show how the subharmonic vibration can be excited. If the forcing frequency is allowed to increase slowly from zero, point B is soon reached, where the subharmonic vibration is excited. This vibration then decreases to zero along path B-C. However, if ω is initially much higher than ω_{min} and is then allowed to decrease, a subharmonic vibration of higher amplitude than A_1 can be achieved since the branch B-D of the hyperbola is then a possible path. It is concluded that the subharmonic vibration results through bifurcation of the harmonic vibration whose amplitude is A_1.

Although the role of damping has not been considered here, its effect is to prohibit the occurrence of subharmonic responses in many nonlinear, single degree of freedom systems. Stoker (1950) discusses the possibility of the nonexistence of subharmonics of frequencies smaller than $\omega/3$ for the system described by (13.11), when the damping term is present.

A dramatic example of the existence of a subharmonic response has been measured on an LST (Landing Ship, Tank) during mooring at a sea berth in the Gulf of Mexico. The details of the mooring were reported by O'Brien and Muga (1964). The seven chain lines were arranged somewhat as shown in Figure 13.1 except that there were four lines from the bow and three from the stern. Figure 13.8 shows time records of the relative water level and the measured tension in the mooring chain at port bow.

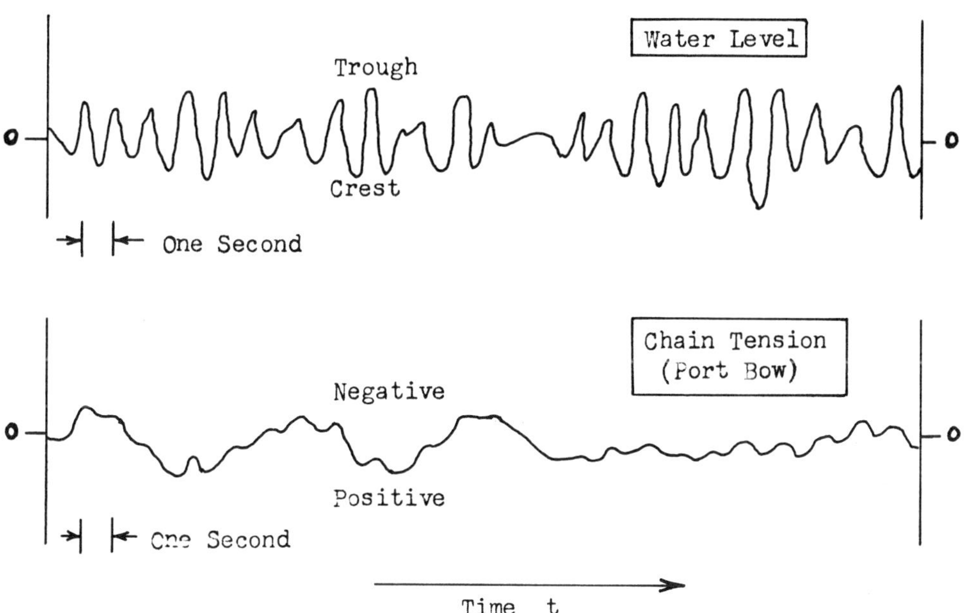

Figure 13.8 Time Histories of Water Level and Mooring Chain Tension

The power spectral densities corresponding to these traces are given in Figure 13.9.

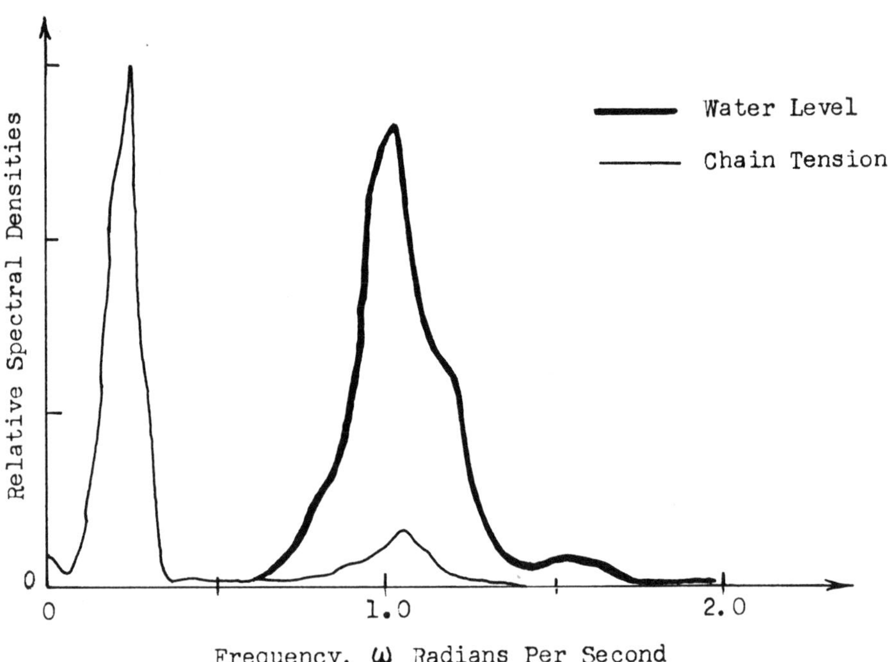

Figure 13.9 Power Spectral Densities for the Water Level and Mooring Chain Tension

Results show that the dominant forcing frequency of the waves is about 1.0 radians per second. It is noticed, however, that there are two frequencies at which the force spectrum of the chain reach maximums. These are at 1 and at about 1/4 radians per second. The interesting result is that the highest force does not occur at the forcing frequency, but at a subharmonic frequency. This shows the importance of considering subharmonic responses in the formulation of analytical models for the design of complex mooring systems.

13.7 REFERENCES

1. Abramson, H. N., and Wilson, B. W. (1955). "A Further Analysis of Moored Vessels to Sea Oscillations," Second Midwestern Conference of Solid Mechanics, p. 236.

2. O'Brien, J. T., and Muga, B. J. (1964). "Sea Tests on a Spread-Moored Landing Craft," Proceedings of 8th Conference on Coastal Engineering, Lisbon, Portugal.

3. Stoker, J. J. (1950). Nonlinear Vibrations in Mechanical and Electrical Systems, Interscience Publishers, Inc., New York.

4. Wilson, B. W. (1951). "Ship Response to Range Action in Harbor Basins," ASCE Transactions, Vol. 116, p. 1129.

Chapter XIV

A SPECIFIC NONLINEAR APPLICATION

In this chapter, nonlinear effects are considered. These effects become more pronounced (1) when wave lengths are short compared with a characteristic length of the vessel (length when waves are head-on and beam when waves are beam-on), (2) when vessel is moored in shallow water so that draft/depth ratio is significant and (3) when natural period of mooring assembly is long relative to dominant wave period.

14.1 PHYSICAL DESCRIPTION OF NONLINEAR MOTION

The sequence of events which takes place in the case of a large vessel moored in shallow water is as follows:

(a) The vessel is initially moored in calm water at a static equilibrium position determined by the mooring configuration.

(b) A train of waves, either regular or irregular, is imposed on the vessel and the vessel acquires (i) linear components of motion corresponding to the period band of the exciting waves in surge, sway, yaw, heave, pitch and roll, and (ii) nonlinear components in surge, sway and yaw. These nonlinear components have frequencies corresponding to the natural frequencies of the mooring system.

(c) As the vessel is displaced, it acquires a velocity (in surge, sway and yaw) and this velocity eventually reaches a maximum and thereafter diminishes. At the same time, the displacements of the vessel cause strain energy to be stored in the mooring lines of the vessel.

(d) As the waves continue to impinge on the vessel, the velocity of the vessel diminishes to zero at which point the strain energy stored in the mooring lines attains its maximum value. The vessel then acquires a velocity in the opposite direction, whose absolute magnitude reaches a maximum and begins to decline thereafter.

(e) Eventually, the vessel attains a zero velocity at which time the displacement reaches its closest point to the initial static equilibrium position.

(f) Thereafter, the cycle is virtually repeated with minor variations depending on the instantaneous position of the vessel with respect to the on-coming waves.

The point at which the vessel attains its maximum velocity is roughly the "dynamic equilibrium position." Thus, the vessel may be considered to oscillate about this new position. A similar position exists which the vessel occupies when currents and/or steady winds are imposed. The unbalanced net force in the moorings is then equal to the imposed current loading or wind loading in the case of currents and winds, respectively. In the case of waves, this unbalanced net force in the moorings results from a combination of (1) the difference in the reflection and transmission characteristics of the vessel to the oncoming waves and (2) the waves generated by the motions of the vessel.

The problem thus consists of finding the time-varying loads F(t) acting on the vessel due to the imposed wave system and then treating the moored-ship system. For very large vessels, some of the coupling terms can be omitted and the resulting equations can then be greatly simplified. The resulting equations can be easily solved using any of the standard methods. For example, the linear acceleration extrapolation method described by Norris, et al (1959) is well suited for this purpose. The actual problem then consists of estimating the time history of the loading pattern (magnitude, direction and point of application). Although the details are proprietary, the method for estimating the loading pattern for a multiple point mooring assumes that waves generated by the vessel may be neglected. The resulting reflecting and transmission characteristics of the vessel are obtained by determining the reflection and transmission properties of an elemental two-dimensional section normal to the oncoming wave direction. This is then corrected for the angle of obliquity using empirically obtained model test data and the result is then integrated along the vessel length. Finally, two types of correction factors are introduced to account for three-dimensional effects, both in the longitudinal and transverse directions relative to the vessel. The result is a composite picture of the effect of the vessel's presence on the incident wave pattern. The loading history itself resembles a surf beat record except that the mean level of the load has been shifted by an amount equal to that force required to displace the vessel from its static to its dynamic equilibrium position.

Figure 14.1.1 illustrates an example of the kind of results obtained by use of the method. This particular solution was obtained by using the phase-plane method of solving differential equations. This method is useful no matter how linear or non-linear the system is or how complicated the original differential equation. The method consists in obtaining numerical estimates of the displacement and velocity at discrete times. These displacement and velocity values are obtained by a step-by-step integration procedure, starting with initial conditions and evaluating the conditions at the end of a discrete time interval. The ratio of the acceleration to the velocity for a given time and instantaneous position gives the corresponding slope of the velocity displacement trajectory in the phase-plane; hence, its descriptive title. To a first approximation, the method requires a prior knowledge of the force-time history but the reader is reminded that the actual force-time history depends on the instantaneous motions of the vessel which are themselves dependent upon the external excitation.

The pattern of motion of a nonlinear moored vessel in surge is well illustrated by Figure 14.1.1. The linear short-term oscillations are clearly indicated by the figure. Additional nonlinear long-term oscillations look similar to the one shown except for minor variations caused by differences in the instantaneous phase position of the vessel with respect to the oncoming waves.

Again, in principle the procedure can be extended to single point moorings but the details are slightly more complex because of the need to account for (i) waves generated by the vessel, and (ii) instantaneous position of the vessel. For large vessels moored at single point type moorings, yaw becomes strongly coupled with sway and surge; thus, since the equations must be solved simultaneously, additional computer time and memory space is required. As a matter of fact, the coupling terms may be said to eventually dominate the behavioral response of the vessel.

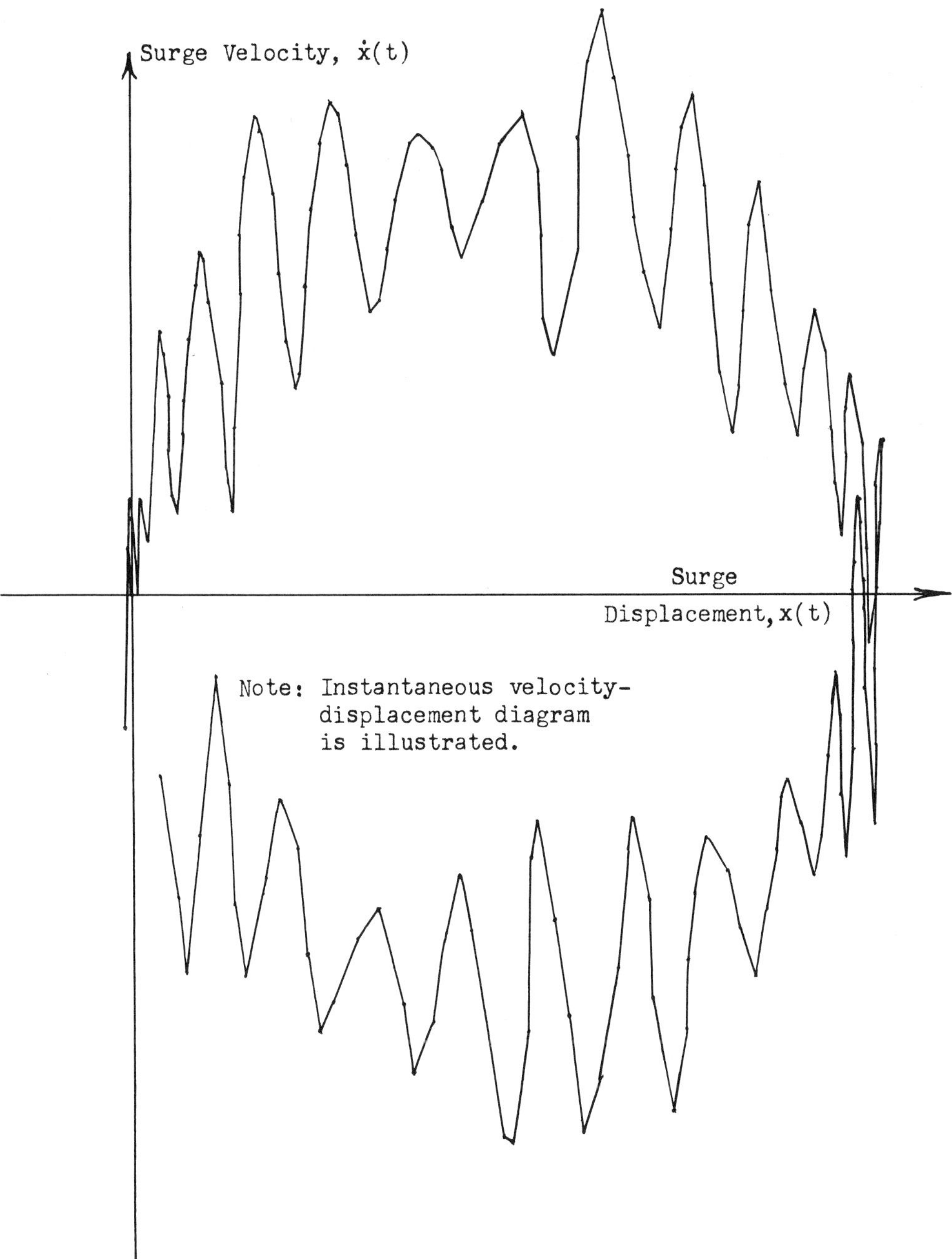

Figure 14.1.1 Phase-plane solution of nonlinear surge response of a moored vessel.

14.2 REFERENCES

1. Norris, C. H., Hansen, R. J., Holley, M. J., Jr., Biggs, J. M., Namyet, S., and Minami, J. K. (1959). Chapter 8, "Introduction to Numerical-Integration Methods and Their Application to Dynamic-Response Calculations," Structural Design for Dynamic Loads, McGraw-Hill Book Company.

Chapter XV

MISCELLANEOUS TOPICS OF CURRENT INTEREST

In addition to the material presented up to now, there are a number of topics of current interest for which it is useful to review the present status. These topics include (i) a summary of the catenary relations and, more specifically, the problem of the double catenary, (ii) dynamic stresses induced in cables by random end loads and (iii) the force required to extract bodies from the ocean bottom. The latter is usually referred to as the bottom breakout force problem.

15.1 DOUBLE CATENARY PROBLEM

The well-known equations for a single catenary (shown in Figure 15.1.1) are given by

$$T_v = ws \qquad \sqrt{L^2 - B^2} = 2c \sinh \frac{k}{2c}$$

$$T_h = wc \qquad \frac{B}{L} = \tanh \frac{x_m}{c}$$

$$T = wy \qquad x_m = x_a + \frac{k}{2}$$

$$y^2 = s^2 + c^2 \qquad\qquad x = x_m + \frac{k}{2}$$

$$s = c \sinh \frac{x}{c}$$

where

T_v = vertical force at point (x,y)

T_h = horizontal force at point (x,y)

T = axial tension at point (x,y)

s = length of curve from point (0,c) to point (x,y)

w = weight of chord per unit length

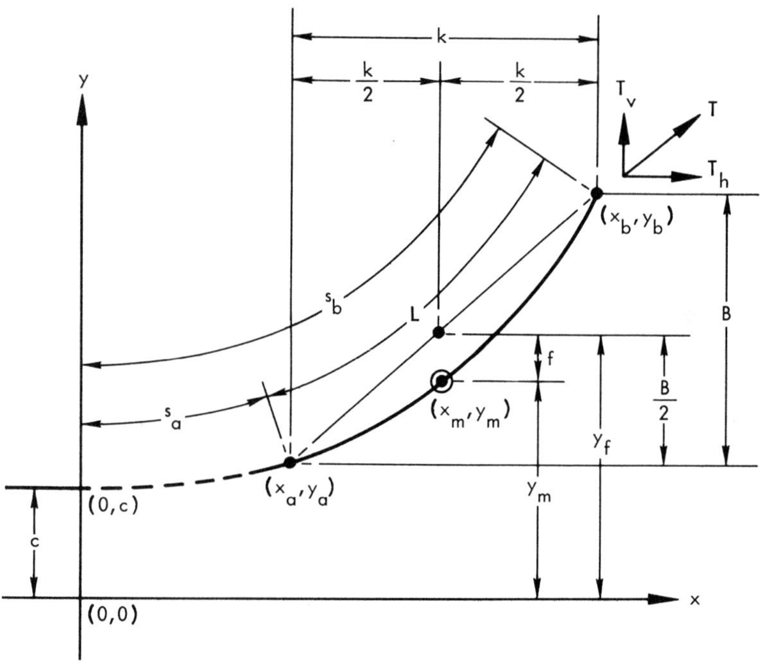

Figure 15.1.1 Definition sketch for single catenary.

When a sinker is placed in the catenary, the profile assumed by the chain is the result of two catenaries as shown in Figure 15.1.2. Here, the sinker lies at the intersection of the two catenary shapes.

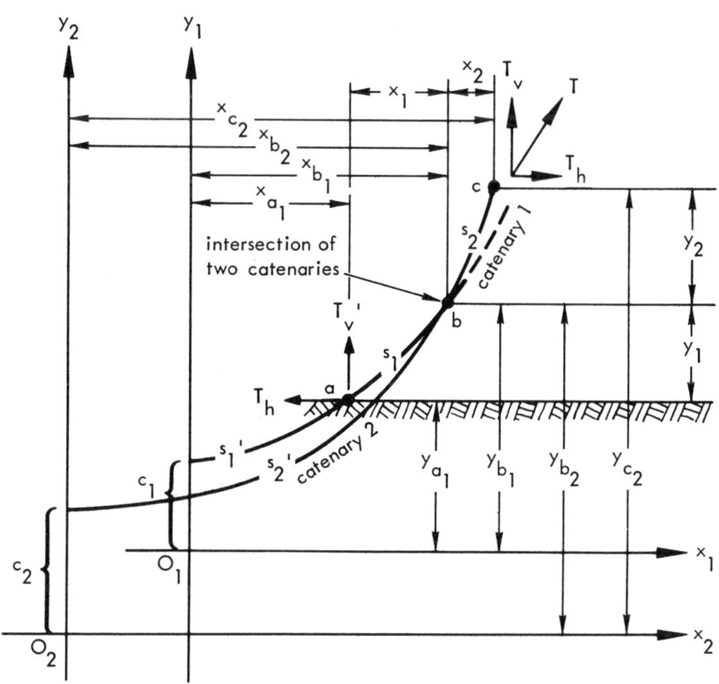

Figure 15.1.2 Definition sketch for double catenary.

The solution of this case which is typical for mooring systems has been solved numerically, but the details are proprietary. However, at least five subcases must be considered. These are:

(1) The sinker and part of the upper chain lie on the bottom.

(2) All of the upper chain is lifted off the bottom but the sinker remains on the bottom.

(3) The upper chain, sinker and part of the lower chain are lifted off of the bottom.

(4) The upper chain, sinker and all of the lower chain are lifted off of the bottom but the anchor does not provide any restraining forces in the vertical direction.

(5) Finally, the entire assemblage excepting the anchor is lifted off of the bottom and the anchor provides some restraining forces in the vertical direction.

Once the force displacement relations for the individual mooring components have been determined, these may be combined to obtain the mooring restoring diagrams (surge, sway and yaw) for the moored vessel. Because of the presence of non-symmetries in the mooring arrangement, displacements in one direction often give rise to reaction forces (or moments) in another direction. For example, displacements in the sway direction may be accompanied by yaw if the mooring arrangement is not symmetrical.

To determine the point about which the moored-ship system rotates, it is often convenient to determine what amounts to the "shear center" (alternatively known as the elastic center). This is the point in space where forces applied in any direction to a restrained body such as a moored vessel, results in translation of the body in the appropriate directions but the body does not rotate. Also, a moment applied to the body results in rotation of the body. It is worth noting that this point may lie outside the body as, for example, in the case of a single point mooring.

15.2 DYNAMIC STRESSES INDUCED IN CABLES BY RANDOM END LOADS

The interest in dynamic stresses induced in cables stems largely from the realization that for sometime to come, major lift capability for load handling operations in the deeper parts of the ocean will have to be provided by surface vessels via long lines. We are thus interested in the propagation of stress waves in cables due to the motions of the floating vessel as illustrated by Figure 15.2.1.

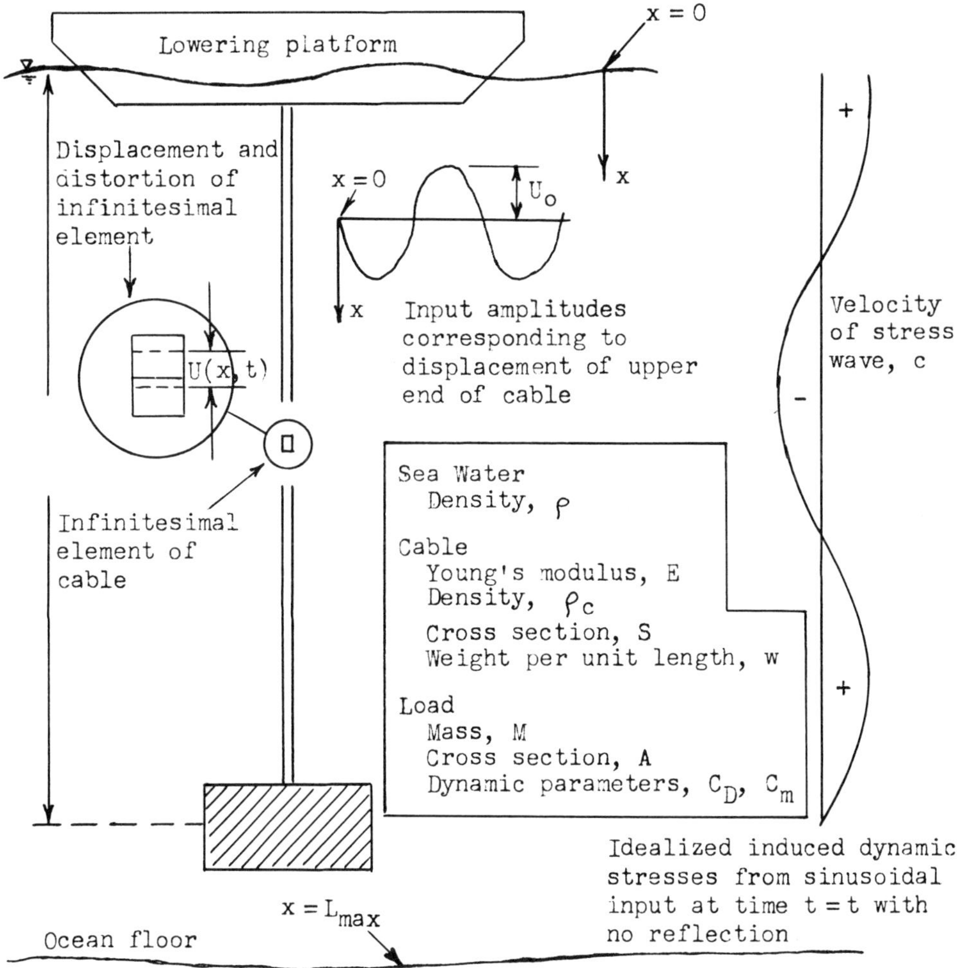

Figure 15.2.1 Definition sketch for raising or lowering operation at sea.

The problem has been studied analytically by Germeles (1963), Walton and Polachek (1959), and experimentally by Muga (1968).

The theory developed by Germeles is based upon the solution of an approximate form of the equation governing the propagation of longitudinal elastic waves. Using the definition sketch illustrated by Figure 15.2.1, the appropriate partial differential equation is

$$\rho_c S \frac{\partial^2 u}{\partial t^2} = S E \frac{\partial^2 u}{\partial x^2} - k \frac{\partial u}{\partial t} \qquad (15.2.1)$$

where

ρ_c = density of the cable

S = material cross-section of the cable

E = modulus of elasticity of the cable

k = constant of friction on the cable due to the surrounding water.

It is assumed that the net vertical motion of the load and cable does not influence the dynamic displacements of the vessel, and that while the length, L, of the cable varies with time, at any given instant of time the length may be considered constant.

By defining the following non-dimensional variables,

$$x' = \frac{x}{L} \qquad t' = \frac{tc}{L}$$

$$c^2 = \frac{E}{\rho_c} \qquad \beta_c = \frac{kL}{\rho_c c S}$$

Equation 15.2.1 may be rewritten (upon substitution) as

$$\frac{\partial^2 u}{\partial t'^2} + \beta_c \frac{\partial u}{\partial t'} = \frac{\partial^2 u}{\partial x'^2} \qquad (15.2.2)$$

The boundary conditions are

(a) At the upper end where $x' = 0$, $u = u(x,t)$ is known for all time, t.

(b) At the lower end where $x' = 1$,

$$M_a \frac{\partial^2 u}{\partial t^2} + E\,S \frac{\partial u}{\partial x} + \frac{1}{2} C_D \rho A \left|\frac{\partial u}{\partial t}\right| \frac{\partial u}{\partial t} = 0 \qquad (15.2.3)$$

where M_a = dynamic mass of the load including added mass effects

C_D = drag coefficient

A = projection of surface area normal to motion direction

By introducing the dimensionless parameters

$$\mu = \frac{\rho_c SL}{M_a}, \qquad \beta = \frac{C_D \rho A}{2M_a}$$

Equation 15.2.3 may be rewritten as

$$\frac{\partial^2 u}{\partial t'^2} + \mu \frac{\partial u}{\partial x'} + \beta \left| \frac{\partial u}{\partial t'} \right| \frac{\partial u}{\partial t'} = 0 \qquad (15.2.4)$$

Germeles has obtained a numerical solution to Equation (15.2.2) by linearizing Equation (15.2.4) and by pointing out that the second term in Equation (15.2.2) may be discarded in some instances. Holmes (1966) has greatly extended the range of available solutions and the limited experimental studies conducted thus far show that the predicted stresses are in excellent agreement with the measured stresses.

15.3 BOTTOM BREAKOUT FORCES

The problem of determining the force required to extract various objects from the ocean floor has not received the attention of investigators which it probably deserves. As a matter of fact, a large class of problems dealing with the marine soil interface has largely gone unnoticed by those most qualified to contribute to their successful solution. For example, the design of ship anchors, which involves the mechanics of both penetration and extraction from a 'soil', has proceeded largely from tradition and limited experience. It seems strange indeed that soil mechanicists as a group have contributed little, if any, to the rational and deliberate design of anchorage systems for marine use. Recently, Vesić (1969) has presented an excellent state-of-the-art review of the breakout problem and the reader is urged to consult this reference as an introduction to the subject. Vesić has also (i) analyzed the basic problem, (ii) discussed the significant features contributing to breakout, and (iii) proposed an analytical approach to solution of the problem which is based on the expansion of cavities near a surface.

One of the difficulties is that of determining the properties of material in which the object is embedded. The benthic ooze found in a number of locations in the ocean has almost no shear strength and thus it is virtually impossible to run the usual laboratory-type tests for soils. On the other hand, the material does not respond consistently to viscometer-type measurements for liquids. At some time or another, it is going to be absolutely necessary to devise means of determining not only the rheological properties of this particular soil-liquid material but also whether the material in situ will behave more as a liquid (such as a non-Newtonian fluid) or more as a soil. As a matter of conjecture, this latter determination almost surely will depend to some extent on the material volume—motion interrelations.

It is obvious that considerably more research should be undertaken in this area and that the most rewarding results are likely to come from high quality model studies under carefully controlled laboratory conditions rather than from large scale prototype field investigations. At this stage, the latter is virtually worthless except for providing gross estimates applicable to specific problems.

15.4 REFERENCES

1. Germeles, P. (1963). "Stress Analysis of Ship-Suspended Heavily Loaded Cables for Deep Underwater Emplacements," Project Trident Technical Report Nobsr-81564, S-7001,0307, Department of the Navy, Bureau of Ships, August.

2. Holmes, P. (1966). "Mechanics of Raising and Lowering Heavy Loads in the Deep Ocean: Cable and Payload Dynamics," U.S. Naval Civil Engineering Laboratory, Report R-433, April.

3. Muga, B. J. (1968). "Experimental Results of Raising and Lowering Heavy Loads in the Ocean," Proceedings of the National Symposium on Science and Engineering of the Atlantic Shelf, Marine Technology Society, Philadelphia, Pennsylvania.

4. Vesić, A. S. (1969). "Breakout Resistance of Objects Embedded in the Ocean Bottom," Proceedings of Second Conference on Civil Engineering in the Oceans, Miami, Florida, December 10-12.

5. Walton, T. S., and Polachek, H. (1959). "Calculation of Non-Linear Transient Motion of Cables," Research and Development Report 1279, David Taylor Model Basin, Applied Mathematics Laboratory, July. (Also appearing in Math. of Computation, Vol. 14, 1960, pp. 27-46.)

AUTHOR INDEX

SUBJECT INDEX